SUSTAINABLE DEVELOPMENT AND ENVIRONMENT MANAGEMENT

INNOVATIONS, SCIENCES AND TECHNOLOGIES

ENVIRONMENTAL REMEDIATION TECHNOLOGIES, REGULATIONS AND SAFETY

Additional books in this series can be found on Nova's website
under the Series tab.

Additional e-books in this series can be found on Nova's website
under the e-book tab.

SUSTAINABLE DEVELOPMENT AND ENVIRONMENT MANAGEMENT

INNOVATIONS, SCIENCES AND TECHNOLOGIES

ABDEEN MUSTAFA OMER

New York

We have partnered with Copyright Clearance Center to make it easy for you to obtain permissions to reuse content from this publication. Simply navigate to this publication's page on Nova's website and locate the "Get Permission" button below the title description. This button is linked directly to the title's permission page on copyright.com. Alternatively, you can visit copyright.com and search by title, ISBN, or ISSN.

For further questions about using the service on copyright.com, please contact:
Copyright Clearance Center
Phone: +1-(978) 750-8400   Fax: +1-(978) 750-4470   E-mail: info@copyright.com.

NOTICE TO THE READER

The Publisher has taken reasonable care in the preparation of this book, but makes no expressed or implied warranty of any kind and assumes no responsibility for any errors or omissions. No liability is assumed for incidental or consequential damages in connection with or arising out of information contained in this book. The Publisher shall not be liable for any special, consequential, or exemplary damages resulting, in whole or in part, from the readers' use of, or reliance upon, this material. Any parts of this book based on government reports are so indicated and copyright is claimed for those parts to the extent applicable to compilations of such works.

Independent verification should be sought for any data, advice or recommendations contained in this book. In addition, no responsibility is assumed by the publisher for any injury and/or damage to persons or property arising from any methods, products, instructions, ideas or otherwise contained in this publication.

This publication is designed to provide accurate and authoritative information with regard to the subject matter covered herein. It is sold with the clear understanding that the Publisher is not engaged in rendering legal or any other professional services. If legal or any other expert assistance is required, the services of a competent person should be sought. FROM A DECLARATION OF PARTICIPANTS JOINTLY ADOPTED BY A COMMITTEE OF THE AMERICAN BAR ASSOCIATION AND A COMMITTEE OF PUBLISHERS.

Additional color graphics may be available in the e-book version of this book.

Library of Congress Cataloging-in-Publication Data

Sustainable development and environment management : innovations, sciences and technologies / editors, Abdeen Mustafa Omer (Associated Researcher, Energy Research Institute (ERI), Nottingham, UK).
    pages cm. -- (Environmental remediation technologies, regulations and safety)
  Includes index.
  ISBN 978-1-63463-973-6 (hardcover)
1. Environmental protection. 2. Sustainable engineering. 3. Sustainable development. 4. Environmental management. I. Omer, Abdeen Mustafa.
  TD170.3.S87 2014
  628--dc23
                                2015010483

Published by Nova Science Publishers, Inc. † New York

CONTENTS

PREFACE

The impacts of the current global financial crisis on sustainable development (SD) have been subject to polarised debates. Interestingly, a common denominator of the debates is that there are various lessons for the global community to learn from the environmental impacts of the crisis, be those impacts positive or negative. While there is a tendency across the board to focus on the economic impacts of the crisis, such as longer unemployment lines, declining individual and corporate assets, mounting toxic loans and silent construction cranes, the view in this address is that the crisis offers a rate but slim window for the world, especially environmental advocates, activists and scholars or analysts, to reflect on, rethink and recalibrate the discourse and strategy of how to ensure and maintain a balanced relationship between growth and society's natural resource base.

New and previously unacknowledged challenges of sustainability, base of pyramid engagement and the management of risk in globalised world, redefine strategic performance for organisations. Based on original research involving senior leadership from a number of countries and organisations 'leading and managing in contemporary uncertainty using a set of three practices taken together and working in harmony'. The first is the practice of "sharing fates and independence", the second is one of "exploring deeper meaning" and the third is the emerging practice of "Zeitgeist" (i.e., integrating cognition, conscience and collective sprit).

In recent decades, the rise of globalisation has increased the prospects of sustaining development through the use of modern information and communication technology (ICT). Most developing countries are still lacking to knowledge and information for increasing productivity and alleviating poverty. The ICT could be used a powerful tool to sustain development by enhancing allocation of resources and management of the environment. To this end, building ICT capacity allows developing countries to absorb, apply and create knowledge and disseminate information. Sustainable development is about learning how to bridge the gap between present and future generations.

During the last decades, energy and energy prices have occupied a substantial space in the literature on development studies and international relations. In modern time, no society can survive without energy use. Currently, conventional energy sources, mainly oil and gas, account for almost 85% of total energy consumption worldwide. These resources are non-renewable and unevenly distributed across regions. Recent increase in oil prices, driven by rapid economic growth, has impacted economic growth in many nations causing unemployment, inflation and global recession. In particular, higher petroleum prices could have negative consequences to development programmes in many developing countries.

Sustainable development is continues process of productivity growth driven by energy production and consumption.

Health research is essential for improvement not only in health but also in social and economic development. During the present era, many challenges face human health (i.e., EBOLA). Among these challenges, communicable, non-communicable diseases and injuries have the highest impact. Making success in these challenges is supported by many needs which should be targeted by all health researchers to ensure a common policy. A triad of success should link policy making with health research and health services.

The aim of this study is to highlight the importance of indigenous knowledge for fostering economic growth and sustainable development. Recent literature on development studies considers knowledge as a key resource that can alleviate poverty, promote innovations, enhance competitiveness and create wealth. Development is a complex process of multidimensional factors involving both local and external forces. Knowledge for development must be appropriate in relation to a country's environmental, social, cultural, spiritual and economic landscape.

BIOGRAPHY

Abdeen Mustafa Omer (BSc, MSc, PhD) is an Associate Researcher at Energy Research Institute (ERI). He obtained both his PhD degree in the Built Environment and Master of Philosophy degree in Renewable Energy Technologies from the University of Nottingham. He is qualified Mechanical Engineer with a proven track record within the water industry and renewable energy technologies. He has been graduated from University of El Menoufia, Egypt, BSc in Mechanical Engineering and Diploma/MSc from University of Khartoum, Sudan. His previous experience involved being a member of the research team at the National Council for Research/Energy Research Institute in Sudan and working director of research and development for National Water Equipment Manufacturing Co. Ltd., Sudan. He has been listed in the book WHO'S WHO in the World 2005, 2006, 2007 and 2010. He has published over 300 papers in peer-reviewed journals, 100 review articles, 5 books and 100 chapters in books.

Contact: Energy Research Institute (ERI), Forest Road West,
Nottingham, United Kingdom
Corresponding author: Email: abdeenomer2@yahoo.co.uk.

ACKNOWLEDGMENTS

It is a pleasure to acknowledge, with gratitude, all those who, at different times and in different ways, have supported Sustainable Development: Process, Challenges and Prospects. This book would not have been possible without the contributions of several people. To all of them, I wish to express my gratitude.

First and foremost, I wish to thank the president, NOVA Science Publishers, Inc., Nadya S. Gotsiridze-Columbus. I also wish to thank Richard Schortemeyer III, Tricia Worthington, Susan Boriotti and Stella Rosa.

Thanks to my wife Kawthar Abdelhai Ali for her warmth and love. Her unwavering faith in me, her intelligence, humour, spontaneity, curiosity and wisdom added to this book and to my life.

I also owe my thanks to the soul of my late parents: Mother Nour El Sham, and Father Mustafa.

And, before all, I would like to thank ALLAH, the Almighty, for making it possible for me to do this book.

KEY TERMS AND DEFINITIONS

Renewable Energy:

Renewable energy is energy generated from natural resources such as sunlight, wind, rain, tides, and geothermal heat, which are renewable (naturally replenished). Energy obtained from sources that are essentially inexhaustible (unlike, for example the fossil fuels, of which there is a finite supply). Energy sources that are, within a short time frame relative to the Earth's natural cycles, sustainable, and include non-carbon technologies such as solar energy, hydropower, and wind, as well as carbon-neutral technologies.

Solar Energy:

Energy from the sun is converted into thermal or electrical energy; "the amount of energy falling on the earth is given by the solar constant, but very little use has been made of solar energy". Energy derived ultimately from the sun. It can be divided into direct and indirect categories. Most energy sources on Earth are forms of indirect solar energy, although we usually do not think of them in that way. Solar energy uses semiconductor material to convert sunlight into electric currents. Although solar energy only provides 0.15% of the world's power and less than 1% of USA energy, experts believe that sunlight has the potential to supply 5,000 times, as much energy as the world currently consumes.

Biomass Energy:

The energy embodied in organic matter ("biomass") that is released when chemical bonds are broken by microbial digestion, combustion, or decomposition. Biofuels are a wide range of fuels, which are in some way derived from biomass. The term covers solid biomass, liquid fuels and various biogases. Biofuels are gaining increased public and scientific attention, driven by factors such as oil price spikes and the need for increased energy security.

Wind Energy:

Kinetic energy present in wind motion that can be converted to mechanical energy for driving pumps, mills, and electric power generators. Wind power is the conversion of wind energy into a useful form of energy, such as using wind turbines to make electricity, wind mills for mechanical power, wind pumps for pumping water or drainage, or sails to propel ships. The conventional and modern designs of wind towers can successfully be used in the hot arid regions to maintain thermal comfort (with or without the use of ceiling fans) during all hours of the cooling season, or a fraction of it.

Hydropower:

Hydropower, hydraulic power or waterpower is power that is derived from the force or energy of moving water, which may be harnessed for useful purposes. Hydropower is using water to power machinery or make electricity. Water constantly moves through a vast global cycle, evaporating from lakes and oceans, forming clouds, precipitating as rain or snow, and then flowing back down to the ocean.

Geothermal Energy:

Geothermal power (from the Greek roots geo, meaning earth, and thermos, meaning heat) is power extracted from heat stored in the earth. This geothermal energy originates from the original formation of the planet, from radioactive decay of minerals, and from solar energy absorbed at the surface. Heat transferred from the earth's molten core to under-ground deposits of dry steam (steam with no water droplets), wet steam (a mixture of steam and water droplets), hot water, or rocks lying fairly close to the earth's surface.

Resource Management:

Efficient incident management requires a system for identifying available resources at all jurisdictional levels to enable timely and unimpeded access to resources needed to prepare for, respond to, or recover from an incident. Resource management is the efficient and effective deployment for an organisation's resources when they are needed. Such resources may include financial resources, inventory, human skills, production resources, or information technology (IT).

Sustainable Development:

The development, which seeks to produce sustainable economic growth, while ensuring future generations' ability to do the same by not exceeding the regenerative capacity of the nature. In other words, it's trying to protect the environment. A process of change in which the resources consumed (both social and ecological) are not depleted to the extent that they cannot be replicated. Environmentally friendly forms of economic growth activities (agriculture, logging, manufacturing, etc.) that allow the continued production of a commodity without damage to the ecosystem (soil, water supplies, biodiversity or other surrounding resources).

Environment:

The natural environment, commonly referred to simply as the environment, encompasses all living and non-living things occurring naturally on Earth or some region thereof. The biophysical environment is the symbiosis between the physical environment and the biological life forms within the environment, and includes all variables that comprise the Earth's biosphere.

Greenhouse Gases:

Greenhouse gases are gases in an atmosphere that absorb and emit radiation within the thermal infrared range. This process is the fundamental cause of the greenhouse effect. The main greenhouse gases in the Earth's atmosphere are water vapour, carbon dioxide, methane, nitrous oxide, and ozone. Changes in the concentration of certain greenhouse gases, due to human activity such as fossil fuel burning, increase the risk of global climate change.

Working fluids:

Working fluids for absorption cycles fall into four categories, each requiring a different approach to cycle modelling and thermodynamic analysis. Liquid absorbents can be non-volatile (i.e., vapour phase is always pure refrigerant, neglecting condensables) or volatile (i.e., vapour concentration varies, so cycle and component modelling must track both vapour and liquid concentration). Solid sorbents can be grouped by whether they are physisorbents (also known as absorbents), for which, as for liquid absorbents, sorbent temperature depends on both pressure and refrigerant loading (bivariance) or chemisorbents, for which sorbent temperature does not vary with loading at least over small ranges. Beyond these distinctions, various other characteristics are either necessary or desirable for suitable liquid absorbent/refrigerant pairs as follows:

Affinity:

The absorbent should have a strong affinity for the refrigerant under conditions in which absorption takes place. Strong affinity allows less absorbent to be circulated for the same refrigeration effect, reducing sensible heat losses, and allows a smaller liquid heat exchanger to transfer heat from the absorbent to the pressurised refrigerant/absorption solution. On the other hand, as affinity increases, extra heat is required in the generators to separate refrigerant from the absorbent and the COP suffers.

Absence of solid phase (solubility field):

The refrigerant/absorbent pair should not solidify over the expected range of composition and temperature. If a solid forms, it will stop flow and shut down equipment. Controls must prevent operation beyond the acceptable solubility range.

Relative volatility:

The refrigerant should be much more volatile than the absorbent so the two can be separated easily. Otherwise, cost and heat requirements may be excessive. Many absorbents are effectively non-volatile.

Pressure:

Operating pressures established by the refrigerant's thermodynamic properties should be moderate. High pressure requires heavy-walled equipment and significant electrical power may be needed to pump fluids from the low-pressure side to the high-pressure side. Vacuum requires large-volume equipment and special means of reducing pressure drop in the refrigerant vapour paths.

Corrosion:

Most absorption fluids corrode materials used in construction. Therefore, corrosion inhibitors are used.

Stability:

High chemical stability is required because fluids are subjects to severe conditions over many years of services. Instability can cause undesirable formation of gases, solids or corrosive substances. Purity of all components charged into the system is critical for high performance and corrosion prevention.

Safety:

Precautions as dictated by code are followed when fluids are toxic, inflammable or at high pressure. Codes vary according to country and region.

Transport properties:

Viscosity, surface tension, thermal diffusivity and mass diffusivity are important characteristics of the refrigerant/absorbent pair. For example, low viscosity promotes heat and mass transfer and reduces pumping power.

Latent heat:

The refrigerant latent heat should be high so the circulation rate of the refrigerant and absorbent can be minimised.

Eco-friendly Natural Refrigerants:

Over the years, all parts of a commercial refrigerator, such as the compressor, heat exchangers, refrigerant, and packaging, have been improved considerably due to the extensive research and development efforts carried out by academia and industry. However, the achieved and anticipated improvement in conventional refrigeration technology are incremental since this technology is already nearing its fundamentals limit of energy efficiency is described is 'magnetic refrigeration' which is an evolving cooling technology. The word 'green' designates more than a colour. It is a way of life, one that is becoming more and more common throughout the world. An interesting topic on 'sustainable technologies for a greener world' details about what each technology is and how it achieves green goals. Recently, conventional chillers using absorption technology consume energy for hot water generator but absorption chillers carry no energy saving. With the aim of providing a single point solution for this dual purpose application, a product is launched but can provide simultaneous chilling and heating using its vapour absorption technology with 40% saving in heating energy. Using energy efficiency and managing customer energy use has become an integral and valuable exercise. The reason for this is green technology helps to sustain life on earth. This not only applies to humans but to plants, animals and the rest of the ecosystem. Energy prices and consumption will always be on an upward trajectory. In fact, energy costs have steadily risen over last decade and are expected to carry on doing so as consumption grows.

Ozone Depletion Potential (ODP):

The ozone layer is damaged by the catalytic action of chlorine, fluorine and bromine in compounds, which reduce ozone to oxygen and thus destroy the ozone layer. The ozone depletion potential (ODP) of a compound is shown as chlorine equivalent (ODP of a chlorine molecule = 1).

Global Warming Potential (GWP):

The greenhouse effect arises from the capacity of materials in the atmosphere to reflect the heat emitted by the earth back onto the earth. The direct global warming potential (GWP) of a compound is shown as a CO_2 equivalent (GWP of a CO_2 molecule = 1)

Refrigeration:

The achievement of a temperature below that of the immediate surroundings.

Latent heat of fusion:

The quantity of heat (Btu/Ib) required changing 1 Ib of material from the solid phase into the liquid phase.

Sensible heat:

Heat that is absorbed/rejected by a material, resulting in a change of temperature.

Latent heat:

Heat that is absorbed/rejected by a material resulting in a change of physical state (occurring at constant temperature).

Saturation temperature:

That temperature at which a liquid starts to boil (or vapour starts to condense). The saturation temperature (boiling temperature) is constant at a given pressure (except for zoetrope refrigerant) and increase as the pressure increases. A liquid cannot be raised above its saturation temperature. Whenever the refrigerant is present in two states (liquid and vapour) the refrigerant mixture will be at the saturation temperature.

Subcooling:

At a given pressure, the difference between a liquid's temperature and its saturation temperature.

Ton of refrigeration:

The amount of cooling required to change (freeze) 1 ton of water at $32^{o}F$ into ice at $32^{o}F$, in a 24 hour period.

Refrigerant circulation rate (RCR):

The amount of refrigerant in Ib/min, which must circulate in the system to meet the demands of the load.

Sustainable Development:

Sustainable development is a road-map, an action plan, for achieving sustainability in any activity that uses resources and where immediate and intergenerational replication is demanded. Sustainable development, although a widely used phrase and idea, has many different meanings and therefore provokes many different responses. In broad terms, the concept of sustainable development is an attempt to combine growing concerns about a range of environmental issues with socio-economic issues. To aid understanding of these different policies this study presents a classification and mapping of different trends of thought on sustainable development, their political and policy frameworks and their attitudes towards change and means of change. Sustainable development has the potential to address fundamental challenges for humanity, now and into the future. However, to do this, it needs more clarity of meaning, concentrating on sustainable livelihoods and well-being rather than well-having and long term environmental sustainability, which requires a strong basis in principles that link the social and environmental to human equity.

Education for Sustainable Development:

Sowing the seeds of tomorrow is important so that future generations can inherit a more sustainable world. Through sustainable development we meet the needs of the present without compromising the ability of future generations to meet their own needs. This vision of development embraces environmental concerns as well as issues such as the fight against poverty, gender equality, human rights, cultural diversity, and education for all. Education is the means through which sustainable development can be achieved. It enables people to develop the knowledge, values and skills to participate in decisions about the way we do things, individually and collectively, locally and globally, that will improve the quality of life now without damaging the planet of the future. Education for Sustainable Development (ESD) is not a separate subject – it is a holistic educational approach. Education for Sustainable Development allows every human being to acquire the knowledge, skills, attitudes and values necessary to shape a sustainable future. Education for Sustainable Development means including key sustainable development issues into teaching and learning; for example, climate change, disaster risk reduction, biodiversity, poverty reduction, and sustainable consumption. It also requires participatory teaching and learning methods that motivate and empower learners to change their behaviour and take action for sustainable development. Education for Sustainable Development consequently promotes competencies like critical thinking, imagining future scenarios and making decisions in a collaborative way. Education for Sustainable Development requires far-reaching changes in the way education is often practised today.

Sustainable Energy:

Energy is central to sustainable development and poverty reduction efforts. It affects all aspects of development -- social, economic, and environmental -- including livelihoods, access to water, agricultural productivity, health, population levels, education, and gender-related issues. None of the Millennium Development Goals (MDGs) can be met without major improvement in the quality and quantity of energy services in developing countries. Access to sustainable sources of clean, reliable and affordable energy has a profound impact on multiple aspects of human development; it relates not only to physical infrastructure (e.g., electricity grids), but also to energy affordability, reliability and commercial viability. In practical terms, this means delivering energy services to households and businesses that are in line with consumers' ability to pay.

Biomass Energy and Sustainability:

Producing heat, power and liquid transport fuels from biomass instead of fossil fuels has the potential to offer a wide range of environmental and socioeconomic benefits. However, these benefits can only be realised if biomass feedback is sourced responsibly and takes into account impacts on life cycle carbon emissions, land use change, soil, water and air quality and living conditions of those involved in the supply chain. A considerable body of work existing on the development, implementation and effectiveness of sustainability and certification scheme exists.

The development of the land and greenhouse gas (GHG) criteria comes from the requirements imposed by the European Community via the Renewable Energy Directive (RED). The RED sets out the sustainability criteria a bioliquid must meet in order to receive support under national incentive schemes. Whilst the RED does not mandate sustainability criteria for solid and gaseous biomass, the government has developed a reporting mechanism for these fuel states based on a recommendation paper published by the European Commission in February 2010.

Advanced technologies such as gasifier/gas turbine systems for electric power generation and fuel cells for transportation make it possible for biomass to provide a substantial share of world energy in the decades ahead, at competitive costs. While biomass energy industries are being launched today using biomass residues of agricultural and forest product industries, the largest potential supplies of biomass will come from plantations dedicated to biomass energy crops. In industrialised countries these plantations will be established primarily on surplus agricultural lands, providing a new source of livelihood for farmers and making it possible eventually to phase out agricultural subsidies. The most promising sites for biomass plantations in developing countries are degraded lands that can be revegetated. For developing countries, biomass energy offers an opportunity to promote rural development. Biomass energy grown sustainably and used to displace fossil fuels can lead to major reductions in carbon dioxide emissions at zero incremental cost, as well as greatly reduced local air pollution through the use of advanced energy conversion and end-use technologies. The growing of biomass energy crops can be either detrimental or beneficial to the environment, depending on how it is done.

Biomass energy systems offer much more flexibility to design plantations that are compatible with environmental goals than is possible with the growing of biomass for food and industrial fiber markets. There is time to develop and put into place environmental guidelines to ensure that the growing of biomass is carried out in environmentally desirable ways, before a biomass energy industry becomes well established.

Solar Energy for Sustainable Development:

1. Originally developed for energy requirement for orbiting earth satellite – Solar Power – have expanded in recent years for our domestic and industrial needs. Solar power is produced by collecting sunlight and converting it into electricity. This is done by using solar panels, which are large flat panels made up of many individual solar cells. It is most often used in remote locations, although it is becoming more popular in urban areas as well.

There is, indeed, enormous amount of advantages lies with use of solar power specially, in the context of environmental impact and self-reliance. However, a few disadvantages such as its initial cost and the effects of weather conditions, make us hesitant to proceed with full vigor. We discuss below the advantages and disadvantages of Solar Power:

2. Advantages of Solar power:

a)   The major advantage of solar power is that no pollution is created in the process of generating electricity. Environmentally it the most Clean and Green energy. Solar Energy is clean, renewable (unlike gas, oil and coal) and sustainable, helping to protect our environment.

b)   Solar energy does not require any fuel.

c) It does not pollute our air by releasing carbon dioxide, nitrogen oxide, sulfur dioxide or mercury into the atmosphere like many traditional forms of electrical generation does.

d) Therefore Solar Energy does not contribute to global warming, acid rain or smog. It actively contributes to the decrease of harmful green house gas emissions.

e) There is no on-going cost for the power it generates – as solar radiation is free everywhere. Once installed, there are no recurring costs.

f) It can be flexibly applied to a variety of stationary or portable applications. Unlike most forms of electrical generation, the panels can be made small enough to fit pocket-size electronic devices, or sufficiently large to charge an automobile battery or supply electricity to entire buildings.

g) It offers much more self-reliance than depending upon a power utility for all electricity.

h) It is quite economical in long run. After the initial investment has been recovered, the energy from the sun is practically free. Solar Energy systems are virtually maintenance free and will last for decades.

i) It is not affected by the supply and demand of fuel and is therefore not subjected to the ever-increasing price of fossil fuel.

j) By not using any fuel, Solar Energy does not contribute to the cost and problems of the recovery and transportation of fuel or the storage of radioactive waste.

k) It is generated where it is needed. Therefore, large scale transmission cost is minimised.

l) Solar Energy can be utilised to offset utility-supplied energy consumption. It does not only reduce your electricity bill, but will also continue to supply your home/ business with electricity in the event of a power outage.

m) A Solar Energy system can operate entirely independently, not requiring a connection to a power or gas grid at all. Systems can therefore be installed in remote locations, making it more practical and cost-effective than the supply of utility electricity to a new site.

n) The use of solar energy indirectly reduces health costs.

o) They operate silently, have no moving parts, do not release offensive smells and do not require you to add any fuel.

p) More solar panels can easily be added in the future when your family's needs grow.

q) Solar Energy supports local job and wealth creation, fuelling local economies.

Although sustainable development is defined in multiple ways, the most often cited definition of the term comes from the Bruntland Report titled, "Our Common Future." According to the report, sustainable development is "development that meets the needs of the present without compromising the ability of future generations to meet their own needs."

From this particular definition, sustainable development can be reduced to two key concepts: needs and limitations. Needs refers to those in need—the world's poor. The limitations are those "imposed by the state of technology and social organisation on the environment's ability to meet present and future needs".

The following are five examples of sustainable development that meet both those needs and limitations.

1. Solar Energy

The greatest advantages of solar energy are that it is completely free and is available in a limitless supply. Both of these factors provide a huge benefit to consumers and help reduce pollution. Replacing non-renewable energy with this type of energy is both environmentally and financially effective.

2. Wind Energy

Wind energy is another readily available energy source. Harnessing the power of wind energy necessitates the use of windmills; however, due to construction cost and finding a suitable location, this kind of energy is meant to service more than just the individual. Wind energy can supplement or even replace the cost of grid power, and therefore may be a good investment and remains a great example of sustainable development.

3. Crop Rotation

The online dictionary defines crop rotation as "the successive planting of different crops on the same land to improve soil fertility and help control insects and diseases". This farming practice is beneficial in several ways; most notably because it is chemical-free. Crop rotation has been proven to maximize the growth potential of land, while also preventing disease and insects in the soil. Not only can this form of development benefit commercial farmers, but it can also aid those who garden at home.

4. Efficient Water Fixtures

Replacing current construction practices and supporting the installation of efficient shower heads, toilets and other water appliances can conserve one of earth's most precious resources: water. Examples of efficient fixtures include products from the EPA's Water Sense programme, as well as dual-flush and composting toilets. According to the EPA, it takes a lot of energy to produce and transport water and to process waste water, and since less than one percent of the Earth's available water supply is fresh water, it is important that sustainable water use is employed at the individual and societal level.

5. Green Space

Green spaces include parks and other areas where plants and wildlife are encouraged to thrive. These spaces also offer the public great opportunities to enjoy outdoor recreation, especially in dense, urban areas. According to the UW-Madison Department of Urban and Regional Planning, advantages of green spaces include, "helping regulate air quality and climate … reducing energy consumption by countering the warming effects of paved surfaces … recharging groundwater supplies and protecting lakes and streams from polluted runoff." Research conducted in the U.K. by the University Of Exeter Medical School also found that moving to a greener area could lead to significant and lasting improvements to an individual's mental health.

Here are many different origins and definitions of the term sustainable development but in 1987 the World Commission on Environment and Development's report called the Bruntland Report is by far the best and is now one of the most widely recognized definitions:

"Sustainable development is development that meets the needs of the present without compromising the ability of future generations to meet their own needs. It contains within it two key concepts:

- The concept of 'needs', in particular the essential needs of the world's poor, to which overriding priority should be given; and
- The idea of limitations imposed by the state of technology and social organization on the environment's ability to meet present and future needs".

So this all sounds great. Nice and scientific but what does this mean and what needs for our current and future life conditions are needed in order to accomplish this? Let's think out loud for a moment shall we? We need clean air to breathe and for plant life to exist. We need transportation as well. For the most part these "needs" will conflict and this is a decision that we have to make. Now take your conflicting needs and multiply them by your town/city, state, country, and world! How about a specific example? Japan's (and most countries) need for energy relies on using nuclear power yet there is a risk to the people's safety (and other countries) of those countries.

How do we as a society decide whose needs are met first? By economic status? Citizens or immigrants? People living in urban or in the rural areas? People first world countries over third world countries? You or your neighbor? The environment or the corporation? This generation or the next generation? When there has to be a trade off, whose needs should go first?

These are very complex questions to answer and really they are just the tip of the iceberg when it comes to the questions that we need

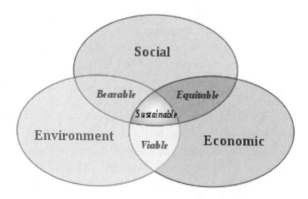

Scheme of sustainable development: at the confluence of three constituent parts. (2006) – en.wikipedia.org/wiki/Sustainable development.

To ask ourselves about sustainable development, people concerned about sustainable development suggest that meeting the needs of the future depends on how well we balance social, economic, and environmental objectives–or needs–when making decisions today.

It is amazing to see how so many things conflict with each other in the short term yet in the long term it works out for the best. For example, third world growth might conflict with preserving natural resources. Yet, in the long run, the responsible use of these natural resources will help ensure that there are resources available for sustained growth of these third world countries into the future.

So if you look at the diagram above you can see that it raises a number of difficult questions. For example, can the long term economic objective of sustained agricultural growth be met if the ecological objective of preserving biodiversity is not? What happens to the environment in the long term if a large number of people cannot afford to meet their basic household needs today? If you did not have access to safe water, and therefore needed wood to boil drinking water so that you and your children would not get sick, would you worry about causing deforestation? Or, if you had to drive a long distance to get to work each day, would you be willing to move or get a new job to avoid polluting the air with your car exhaust? If we do not balance our social, economic, and environmental objectives in the short term, how can we expect to sustain our development in the long term?

INTRODUCTION

This is an introduction to present sustainable development: processes, challenges and prospects. We are living in an era of rapid technological changes where innovation has increased exponentially. Accordingly companies with cutting edge research are facing lots of challenges in creating new and creative products to enable a sustainable human existence and to ensure their acceptance by the society especially when we heat of reports like tomorrow's market signaling that the current trends are leading to an unsustainable human society. The relevance and importance of the study is discussed in the communication, which also highlights the objectives of the study, and the scope of the book.

1.1. BACKGROUND

The move towards a low-carbon world, driven partly by climate science and partly by the business opportunities it offers, will need the promotion of environmentally friendly alternatives if an acceptable stabilisation level of atmospheric carbon dioxide is to be achieved. This requires the harnessing and use of natural resources that produce no air pollution or greenhouse gases (GHGs) and provides comfortable coexistence of humans, livestock , and plants.

Today, the challenge before many cities is to support large numbers of people while limiting their impact on the natural environment. Buildings are significant users of energy and materials in a modern society and, hence, energy conservation in buildings plays an important role in urban environmental sustainability. A challenging task of architects and other building professionals, therefore, is to design and promote low energy buildings in a cost effective and environmentally responsive way. Passive and low energy architecture has been proposed and investigated in different locations of the world; design guides and handbooks were produced for promoting energy efficient buildings. However, at present, little information is available for studying low energy building design in densely populated areas.

Designing low energy buildings in high-density areas requires special treatment to the planning of urban structure, co-ordination of energy systems, integration of architectural elements, and utilisation of space. At the same time, the study of low energy buildings will lead to a better understanding of the environmental conditions and improved design practices. This may help people study and improve the quality of the built environment and living conditions.

1.2. OBJECTIVE OF THE STUDY

The key factors to reducing and controlling CO2 , which is the major contributor to global warming , are the use of alternative approaches to energy generation and the exploration of how these alternatives are used today and may be used in the future as green energy sources. Even with modest assumptions about the availability of land, comprehensive fuel-wood farming programmes offer significant energy, economic and environmental benefits. These benefits would be dispersed in rural areas where they are greatly needed and can serve as linkages for further rural economic development. There is strong scientific evidence that the average temperature of the earth's surface is rising. This was a result of the increased concentration of carbon dioxide (CO2), and other greenhouse gas es (GHGs) in the atmosphere as released by burning fossil fuels. This global warming will eventually lead to substantial changes in the world's climate , which will, in turn, have a major impact on human life and the environment . Energy use reductions can be achieved by minimising the energy demand, by rational energy use, by recovering heat and the use of more green energies. This study was a step towards achieving this goal. The adoption of green or sustainable approaches to the way in which society is run is seen as an important strategy in finding a solution to the energy problem. The main purpose of the session is to:

1) Bring together scientists and technologists from the globe to present, discuss and further develop views on various fields of renewable energy technologies.
2) Identify the most feasible and cost -effective applications of both technologies.
3) Review the conversion and efficient utilisation methods of traditional energy sources and discuss implications of energy efficiency on the development of renewable energy technologies.
4) Ensure that renewable energy takes its proper place in the sustainable supply and use of energy, taking due account of research requirements, energy efficiency, conversion and cost criteria for renewable energy technologies.

1.3. SCOPE OF THE BOOK

Chapter 1 - this chapter presents a comprehensive review of energy sources, and the development of sustainable technologies to explore these energy sources. The move towards a de-carbonised world, driven partly by climate science and partly by the business opportunities it offers, will need the promotion of environmentally friendly alternatives, if an acceptable stabilisation level of atmospheric carbon dioxide is to be achieved. This requires the harnessing and use of natural resources that produce no air pollution or greenhouse gases and provides comfortable coexistence of human, livestock, and plants. This communication also includes potential renewable energy technologies, efficient energy systems, energy savings techniques and other mitigation measures necessary to reduce climate changes. The approach concludes with the technical status of the ground source heat pumps (GSHP) technologies.

Chapter 2 discusses and reviews some interactions between buildings and environment. The correct assessment of climate helps to create buildings, which are successful in their external environment, while knowledge of sick buildings helps to avoid unsuccessful internal environments. Geothermal energy is the natural heat that exists within the earth and that can be absorbed by fluids occurring within, or introduced into, the crystal rocks. Although, geographically, this energy has local concentrations, its distribution globally is widespread. The amount of heat that is, theoretically, available between the earth's surface and a depth of 5 km is around 140×10^{24} joules. Of this, only a fraction (5×10^{21} joules) can be regarded as having economic prospects within the next five decades, and only about 500×10^{18} joules is likely to be exploited by the year 2020. Three main techniques used to exploit the heat available are: geothermal aquifers, hot dry rocks and ground source heat pumps (GSHPs). The GSHPs play a key role in geothermal development in Central and Northern Europe. With borehole heat exchangers as heat source, they offer de-central geothermal heating at virtually any location, with great flexibility to meet given demands. In the vast majority of systems, no space cooling is included, leaving the GSHPs with some economic constraints. Nevertheless, a promising market development first occurred in Switzerland and Sweden, and now also is obvious in Austria and Germany. Approximately 20 years of R&D focusing on borehole heat exchangers resulted in a well-established concept of sustainability for this technology, as well as in sound design and installation criteria. The market success brought Switzerland to the third rank worldwide in geothermal direct use. The future prospects are good, with an increasing range of applications including large systems with thermal energy storage for heating and cooling, The GSHPs in densely populated development areas, borehole heat exchangers for cooling of telecommunication equipment, etc. The sections on energy conservation and green buildings suggest how the correct design and use of buildings can help to improve total environment.

Chapter 3 gives a comprehensive review of biomass energy sources, environment and sustainable development. This includes all the biomass energy technologies, energy efficiency systems, energy conservation scenarios, energy savings and other mitigation measures necessary to reduce emissions globally. The current literature is reviewed regarding the ecological, social, cultural and economic impacts of biomass technology. This study gives an overview of present and future use of biomass as an industrial feedstock for production of fuels, chemicals and other materials. However, to be truly competitive in an open market situation, higher value products are required. Results suggest that biomass technology must be encouraged, promoted, invested, implemented, and demonstrated, but especially in remote rural areas.

Chapter 4 discusses the overall problem and identifies possible solutions. For the thirty-nine million, who live in Sudan, environmental pollution is a major concern; therefore industry, communities, local authorities and central government, to deal with pollution issues, should adopt an integrated approach. Most polluters pay little or no attention to the control and proper management of polluting effluents. This may be due to a lack of enforceable legislation and/or the fear of spending money on the treatment of their effluent prior to discharge. Furthermore, the imposed fines are generally low and therefore do not deter potential offenders. The present problems that are related to water and sanitation in Sudan are many and varied, and the disparity between water supply and demand is growing with time due to the rapid population growth and aridity. The situation of the sewerage system in the cities is extremely critical, and there are no sewerage systems in the rural areas. There is an urgent need for substantial improvements and extensions to the sewerage systems treatment plants. The further development of water resources for agriculture and domestic use is one of the priorities to improve the agricultural yield of the country, and the domestic and industrial demands for water.

Chapter 5 this chapter summaries the renewable energy technologies, energy efficiency systems, energy conservation scenarios, energy savings in greenhouses environment and other mitigation measures necessary to reduce climate change. People are relying upon oil for primary energy and this for a few more decades. Other conventional sources may be more enduring, but are not without serious disadvantages. The renewable energy resources are particularly suited for the provision of rural power supplies and a major advantage is that equipment such as flat plate solar driers, wind machines, etc., can be constructed using local resources and without the advantage results from the feasibility of local maintenance and the general encouragement such local manufacture gives to the build up of small-scale rural based industry. This chapter comprises a comprehensive review of energy sources, the environment and sustainable development. This study gives some examples of small-scale energy converters, nevertheless it should be noted that small conventional, i.e., engines are currently the major source of power in rural areas and will continue to be so for a long time to come. There is a need for some further development to suit local conditions, to minimise spares holdings, to maximise interchangeability both of engine parts and of the engine application. Emphasis should be placed on full local manufacture. It is concluded that renewable environmentally friendly energy must be encouraged, promoted, implemented and demonstrated by full-scale plant (device) especially for use in remote rural areas.

Chapter 6, this study reveals the low incentives, poor working conditions, job dissatisfaction and lack of professional development programmes as main reasons for the immigration to the private-sector. Worldwide there are different systems for providing pharmacy services. Most countries have some element of state assistance, either for all patients or selected groups such as children, and some private provisions. Medicines are financed either through cost sharing or full private. The role of the private services is therefore much more significant. Nationally, there is a mismatch between the numbers of pharmacists and where are they worked, and the demand for pharmacy services. The position is exacerbated locally where in some areas of poor; there is a real need for pharmacy services, which is not being met and where pharmacists have little spare capacity. Various changes within the health-care system require serious attention be given to the pharmacy human resources need. In order to stem the brain drain of pharmacists, it is, however, necessary to have accurate information regarding the reasons that make the pharmacists emigrate to the private sector. Such knowledge is an essential in making of informed decisions regarding the retention of qualified, skilled pharmacists in the public sector for long time. There are currently 3000 pharmacists registered with the Sudan Medical Council of whom only 10% are working with the government. The pharmacist: population ratio indicates there is one pharmacist for every 11,433 inhabitants in Sudan, compared to the World Health Organisation (WHO) average for industrialised countries of one pharmacist for 2,300 inhabitants. The situation is particularly problematic in the Southern states where there is no pharmacist at all. The distribution of pharmacists indicates the majority are concentrated in Khartoum state. When population figures are taken into consideration all states except Khartoum and Gezira states are under served compared to the WHO average. This mal-distribution requires serious action as majority of the population is served in the public sector. The objective of this communication is to highlight and provide an overview of the reasons that lead to the immigration of the public sector pharmacists to the private-sector in Sudan.

The survey has been carried out in September 2004. Data gathered by the questionnaires were analysed using Statistical Package for Social Sciences (SPSS) version 12.0 for windows. The result have been evaluated and tabulated in this study. The data presented in this theme can be considered as nucleus information for executing research and development for pharmacists and pharmacy. More measures must be introduced to attract pharmacists into the public sector. The emerging crisis in pharmacy human resources requires significant additional effort to gather knowledge and dependable data that can inform reasonable, effective, and coordinated responses from government, industry, and professional associations.

Chapter 7 this chapter comprises and gives some examples of small-scale energy converters, nevertheless it should be noted that small conventional, i.e., engines are currently the major source of power in rural areas and will continue to be so for a long time to come. People are relying upon oil for primary energy and this for a few more decades. Other conventional sources may be more enduring, but are not without serious disadvantages. The renewable energy resources are particularly suited for the provision of rural power supplies and a major advantage is that equipment such as flat plate solar driers, wind machines, etc., can be constructed using local resources and without the advantage results from the feasibility of local maintenance and the general encouragement such local manufacture gives to the build up of small-scale rural based industry. This chapter comprises a comprehensive review of energy sources, the environment and sustainable development. It includes the renewable energy technologies, energy efficiency systems, energy conservation scenarios, energy savings in greenhouses environment and other mitigation measures necessary to reduce climate change. There is a need for some further development to suit local conditions, to minimise spares holdings, to maximise interchangeability both of engine parts and of the engine application. Emphasis should be placed on full local manufacture. It is concluded that renewable environmentally friendly energy must be encouraged, promoted, implemented and demonstrated by full-scale plant (device) especially for use in remote rural areas.

Chapter 8 presents and reviews the development of scientific and technological research highlights the efforts made sporadic in this area. A booming economy, high population, land-locked locations, vast area, remote separated and poorly accessible rural areas, large reserves of oil, excellent sunshine, large mining sector and cattle farming on a large-scale, are factors which are most influential to the total water scene in Africa. In this study I try to put some solutions and recommendations that help the advancement of scientific research to resolve issues in the Sudanese society. Despite the obstacles, the movement of scientific research did not stop completely, because a number of researchers still believe in the inevitability of continued scientific research to benefit the maximum of what is available (and the efforts of individual) to attain the objectives of development, prosperity and keep pace with scientific development. Research has become a pedestal to build a modern state in today's world, and became the backbone for all plans developed nations and even developing countries. And enter the world in the era of World Trade Organisation (WTO) and intellectual property Guanyin and the demands of globalisation for the next century that followed, we are in need to change. We should employ scientific research to address the backlog of cases over the years such as the issue of poverty and human capacity development and exploitation of natural resources of the country and the fight against desertification and settling of scientific technologies for the stability of the pastoral communities, and others.

Chapter 9 presents and discusses the potential for such integrated systems in the stationary and portable power market in response to the critical need for a cleaner energy technology. The massive increases in fuel prices over the last years have however, made any scheme not requiring fuel appear to be more attractive and to be worth reinvestigation. In considering the atmosphere and the oceans as energy sources the four main contenders are wind power, wave power, tidal and power from ocean thermal gradients. The renewable energy resources are particularly suited for the provision of rural power supplies and a major advantage is that equipment such as flat plate solar driers, wind machines, etc., can be constructed using local resources and without the advantage results from the feasibility of local maintenance and the general encouragement such local manufacture gives to the build up of small-scale rural based industry. The key factors to reducing and controlling CO_2, which is the major contributor to global warming, are the use of alternative approaches to energy generation and the exploration of how these alternatives are used today and may be used in the future as green energy sources. Even with modest assumptions about the availability of land, comprehensive fuel-wood farming programmes offer significant energy, economic and environmental benefits. These benefits would be dispersed in rural areas where they are greatly needed and can serve as linkages for further rural economic development. Self-renewing resources such as wind, sun, plants and heat from the earth can provide clean abundant energy through the development of renewable technologies. Virtually all regions of the world have renewable resources of one type or another. Research and development investments in the past 25 years in renewable technologies development has lead to important advances in performance and resulting cost effectiveness. Renewable resources currently account for about 9%-10% of the energy consumed in the world; most of this is from hydropower and traditional biomass sources. Wind, solar, biomass and geothermal technologies are cost effective today in an increasing number of markets and are making important steps to broader commercialisation. The present situation is best characterised as one of very rapid growth for wind and solar technologies and of significant promise for biomass and geothermal technologies. Each of the renewable energy technologies is in a different stage of research, development and commercialisation and all have differences in current and future expected costs, current industrial base, resource availability and potential impact on energy supply. Anticipated patterns of future energy use and consequent environmental impacts (acid precipitation, ozone depletion and the greenhouse effect or global warming) are comprehensively discussed in this paper. Throughout the theme several issues relating to renewable energies, environment and sustainable development are examined from both current and future perspectives.

Chapter 10 this study gives a comprehensive review of energy sources, the environment and sustainable development. It reviews the renewable energy technologies, energy efficiency systems, energy conservation scenarios, energy savings in greenhouses environment and other mitigation measures necessary to reduce climate change. People will have to rely upon mineral oil for primary energy and this will go on for a few more decades. Other conventional sources of energy may be more enduring, but are not without serious disadvantages. The renewable energy resources are particularly suited for the provision of rural energy supplies. A major advantage of using the renewable energy sources is that equipment such as flat plate solar driers, wind machines, etc., can be constructed using local resources and with the advantage of local maintenance which can encourage local manufacturing that can give a boost to the building of small-scale rural based industries. This communication gives some examples of small-scale energy converters, nevertheless it should be noted that small conventional, i.e., engines are currently the major source of power in rural areas and will continue to be so for a long time to come. There is a need for some further development to suit local conditions, to minimise spares holdings, to maximise interchangeability both of engine parts and of the engine application. Emphasis should be placed on full local manufacturing of some of the energy systems. It is concluded that renewable environmentally friendly energy must be encouraged, promoted, implemented and demonstrated by full-scale plant (device) especially for use in remote rural areas of many developing nations.

The conclusions drawn, lessons learnt and recommendations for future work are summarised in a closing chapter.

Chapter 1

SUSTAINABLE ENERGY TECHNOLOGIES: CLEANER AND GREENER ENVIRONMENT

ABSTRACT

The move towards a de-carbonised world, driven partly by climate science and partly by the business opportunities it offers, will need the promotion of environmentally friendly alternatives, if an acceptable stabilisation level of atmospheric carbon dioxide is to be achieved. This requires the harnessing and use of natural resources that produce no air pollution or greenhouse gases and provides comfortable coexistence of human, livestock, and plants. This chapter presents a comprehensive review of energy sources, and the development of sustainable technologies to explore these energy sources. It also includes potential renewable energy technologies, efficient energy systems, energy savings techniques and other mitigation measures necessary to reduce climate changes. The approach concludes with the technical status of the ground source heat pumps (GSHP) technologies.

Keywords: renewable energy resources, technologies, sustainable development

1. INTRODUCTION

Over millions of years ago, plants have covered the earth converting the energy of sunlight into living plants and animals, some of which was buried in the depths of the earth to produce deposits of coal, oil and natural gas (Lin and Chang, 2013; Glaas and Juhola, 2013; Gerald, 2012). The past few decades, however, have experienced many valuable uses for these complex chemical substances and manufacturing from them plastics, textiles, fertiliser and the various end products of the petrochemical industry. Indeed, each decade sees increasing uses for these products. Coal, oil and gas, which will certainly be of great value to future generations, as they are to ours, are however non-renewable natural resources. The rapid depletion of these non-renewable fossil resources need not continue. This is particularly true now as it is, or soon will be, technically and economically feasible to supply all of man's needs from the most abundant energy source of all, the sun. The sunlight is not only inexhaustible, but, moreover, it is the only energy source, which is completely non-polluting (Bendewald and Zhai, 2013).

Industry's use of fossil fuels has been largely blamed for warming the climate. When coal, gas and oil are burnt, they release harmful gases, which trap heat in the atmosphere and cause global warming. However, there had been an ongoing debate on this subject, as scientists have struggled to distinguish between changes, which are human induced, and those, which could be put down to natural climate variability. Notably, human activities that emit carbon dioxide (CO_2), the most significant contributor to potential climate change, occur primarily from fossil fuel production. Consequently, efforts to control CO_2 emissions could have serious, negative consequences for economic growth, employment, investment, trade and the standard of living of individuals everywhere.

2. ENERGY SOURCES AND USE

Scientifically, it is difficult to predict the relationship between global temperature and greenhouse gas (GHG) concentrations. The climate system contains many processes that will change if warming occurs. Critical processes include heat transfer by winds and tides, the hydrological cycle involving evaporation, precipitation, runoff and groundwater and the formation of clouds, snow, and ice, all of which display enormous natural variability. The equipment and infrastructure for energy supply and use are designed with long lifetimes, and the premature turnover of capital stock involves significant costs. Economic benefits occur if capital stock is replaced with more efficient equipment in step with its normal replacement cycle. Likewise, if opportunities to reduce future emissions are taken in a timely manner, they should be less costly. Such a flexible approach would allow society to take account of evolving scientific and technological knowledge, while gaining experience in designing policies to address climate change (Bendewald and Zhai, 2013).

The World Summit on Sustainable Development in Johannesburg in 2002 (Bendewald and Zhai, 2013) committed itself to ''encourage and promote the development of renewable energy sources to accelerate the shift towards sustainable consumption and production''. Accordingly, it aimed at breaking the link between resource use and productivity. This can be achieved by the following:

- Trying to ensure economic growth does not cause environmental pollution.
- Improving resource efficiency.
- Examining the whole life-cycle of a product.
- Enabling consumers to receive more information on products and services.
- Examining how taxes, voluntary agreements, subsidies, regulation and information campaigns, can best stimulate innovation and investment to provide cleaner technology.

The energy conservation scenarios include rational use of energy policies in all economy sectors and the use of combined heat and power systems, which are able to add to energy savings from the autonomous power plants. Electricity from renewable energy sources is by definition the environmental green product. Hence, a renewable energy certificate system, as recommended by the World Summit, is an essential basis for all policy systems, independent of the renewable energy support scheme. It is, therefore, important that all parties involved support the renewable energy certificate system in place if it is to work as planned. Moreover, existing renewable energy technologies (RETs) could play a significant mitigating role, but the economic and political climate will have to change first. It is now universally accepted that climate change is real. It is happening now, and GHGs produced by human activities are significantly contributing to it. The predicted global temperature increase of between 1.5 and 4.5°C could lead to potentially catastrophic environmental impacts (Morrow, 2012). These include sea level rise, increased frequency of extreme weather events, floods, droughts, disease migration from various places and possible stalling of the Gulf Stream. This has led scientists to argue that climate change issues are not ones that politicians can afford to ignore, and policy makers tend to agree (Morrow, 2012). However, reaching international agreements on climate change policies is no trivial task as the difficulty in ratifying the Kyoto Protocol and reaching agreement at Copenhagen have proved.

Therefore, the use of renewable energy sources and the rational use of energy, in general, are the fundamental inputs for any responsible energy policy. However, the energy sector is encountering difficulties because increased production and consumption levels entail higher levels of pollution and eventually climate change, with possibly disastrous consequences. At the same time, it is important to secure energy at an acceptable cost in order to avoid negative impacts on economic growth. To date, renewable energy contributes only as much as 20% of the global energy supplies worldwide (Morrow, 2012). Over two thirds of this comes from biomass use, mostly in developing countries, and some of this is unsustainable. However, the potential for energy from sustainable technologies is huge. On the technological side, renewables have an obvious role to play. In general, there is no problem in terms of the technical potential of renewables to deliver energy. Moreover, there are very good opportunities for the RETs to play an important role in reducing emissions of the GHGs into the atmosphere, certainly far more than have been exploited so far. However, there are still some technical issues to address in order to cope with the intermittency of some renewables, particularly wind and solar. Nevertheless, the biggest problem with relying on renewables to deliver the necessary cuts in the GHG emissions is more to do with politics and policy issues than with technical ones (Cantrell and Wepfer, 1984). For example, the single most important step governments could take to promote and increase the use of renewables is to improve access for renewables to the energy market. This access to the market needs to be under favourable conditions and, possibly, under favourable economic rates as well. One move that could help, or at least justify, better market access would be to acknowledge that there are environmental costs associated with other energy supply options and that these costs are not currently internalised within the market price of electricity or fuels. This could make a significant difference, particularly if appropriate subsidies were applied to renewable energy in recognition of the environmental benefits it offers. Similarly, cutting energy consumption through end-use efficiency is absolutely essential. This suggests that issues of end-use consumption of energy will have to come into the discussion in the foreseeable future (ASHRAE, 2005).

However, the RETs have the benefit of being environmentally benign when developed in a sensitive and appropriate way with the full involvement of local communities. In addition, they are diverse, secure, locally based and abundant. In spite of the enormous potential and the multiple benefits, the contribution from renewable energy still lags behind the ambitious claims for it due to the initially high development costs, concerns about local impacts, lack of research funding and poor institutional and economic arrangements (ASHRAE, 2005). Hence, an approach is needed to integrate renewable energies in a way that meets the rising demand in a cost-effective way.

2.1. Role of Energy Efficiency System

The prospects for development in power engineering are, at present, closely related to ecological problems. Power engineering has harmful effects on the environment, as it discharges toxic gases into atmosphere and also oil-contaminated and saline waters into rivers, as well as polluting the soil with ash and slag and having adverse effects on living things on account of electromagnetic fields and so on. Thus there is an urgent need for new approaches to provide an ecologically safe strategy. Substantial economic and ecological effects for thermal power projects (TPPs) can be achieved by improvement, upgrading the efficiency of the existing equipment, reduction of electricity loss, saving of fuel, and optimisation of its operating conditions and service life leading to improved access for rural and urban low-income areas in developing countries through energy efficiency and renewable energies.

Sustainable energy is a prerequisite for development. Energy-based living standards in developing countries, however, are clearly below standards in developed countries. Low levels of access to affordable and environmentally sound energy in both rural and urban low-income areas are therefore a predominant issue in developing countries. In recent years many programmes for development aid or technical assistance have been focusing on improving access to sustainable energy, many of them with impressive results. Apart from success stories, however, experience also shows that positive appraisals of many projects evaporate after completion and vanishing of the implementation expert team. Altogether, the diffusion of sustainable technologies such as energy efficiency and renewable energy for cooking, heating, lighting, electrical appliances and building insulation in developing countries has been slow. Energy efficiency and renewable energy programmes could be more sustainable and pilot studies more effective and pulse releasing if the entire policy and implementation process was considered and redesigned from the outset (Kavanaugh and Rafferty, 1997).

New financing and implementation processes, which allow reallocating financial resources and thus enabling countries themselves to achieve a sustainable energy infrastructure, are also needed. The links between the energy policy framework, financing and implementation of renewable energy and energy efficiency projects have to be strengthened and as well as efforts made to increase people's knowledge through training.

2.2. Energy Use in Buildings

Buildings consume energy mainly for cooling, heating and lighting. The energy consumption was based on the assumption that the building operates within ASHRAE-thermal comfort zone during the cooling and heating periods (UN, 2003; UNFCCC, 2009). Most of the buildings incorporate energy efficient passive cooling, solar control, photovoltaic, lighting and day lighting, and integrated energy systems. It is well known that thermal mass with night ventilation can reduce the maximum indoor temperature in buildings in summer (Rees, 1999). Hence, comfort temperatures may be achieved by proper application of passive cooling systems. However, energy can also be saved if an air conditioning unit is used (Bos et al., 1994). The reason for this is that in summer, heavy external walls delay the heat transfer from the outside into the inside spaces. Moreover, if the building has a lot of internal mass the increase in the air temperature is slow. This is because the penetrating heat raises the air temperature as well as the temperature of the heavy thermal mass. The result is a slow heating of the building in summer as the maximal inside temperature is reached only during the late hours when the outside air temperature is already low. The heat flowing from the inside heavy walls could be reduced with good ventilation in the evening and night. The capacity to store energy also helps in winter, since energy can be stored in walls from one sunny winter day to the next cloudy one. However, the admission of daylight into buildings alone does not guarantee that the design will be energy efficient in terms of lighting. In fact, the design for increased daylight can often raise concerns relating to visual comfort (glare) and thermal comfort (increased solar gain in the summer and heat losses in the winter from larger apertures).

Such issues will clearly need to be addressed in the design of the window openings, blinds, shading devices, heating system, etc. In order for a building to benefit from daylight energy terms, it is a prerequisite that lights are switched off when sufficient daylight is available. The nature of the switching regime; manual or automated, centralised or local, switched, stepped or dimmed, will determine the energy performance. Simple techniques can be implemented to increase the probability that lights are switched off (Duchin, 1995). These include:

- Making switches conspicuous and switching banks of lights independently.
- Loading switches appropriately in relation to the lights.
- Switching banks of lights parallel to the main window wall.

There are also a number of methods, which help reduce the lighting energy use, which, in turn, relate to the type of occupancy pattern of the building (Givoni, 1998). The light switching options include:

- Centralised timed off (or stepped)/manual on.
- Photoelectric off (or stepped)/manual on.
- Photoelectric and on (or stepped), and photoelectric dimming.
- Occupant sensor (stepped) on/off (movement or noise sensor).

Likewise, energy savings from the avoidance of air conditioning can be very substantial. Whilst day-lighting strategies need to be integrated with artificial lighting systems in order to become beneficial in terms of energy use, reductions in overall energy consumption levels by employment of a sustained programme of energy consumption strategies and measures would have considerable benefits within the buildings sector. The perception is often given however is that rigorous energy conservation as an end in itself imposes a style on building design resulting in a restricted aesthetic solution.

It would perhaps be better to support a climate sensitive design approach that encompasses some elements of the pure conservation strategy together with strategies, which work with the local ambient conditions making use of energy technology systems, such as solar energy, where feasible. In practice, low energy environments are achieved through a combination of measures that include:

- Application of environmental regulations and policy.
- Application of environmental science and best practice.
- Mathematical modelling and simulation.
- Environmental design and engineering.
- Construction and commissioning.
- Management and modifications of environments in use.

While the overriding intention of passive solar energy design of buildings is to achieve a reduction in purchased energy consumption, the attainment of significant savings is in doubt. The non-realisation of potential energy benefits is mainly due to the neglect of the consideration of post-occupancy user and management behaviour by energy scientists and designers alike. Calculating energy inputs in agricultural production is more difficult in comparison to the industry sector due to the high number of factors affecting agricultural production, as it is shown in Table 1. However, considerable studies have been conducted in different countries on energy use in agriculture (ASHRAE, 2003; Kammerud et al., 1984; Shaviv, 1989; Singh, 2000) in order to quantify the influence of these factors.

Table 1. Energy equivalent of inputs and outputs

Nr.	Energy source	Unit	Equivalent energy (MJ)
	Inputs		
1.	Human labour	h	2.3
2.	Animal labour		
	Horse	h	10.10
	Mule	h	4.04
	Donkey	h	4.04
	Cattle	h	5.05
	Water buffalo	h	7.58
3.	Electricity	kWh	11.93
4.	Diesel	Litre	56.31
5.	Chemicals fertilisers		
	Nitrogen	kg	64.4
	P_2O_5	kg	11.96
	K_2O	kg	6.7
6.	Seeds		
	Cereals and pulses	kg	25
	Oil seed	kg	3.6
	Tuber	kg	14.7
	Total input	**kg**	**227.71**
	Outputs		
7	Major products		
	Cereal and pulses	kg	14.7
	Sugar beet	kg	5.04
	Tobacco	kg	0.8
	Cotton	kg	11.8

Nr.	Energy source	Unit	Equivalent energy (MJ)
	Oil seed	kg	25
	Fruits	kg	1.9
	Vegetables	kg	0.8
	Water melon	kg	1.9
	Onion	kg	1.6
	Potatoes	kg	3.6
	Olive	kg	11.8
	Tea	kg	0.8
8.	By products		
	Husk	kg	13.8
	Straw	kg	12.5
	Cob	kg	18.0
	Seed cotton	kg	25.0
	Total output	**kg**	**149.04**

2.3. Renewable Energy Technologies

Sustainable energy is the energy that, in its production or consumption, has minimal negative impacts on human health and the healthy functioning of vital ecological systems, including the global environment (CAEEDAC, 2000; Yaldiz et al., 1993). It is an accepted fact that renewable energy is a sustainable form of energy, which has attracted more attention during recent years.

Increasing environmental interest, as well as economic consideration of fossil fuel consumption and high emphasis of sustainable development for the future helped to bring the great potential of renewable energy into focus.

Nearly a fifth of all global power is generated by renewable energy sources, according to a book published by the OECD/IEA (Dutt, 1982). "Renewables for power generation: status and prospects" claims that, at approximately 20%, renewables are the second largest power source after coal (39%) and ahead of nuclear (17%), natural gas (17%) and oil (8%) respectively. From 1973-2000 renewables grew at 9.3% a year and it is predicted that this will increase by 10.4% a year to 2020. Wind power grew fastest at 52% and will multiply seven times by 2010, overtaking bio-power and hence help reducing greenhouse gases (GHGs) emissions to the environment.

Table 2 shows some applications of different renewable energy sources. The challenge is to match leadership in GHG reduction and production of renewable energy with developing a major research and manufacturing capacity in environmental technologies (wind, solar, fuel cells, etc.).

More than 50% of the world's area is classified as arid, representing the rural and desert part, which lack electricity and water networks. The inhabitants of such areas obtain water from borehole wells by means of water pumps, which are mostly driven by diesel engines. The diesel motors are associated with maintenance problems, high running cost, and environmental pollution. Alternative methods are pumping by photovoltaic (PV) or wind systems.

At present, renewable sources of energy are regional and site specific. It has to be integrated in the regional development plans (Baruah, 1995).

Table 2. Sources of renewable energy

Energy source	Technology	Size
Solar energy	Domestic solar water heaters	Small
	Solar water heating for large demands	Medium-large
	PV roofs: grid connected systems generating electric energy	Medium-large
Wind energy	Wind turbines (grid connected)	Medium-large
Hydraulic energy	Hydro plants in derivation schemes	Medium-small
	Hydro plants in existing water distribution networks	Medium-small
Biomass	High efficiency wood boilers	Small
	CHP plants fed by agricultural wastes or energy crops	Medium
Animal manure	CHP plants fed by biogas	Small
CHP	High efficiency lighting	Wide
	High efficiency electric	Wide
	Householders appliances	Wide
	High efficiency boilers	Small-medium
	Plants coupled with refrigerating absorption machines	Medium-large

2.4. Solar Energy

The availability of data on solar radiation is a critical problem. Even in developed countries, very few weather stations have been recording detailed solar radiation data for a period of time long enough to have statistical significance. Solar radiation arriving on earth is the most fundamental renewable energy source in nature. It powers the bio-system, the ocean and atmospheric current system and affects the global climate. Reliable radiation information is needed to provide input data in modelling solar energy devices and a good database is required in the work of energy planners, engineers, and agricultural scientists. In general, it is not easy to design solar energy conversion systems when they have to be installed in remote locations. First, in most cases, solar radiation measurements are not available for these sites. Second, the radiation nature of solar radiation makes the computation of the size of such systems difficult. While solar energy data are recognised as very important, their acquisition is by no means straightforward. The measurement of solar radiation requires the use of costly equipment such as pyrheliometers and pyranometers. Consequently, adequate facilities are often not available in developing countries to mount viable monitoring programmes. This is partly due to the equipment cost as well as the cost of technical manpower. Several attempts have, however, been made to estimate solar radiation through the use of meteorological and other physical parameter in order to avoid the use of expensive network of measuring instruments (Sivkov, 1964).

Two of the most essential natural resources for all life on the earth and for man's survival are sunlight and water. Sunlight is the driving force behind many of the RETs. The worldwide potential for utilising this resource, both directly by means of the solar technologies and indirectly by means of biofuels, wind and hydro technologies, is vast. During the last decade interest has been refocused on renewable energy sources due to the increasing prices and fore-seeable exhaustion of presently used commercial energy sources.

The most promising solar energy technology are related to thermal systems; industrial solar water heaters, solar cookers, solar dryers for peanut crops, solar stills, solar driven cold stores to store fruits and vegetables, solar collectors, solar water desalination, solar ovens, and solar commercial bakers. Solar PV system: solar PV for lighting, solar refrigeration to store vaccines for human and animal use, solar PV for water pumping, solar PV for battery chargers, solar PV for communication network, microwave, receiver stations, radio systems in airports, VHF and beacon radio systems in airports, and educational solar TV posts in villages. Solar pumps are most cost effective for low power requirement (up to 5 kW) in remote places. Applications include domestic and livestock drinking water supplies, for which the demand is constant throughout the year, and irrigation. However, the suitability of solar pumping for irrigation, though possible, is uncertain because the demand may vary greatly with seasons. Solar systems may be able to provide trickle irrigation for fruit farming, but not usually the large volumes of water needed for wheat growing (Thakur and Mistra, 1993; Wu and Boggess, 1999).

The hydraulic energy required to deliver a volume of water is given by the formula (1):

$$E_w = \rho_w \, g \, V \, H \tag{1},$$

where:
  E_w is the required hydraulic energy (kWh day^{-1})
  ρ_w is the water density (kg m^{-3})
  g is the gravitational acceleration (ms^{-2})
  V is the required volume of water (m^3 day^{-1}), and
  H is the head of water (m).

The solar array power required is given by the formula (2):

$$P_{sa} = E_w \, / \, E_{sr} \, \eta \, F \tag{2}$$

where:
  P_{sa} is the solar array power (kW$_p$)
  E_{sr} is the average daily solar radiation (kWhm^{-2} day^{-1})
  F is the array mismatch factor, and
  η is the daily subsystem efficiency.

Substituting Eq. (1) in Eq. (2), the following equation is obtained for the amount of water that can be pumped:

$$V = P_{sa} \, E_{sr} \, \eta \, F / \, \rho_w \, g \, H \tag{3}$$
$$P_{sa} = 1.6 \, kW_p$$
$$F = 0.85$$
$$\eta = 40\%.$$

A further increase of PV depends on the ability to improve the durability, performance and the local manufacturing capabilities of PV.

Table 3. Classifications of data requirements

Criteria	Plant data	System data
Existing data	Size Life Cost (fixed and variation operation and maintenance) Forced outage Maintenance Efficiency Fuel Emissions	Peak load Load shape Capital costs Fuel costs Depreciation Rate of return Taxes
Future data	All of above, plus Capital costs Construction trajectory Date in service	System lead growth Fuel price growth Fuel import limits Inflation

2.5. Biomass Energy

The data required to perform the trade-off analysis simulation of bio-energy resources can be classified according to the divisions given in Table 3, namely the overall system or individual plants, and the existing situation or future development. The effective economical utilisations of these resources are shown in Table 4, but their use is hindered by many problems such as those related to harvesting, collection, and transportation, besides the photo-sanitary control regulations.

Table 4. Effective biomass resource utilisation

Subject	Tools	Constraints
Utilisation and land clearance for agriculture expansion	Stumpage fees Control Extension Conversion Technology	Policy Fuel-wood planning Lack of extension Institutional
Utilisation of agricultural residues	Briquetting Carbonisation Carbonisation and briquetting Fermentation Gasification	Capital Pricing Policy and legislation Social acceptability

Biomass energy is experiencing a surge in interest stemming from a combination of factors, e.g., greater recognition of its current role and future potential contribution as a modern fuel, global environmental benefits, its development and entrepreneurial opportunities, etc. Possible routes of biomass energy development are shown in Table 5.

Table 5. Agricultural residues routes for development

Source	Process	Product	End use
Agricultural residues	Direct	Combustion	Rural poor Urban household Industrial use
	Processing	Briquettes	Industrial use Limited household use
	Processing	Carbonisation (small scale)	Rural household (self-sufficiency)
	Carbonisation	Briquettes Carbonised	Urban fuel Energy services
	Fermentation	Biogas	Household and industry
Agricultural and animal residues	Direct	Combustion	(Save or less efficiency as wood)
	Briquettes	Direct combustion	(Similar end use devices or improved)
	Carbonisation	Carbonised	Use
	Carbonisation	Briquettes	Briquettes use
	Fermentation	Biogas	Use

However, biomass usage and application can generally be divided into the following three categories:

(a) Biomass energy for petroleum substitution, driven by the following factors:
    (1) Oil price increase
    (2) Balance of payment problems, and economic crisis
    (3) Fuel-wood plantations and residue utilisation
    (4) Wood based heat and electricity
    (5) Liquid fuels from biomass
    (6) Producer gas technology
(b) Biomass energy for domestic needs driven by:
    (1) Population increase
    (2) Urbanisation
    (3) Agricultural expansion
    (4) Fuel-wood crisis
    (5) Ecological crisis
    (6) Fuel-wood plantations, agro-forestry
    (7) Community forestry and residue utilisation
    (8) Improved stoves, and improved charcoal production.
(c) Biomass energy for development driven by:
    (1) Electrification
    (2) Irrigation and water supply
    (3) Economic and social development
    (4) Fuel-wood plantations and community forestry

(5) Agro-forestry
(6) Briquettes
(7) Producer gas technology.

The use of biomass through direct combustion has long been, and still is, the most common mode of biomass utilisation (Table 5). Examples for dry (thermo-chemical) conversion processes are charcoal making from wood (slow pyrolysis), gasification of forest and agricultural residues (fast pyrolysis – this is still in demonstration phase), and of course, direct combustion in stoves, furnaces, etc. Wet processes require substantial amount of water to be mixed with the biomass (OECD/IEA, 2004; Duffie and Beckman, 1980). Biomass technologies include:

- Carbonisation and briquetting.
- Improved stoves
- Biogas
- Improved charcoal
- Gasification.

2.5.1. Briquetting and Carbonisation

Briquetting is the formation of a charcoal (an energy-dense solid fuel source) from otherwise wasted agricultural and forestry residues. One of the disadvantages of wood fuel is that it is bulky with a low energy density and therefore requires transport. Briquette formation allows for a more energy-dense fuel to be delivered, thus reducing the transportation cost and making the resource more competitive. It also adds some uniformity, which makes the fuel more compatible with systems that are sensitive to the specific fuel input. Charcoal stoves are very familiar to African societies. As for the stove technology, the present charcoal stove can be used, and can be improved upon for better efficiency. This energy term will be of particular interest to both urban and rural households and all the income groups due to its simplicity, convenience, and lower air polluting characteristics. However, the market price of the fuel together with that of its end-use technology may not enhance its early high market penetration especially in the urban low income and rural households.

Charcoal is produced by slow heating wood (carbonisation) in airtight ovens or retorts, in chambers with various gases, or in kilns supplied with limited and controlled amounts of air. The charcoal yield decreased gradually from 42.6 to 30.7% for the hazelnut shell and from 35.6 to 22.7% for the beech wood with an increase of temperature from 550 to 1.150 K while the charcoal yield from the lignin content decreases sharply from 42.5 to 21.7% until it was at 850 K during the carbonisation procedures (Sivkov, 1964). The charcoal yield decreases as the temperature increases, while the ignition temperature of charcoal increases as the carbonisation temperature increases. The charcoal briquettes that are sold on the commercial market are typically made from a binder and filler.

2.5.2. Improved Cook Stoves

Traditional wood stoves are commonly used in many rural areas. These can be classified into four types: three stone, metal cylindrical shaped, metal tripod and clay type. Indeed, improvements of traditional cookers and ovens to raise the efficiency of fuel saving can secure rural energy availability, where woody fuels have become scarce. However, planting fast growing trees to provide a constant fuel supply should also be considered. The rural development is essential and economically important since it will eventually lead to a better standard of living, people's settlement, and self-sufficiency.

2.5.3. Biogas Technology

Biogas technology cannot only provide fuel, but is also important for comprehensive utilisation of biomass forestry, animal husbandry, fishery, agricultural economy, protecting the environment, realising agricultural recycling as well as improving the sanitary conditions, in rural areas. However, the introduction of biogas technology on a wide scale has implications for macro planning such as the allocation of government investment and effects on the balance of payments. Hence, factors that determine the rate of acceptance of biogas plants, such as credit facilities and technical backup services, are likely to have to be planned as part of general macro-policy, as do the allocation of research and development funds (Barabaro et al., 1978).

2.5.4. Improved Charcoal

Dry cell batteries are a practical but expensive form of mobile fuel that is used by rural people when moving around at night and for powering radios and other small appliances. The high cost of dry cell batteries is financially constraining for rural households, but their popularity gives a good indication of how valuable a versatile fuel like electricity is in rural areas (Table 6). However, dry cell batteries can constitute an environmental hazard unless they are recycled in a proper fashion.

Table 6. Energy carrier and energy services in rural areas

Energy carrier	Energy end-use
Fuel-wood	Cooking Water heating Building materials Animal fodder preparation
Kerosene	Lighting Ignition fires
Dry cell batteries	Lighting Small appliances
Animal power	Transport Land preparation for farming Food preparation (threshing)
Human power	Transport Land preparation for farming Food preparation (threshing)

Tables 6-7 further show that direct burning of fuel-wood and crop residues constitute the main usage of biomass, as is the case with many developing countries. In fact, biomass resources play a significant role in energy supply in all developing countries. However, the direct burning of biomass in an inefficient manner causes economic loss and adversely affects human health. In order to address the problem of inefficiency, research centres around the world (Hall and Scrase, 1998) have investigated the viability of converting the resource to a more useful form of improved charcoal, namely solid briquettes and fuel gas. Accordingly, biomass resources should be divided into residues or dedicated resources, the latter including firewood and charcoal can also be produced from forest residues (Table 7). Whichever form of biomass resource used, its sustainability would primarily depend on improved forest and tree management.

Table 7. Biomass residues and current use

Type of residue	Current use
Wood industry waste	Residues available
Vegetable crop residues	Animal feed
Food processing residue	Energy needs
Sorghum, millet, and wheat residues	Fodder, and building materials
Groundnut shells	Fodder, brick making, and direct fining oil mills
Cotton stalks	Domestic fuel considerable amounts available for short period
Sugar, bagasse, and molasses	Fodder, energy need, and ethanol production (surplus available)
Manure	Fertiliser, brick making, and plastering

2.5.5. Gasification

Gasification is based on the formation of a fuel gas (mostly CO_2 and H_2) by partially oxidising raw solid fuel at high temperatures in the presence of steam or air. The technology can use wood chips, groundnut shells, sugar cane bagasse, and other similar fuels to generate capacities from 3 kW to 100 kW. Many types of gasifier designs have been developed to make use of the diversity of fuel inputs and to meet the requirements of the product gas output (degree of cleanliness, composition, heating value, etc.) (Pernille, 2004).

2.5.6. Biomass Potential and Sustainability

A sustainable energy system includes energy efficiency, energy reliability, energy flexibility, fuel poverty, and environmental impacts. A sustainable biofuel has two favourable properties, which are availability from renewable raw material, and its lower negative environmental impact than that of fossil fuels. Global warming, caused by CO_2 and other substances, has become an international concern in recent years. To protect forestry resources, which act as major absorbers of CO_2, by controlling the ever-increasing deforestation and the increase in the consumption of wood fuels, such as firewood and charcoal, is therefore an urgent issue. Given this, the development of a substitute fuel for charcoal is necessary.

Briquette production technology, a type of clean coal technology, can help prevent flooding and serve as a global warming countermeasure by conserving forestry resources through the provision of a stable supply of briquettes as a substitute for charcoal and firewood.

There are many emerging biomass technologies with large and immediate potential applications, e.g., biomass gasifier/gas turbine (BGST) systems for power generation with pilot plants, improved techniques for biomass harvesting, transportation and storage. Gasification of crop residues such as rice husks, groundnut shells, etc., with plants already operating in China, India, and Thailand. Treatment of cellulosic materials by steam explosion which may be followed by biological or chemical hydrolysis to produce ethanol or other fuels, cogeneration technologies, hydrogen from biomass, striling energies capable of using biomass fuels efficiently, etc. Table 8 gives a view of the use of biomass and its projection worldwide.

Table 8. Final energy projections including biomass (Mtoe)

Region 1995				
	Biomass	Conventional Energy	Total	Share of Biomass (%)
Africa	**205**	**136**	**341**	**60**
China	206	649	855	24
East Asia	106	316	422	25
Latin America	73	342	416	18
South Asia	235	188	423	56
Total developing countries	825	1632	2456	34
Other non-OECD countries	24	1037	1061	1
Total non-OECD countries	849	2669	3518	24
OECD countries	81	3044	3125	3
World	**930**	**5713**	**6643**	**14**
Region 2020				
	Biomass	Conventional Energy	Total	Share of Biomass (%)
Africa	**371**	**266**	**631**	**59**
China	224	1524	1748	13
East Asia	118	813	931	13
Latin America	81	706	787	10
South Asia	276	523	799	35
Total developing countries	1071	3825	4896	22
Other non-OECD countries	26	1669	1695	1
Total non-OECD countries	1097	5494	6591	17
OECD countries	96	3872	3968	2
World	**1193**	**9365**	**10558**	**11**

However, a major gap with biomass energy is that research has usually been aimed at obtaining supply and consumption data, with insufficient attention and resources being allocated to basic research, to production, harvesting and conservation processes. Biomass has not been closely examined in terms of a substitute for fossil fuels compared to carbon sequestration and overall environmental benefits related to these different approaches. To achieve the full potential of biomass as a feedstock for energy, food, or any other use, requires the application of considerable scientific and technological inputs (D'Apote, 1998). However, the aim of any modern biomass energy systems must be:

(1) To maximise yields with minimum inputs.
(2) Utilise and select adequate plant materials and processes.
(3) Optimise use of land, water, and fertiliser.
(4) Create an adequate infrastructure and strong (R & D) base.

An afforestation programme appears an attractive option for any country to pursue in order to reduce the level of atmospheric carbon by enhancing carbon sequestration in the nation's forests, which would consequently mitigate climate change. However, it is acknowledged that certain barriers need to be overcome if the objectives are to be fully achieved. These include the followings.

- Low level of public awareness of the economic/environmental benefits of forestry.
- The generally low levels of individuals' income.
- Pressures from population growth.
- The land tenural system, which makes it difficult (if at all possible) for individuals to own or establish forest plantations.
- Poor pricing of forest products especially in the local market.
- Inadequate financial support on the part of governments.
- Weak institutional capabilities of the various Forestry Departments as regards technical manpower to effectively manage tree plantations.

However, social policy conditions are also critical. This is still very much lacking particularly under developing countries conditions. During the 1970s and 1980s different biomass energy technologies were perceived in sub-Saharan Africa as a panacea for solving acute problems. On the account of these expectations, a wide range of activities and projects were initiated. However, despite considerable financial and human efforts, most of these initiatives have unfortunately been a failure.

Therefore, future research efforts should concentrate on the following areas.

- Directed R and D in the most promising areas of biomass to increase energy supply and to improve the technological base.
- Formulate a policy framework to encourage entrepreneurial and integrated process.
- Pay more attention to sustainable production and use of biomass energy feedstocks, methodology of conservation and efficient energy flows.
- More research aimed at pollution abatement.
- Greater attentions to interrelated socio-economic aspects.
- Support R and D on energy efficiency in production and use.
- Improve energy management skills and take maximum advantage of existing local knowledge.
- Closely examine past successes and failures to assist policy makers with well-informed recommendations.

2.5.7. Combined Heat and Power (CHP)

District Heating (DH), also known as community heating can be a key factor to achieve energy savings, reduce CO_2 emissions and at the same time provide consumers with a high quality heat supply at a competitive price. Generally, DH should only be considered for areas where the heat density is sufficiently high to make DH economical. In countries like Denmark for example, DH may today be economical even to new developments with lower density areas, due to the high level of taxation on oil and gas fuels combined with the efficient production of DH.

Most of the heat used for the DH can be produced by large CHP plants (gas-fired combined cycle plants using natural gas, biomass, waste or biogas) as shown in Table 2. The DH is energy efficient because of the way the heat is produced and the required temperature level is an important factor. Buildings can be heated to a temperature of 21°C and domestic hot water (DHW) can be supplied at a temperature of 55°C using energy sources other than DH that are most efficient when producing low temperature levels (<95°C) for the DH water (David, 2000). Most of these heat sources are CO_2 neutral or emit low levels. However, only a few of these sources are available to small individual systems at a reasonable cost, whereas DH schemes because of the plant's size and location can have access to most of the heat sources and at a low cost. Low temperature DH, with return temperatures of around 30-40°C can utilise the following heat sources:

- Efficient use of the CHP by extracting heat at low calorific value (CV).
- Efficient use of biomass or gas boilers by condensing heat in economisers.
- Efficient utilisation of geothermal energy.
- Direct utilisation of excess low temperature heat from industrial processes.
- Efficient use of large-scale solar heating plants.

Heat tariffs may include a number of components such as a connection charge, a fixed charge and a variable energy charge. Also, consumers may be incentivised to lower the return temperature. Hence, it is difficult to generalise but the heat practice for any DH company, no matter what the ownership structure is, can be highlighted as follows:

- To develop and maintain a development plan for the connection of new consumers.
- To evaluate the options for least cost production of heat.
- To implement the most competitive solutions by signing agreements with other companies or by implementing own investment projects.
- To monitor all internal costs and with the help of benchmarking, and improve the efficiency of the company.
- To maintain a good relationship with the consumer and deliver heat supply services at a sufficient quality.

Also, installing DH should be pursued to meet the objectives for improving the environment through the improvement of energy efficiency in the heating sector. At the same time DH can serve the consumer with a reasonable quality of heat at the lowest possible cost. The variety of possible solutions combined with the collaboration between individual companies, the district heating association, the suppliers and consultants can, as it has been in Denmark, be the way forward for developing DH in the United Kingdom.

3. RESULTS AND DISSCUSIONS

Aims/Purpose: The purpose of this study, however, is to contribute to the reduction of energy consumption in buildings, identify GSHPs as an environmental friendly technology able to provide efficient utilisation of energy in the buildings sector, promote using GSHPs applications as an optimum means of heating and cooling, and to present typical applications and recent advances of DX GSHPs.

Study design: The main concept of this technology is that it utilises the lower temperature of the ground (approximately <32°C), which remains relatively stable throughout the year, to provide space heating, cooling and domestic hot water inside the building area. The main goal of this study is to stimulate the uptake of the GSHPs. Recent attempts to stimulate alternative energy sources for heating and cooling of buildings has emphasised the utilisation of the ambient energy from ground source and other renewable energy sources.

Place and Duration of Study: Energy Research Institute (ERI), between November 2011 and March 2012.

Methodology/Approach: This chapter highlights the potential energy saving that could be achieved through use of ground energy source. It also focuses on the optimisation and improvement of the operation conditions of the heat cycles and performances of the direct expansion (DX) GSHP.

Results/Findings: It is concluded that the direct expansion of the GSHP are extendable to more comprehensive applications combined with the ground heat exchanger in foundation piles and the seasonal thermal energy storage from solar thermal collectors.

Originality/Value: The study highlighted the potential energy saving that could be achieved through the use of ground energy sources. It also focuses on the optimisation and improvement of the operation conditions of the heat cycle and performance of the DX GSHP. It is concluded that the direct expansion of the GSHP, combined with the ground heat exchanger in foundation piles and the seasonal thermal energy storage from solar thermal collectors, is extendable to more comprehensive applications.

3.1. Fuel Cells

Platinum is a catalyst for fuel cells and hydrogen-fuelled cars presently use about two ounces of the metal. There is currently no practicable alternative. Reserves are in South Africa (70%), and Russia (22%). Although there are sufficient accessible reserves in South Africa to increase supply by up to 5% per year for the next 50 years, there are significant environmental impacts associated with its mining and refining, such as groundwater pollution and atmospheric emissions of sulphur dioxide ammonia, chlorine and hydrogen chloride. The carbon cost of platinum use equates to 360 kg for a current fuel cell car, or 36 kg for a future car, with the target platinum loading of 0.2 oz, which is negligible compared to the CO_2 currently emitted by vehicles (IHA, 2003). Furthermore, Platinum is almost completely recyclable. At current prices and loading, platinum would cost 3% of the total cost of a fuel cell engine. Also, the likely resource costs of hydrogen as a transport fuel are apparently cheapest if it is reformed from natural gas with pipeline distribution, with or without carbon sequestration. However, this is not as sustainable as using renewable energy sources. Substituting hydrogen for fossils fuels will have a positive environmental impact in reducing both photochemical smog and climate change. There could also be an adverse impact on the ozone layer but this is likely to be small, though potentially more significant if hydrogen was to be used as aviation fuel.

3.2. Hydrogen Production

Hydrogen is now beginning to be accepted as a useful form for storing energy for reuse on, or for export off, the grid. Clean electrical power harvested from wind and wave power projects can be used to produce hydrogen by electrolysis of water. Electrolysers split water molecules into its constituent parts: hydrogen and oxygen. These are collected as gases; hydrogen at the cathode and oxygen at the anode. The process is quite simple. Direct current is applied to the electrodes to initiate the electrolysis process. Production of hydrogen is an elegant environmental solution. Hydrogen is the most abundant element on the planet, it cannot be destroyed (unlike hydrocarbons) it simply changes state (water to hydrogen and back to water) during consumption. There is no CO or CO_2 generation in its production and consumption and, depending upon methods of consumption, even the production of oxides of nitrogen can be avoided too. However, the transition will be very messy, and will take many technological paths to convert fossil fuels and methanol to hydrogen, building hybrid engines and so on. Nevertheless, the future of hydrogen fuel cells is promising. Hydrogen can be used in internal combustion engines, fuel cells, turbines, cookers gas boilers, road-side emergency lighting, traffic lights or signalling where noise and pollution can be a considerable nuisance, but where traffic and pedestrian safety cannot be compromised.

Hydrogen is already produced in huge volumes and used in a variety of industries. Current worldwide production is around 500 billion Nm^3 per year (EWEA, 2003). Most of the hydrogen produced today is consumed on-site, such as at oil refineries, at a cost of around $0.70/kg and is not sold on the market (Steele, 1997). When hydrogen is sold on the market, the cost of liquefying the hydrogen and transporting it to the user adds considerably to the production cost. The energy required to produce hydrogen via electrolysis (assuming 1.23 V) is about (33 kWh/kg). For 1 mole (2 g) of hydrogen the energy is about (0.066 kWh/mole) (Sitarz, 1992). The achieved efficiencies are over 80% and on this basis electrolytic hydrogen can be regarded as a storable form of electricity. Hydrogen can be stored in a variety of forms:

- Cryogenic; this has the highest gravimetric energy density.
- High-pressure cylinders; pressures of 10 000 psi are quite normal.
- Metal hydride absorbs hydrogen, providing a very low pressure and extremely safe mechanism, but is heavy and more expensive than cylinders.
- Chemical carriers offer an alternative, with anhydrous ammonia offering similar gravimetric and volumetric energy densities to ethanol and methanol.

3.3. Hydropower Generation

Hydropower has a valuable role as a clean and renewable source of energy in meeting a variety of vital human needs. The recognition of the role of hydropower as one of the renewable and clean energy sources and that its potential should be realised in an environmentally sustainable and socially acceptable manner. Water is a basic requirement for survival: for drinking, for food, energy production and for good health. As water is a commodity, which is finite and cannot be created, and in view of the increasing requirements as the world population grows, there is no alternative but to store water for use when it is needed. However, the major challenges are to feed the increasing world population, to improve the standards of living in rural areas and to develop and manage land and water in a sustainable way. Hydropower plants are classified by their rated capacity into one of four regimes: micro (<50kW), mini (50-500 kW), small (500 kW-5 MW), and large (>5 MW)

(John and James, 1989). The total world installed hydro capacity today is around 1000 GW and a lot more are currently planned, principally in developing countries in Asia, Africa and South America as shown in Table 9, which is reproduced from (Okkan, 1993). However, the present production of hydroelectricity is only about 18 per cent of the technically feasible potential (and 32 per cent of the economically feasible potential); there is no doubt that a large amount of hydropower development lays ahead (Okkan, 1993).

Table 9. World hydro potential and development

Continent	Africa	Asia	Australia & Oceania	Europe	North & Central America	South America
Gross theoretical hydropower potential (GWhy^{-1})	4×10^6	19.4×10^6	59.4×10^6	3.2×10^6	6×10^6	6.2×10^6
Technically feasible hydropower potential (GWhy^{-1})	1.75×10^6	6.8×10^6	2×10^6	10^6	1.66×10^6	2.7×10^6
Economically feasible hydropower potential (GWhy^{-1})	1.1×10^5	3.6×10^6	90×10^4	79×10^4	10^6	1.6×10^6
Installed hydro capacity (MW)	21×10^3	24.5×10^4	13.3×10^4	17.7×10^4	15.8×10^4	11.4×10^4
Production by hydro plants in 2002 or average (GWhy^{-1})	83.4×10^3	80×10^4	43×10^3	568×10^3	694×10^3	55×10^4
Hydro capacity under construction (MW)	> 3024	>72.7 $\times 10^3$	>177	>23 $\times 10^2$	58×10^2	>17 $\times 10^3$
Planned hydro capacity (MW)	77.5×10^3	>17.5 $\times 10^4$	>647	>10^3	>15 $\times 10^3$	>59 $\times 10^3$

3.4. Wind Energy

Water is the most natural commodity for the existence of life in the remote desert areas. However, as a condition for settling and growing, the supply of energy is the close second priority. The high cost and the difficulties of mains power line extensions, especially to a low populated region can focus attention on the utilisation of different and more reliable and independent sources of energy like renewable wind energy. Accordingly, the utilisation of wind energy, as a form of energy, is becoming increasingly attractive and is being widely used for the substitution of oil-produced energy, and eventually to minimise atmospheric degradation, particularly in remote areas. Indeed, utilisation of renewables, such as wind energy, has gained considerable momentum since the oil crises of the 1970s. Wind energy, though site-dependent, is non-depleting, non-polluting, and a potential option of the alternative energy source. Wind power could supply 12% of global electricity demand by 2020, according to a report by the European Wind Energy Association and Greenpeace (Njeru, 2013).

Wind energy can and will constitute a significant energy resource when converted into a usable form. As Figure 1 illustrates, information sharing is a four-stage process and effective collaboration must also provide ways in which the other three stages of the 'renewable' cycle:

gather, convert and utilise, can be integrated. Efficiency in the renewable energy sector translates into lower gathering, conversion and utilisation (electricity) costs. A great level of installed capacity has already been achieved.

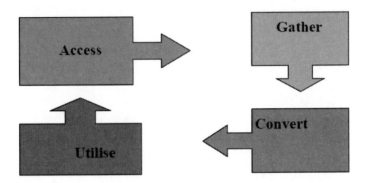

Figure 1. The renewable cycle.

Figure 2 clearly shows that the offshore wind sector is developing fast, and this indicates that wind is becoming a major factor in electricity supply with a range of significant technical, commercial and financial hurdles to be overcome.

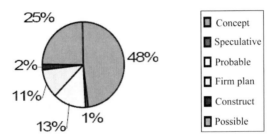

Figure 2. Global prospects of wind energy utilisation by 2003-2010.

The offshore wind industry has the potential for a very bright future and to emerge as a new industrial sector, as Figure 3 implies. The speed of turbine development is such that more powerful models would supersede the original specification turbines in the time from concept to turbine order.

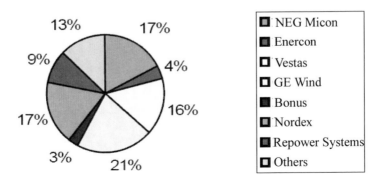

Figure 3. Prospect turbines share for 2003-2010.

Levels of activities are growing at a phenomenal rate (Figure 4), new prospects developing, new players entering, existing players growing in experience; technology evolving and, quite significantly, politics appear to support the sector.

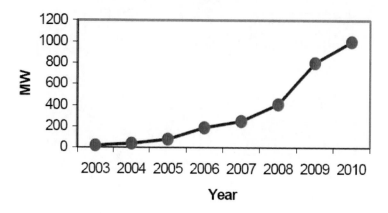

Figure 4. Average wind farm capacity 2003-2010.

3.5. Energy and Sustainable Development

Sustainability is defined as the extent to which progress and development should meet the need of the present without compromising the ability of the future generations to meet their own needs (Odeku et al., 2013). This encompasses a variety of levels and scales ranging from economic development and agriculture, to the management of human settlements and building practices. Tables 10-12 indicate the relationship between energy conservation, sustainable development and environment.

Table 10. Energy and sustainable environment

Technological criteria	Energy and environment criteria	Social and economic criteria
Primary energy saving in regional scale	Sustainability according to greenhouse gas pollutant emissions	Labour impact
Technical maturity, reliability	Sustainable according to other pollutant emissions	Market maturity
Consistence of installation and maintenance requirements with local technical known-how	Land requirement	Compatibility with political, legislative and administrative situation
Continuity and predictability of performance	Sustainability according to other environmental impacts	Cost of saved primary energy

In some countries, a wide range of economic incentives and other measures are already helping to protect the environment. These include: (1) Taxes and user charges that reflect the costs of using the environment, e.g., pollution taxes and waste disposal charges. (2) Subsidies, credits and grants that encourage environmental protection. (3) Deposit-refund systems that prevent pollution on resource misuse and promote product reuse or recycling. (4) Financial enforcement incentives, e.g., fines for non-compliance with environmental regulations. (5) Tradable permits for activities that harm the environment.

Table 11. Classification of key variables defining facility sustainability

Criteria	Intra-system impacts	Extra-system impacts
Stakeholder satisfaction	Standard expectations met Relative importance of standard expectations	Covered by attending to extra-system resource base and ecosystem impacts
Resource base impacts	Change in intra-system resource bases Significance of change	Resource flow into/out of facility system Unit impact exerted by flow on source/sink system Significance of unit impact
Ecosystem impacts	Change in intra-system ecosystems Significance of change	Resource flows into/out of facility system Unit impact exerted by how on source/sink system Significance of unit impact

Table 12. Positive impact of durability, adaptability and energy conservation on economic, social and environment systems

Economic system	Social system	Environmental system
Durability	Preservation of cultural values	Preservation of resources
Meeting changing needs of economic development	Meeting changing needs of individuals and society	Reuse, recycling and preservation of resources
Energy conservation and saving	Savings directed to meet other social needs	Preservation of resources, reduction of pollution and global warming

And, the following action areas for producers were recommended (Abdeen, 2009):

- Management and measurement tools - adopting environmental management systems appropriate for the business.
- Performance assessment tools - making use of benchmarking to identify scope for impact reduction and greater eco-efficiency in all aspects of the business.
- Best practice tools - making use of free help and advice from government best practice programmes (energy efficiency, environmental technology, resource savings).
- Innovation and ecodesign - rethinking the delivery of 'value added' by the business, so that impact reduction and resource efficiency are firmly built in at the design stage.

- Cleaner, leaner production processes - pursuing improvements and savings in waste minimisation, energy and water consumption, transport and distribution, as well as reduced emissions.
- Supply chain management - specifying more demanding standards of sustainability from "upstream" suppliers, while supporting smaller firms to meet those higher standards.
- Product stewardship - taking the broadest view of 'producer responsibility' and working to reduce all the 'downstream' effects of products after they have been sold on to customers.
- Openness and transparency - publicly reporting on environmental performance against meaningful targets; actively using clear labels and declarations so that customers are fully informed; building stakeholder confidence by communicating sustainability aims to the workforce, the shareholders and the local community (Figure 5).

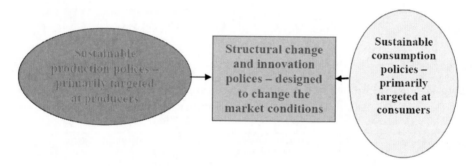

Figure 5. Link between resources and productivity.

The following issues were addressed during the Rio Earth Summit in 1992 (Abdeen, 2009):

- The use of local materials and indigenous building sources.
- Incentive to promote the continuation of traditional techniques, with regional resources and self-help strategies.
- Regulation of energy-efficient design principles.
- International information exchange on all aspects of construction related to the environment, among architects and contractors, particularly non-conventional resources.
- Exploration of methods to encourage and facilitate the recycling and reuse of building materials, especially those requiring intensive energy use during manufacturing and the use of clean technologies.

This is the step in a long journey to encourage progressive economy, which continues to provide people with high living standards, but, at the same time helps reduce pollution, waste mountains, other environmental degradation, and environmental rationale for future policy-making and intervention to improve market mechanisms. This vision will be accomplished by:

- "Decoupling" economic growth and environmental degradation. The basket of indicators illustrated in Table 13 shows the progress being made. Decoupling air and water pollution from growth, making good headway with CO_2 emissions from energy, and transport. The environmental impact of our own individual behaviour is more closely linked to consumption expenditure than the economy as a whole.
- Focusing policy on the most important environmental impacts associated with the use of particular resources, rather than on the total level of all resource use.
- Increasing the productivity of material and energy use that are economically efficient by encouraging patterns of supply and demand, which are more efficient in the use of natural resources. The aim is to promote innovation and competitiveness. Investment in areas like energy efficiency, water efficiency and waste minimisation.
- Encouraging and enabling active and informed individual and corporate consumers.

Table 13. The basket of indicators for sustainable consumption and production

Item	Economy-wide decoupling indicators
1.	Greenhouse gas emissions
2.	Air pollution
3.	Water pollution (river water quality)
4.	Commercial and industrial waste raisings and household waste not cycled
	Resource use indicators
5.	Material use
6.	Water abstraction
7.	Homes built on land not previously developed, and number of households
	Decoupling indicators for specific sectors
8.	Emissions from electricity generation
9.	Motor vehicle kilometers and related emissions
10.	Agricultural output, fertilizer use, methane emissions and farmland bird populations
11.	Manufacturing output, energy consumption and related emissions
12.	Household consumption, expenditure energy, water consumption and waste generated

3.6. Chemicals Compounds

Humans and wildlife are being contaminated by a host of commonly used chemicals in food packaging and furniture, according to the World Wildlife Federation (WWF) and European Union (Abdeen, 2008). Currently, the chemical industry has been under no obligation to make the information public. However, the new proposed rules would change this. Future dangers will only be averted if the effects of chemicals are exposed and then the dangerous ones are never used.

Indeed, chemicals used for jacket waterproofing, food packaging and non-stick coatings have been found in dolphins, whales, cormorants, seals, sea eagles and polar bears from the Mediterranean to the Baltic. The European Commission has adopted an ambitious action plan to improve the development and wider use of environmental technologies such as recycling systems for wastewater in industrial processes, energy-saving car engines and soil remediation techniques, using hydrogen and fuel cells (Abdeen, 2012).

The legislation, which has not been implemented in time, concerns the incineration of waste, air quality limit, values for benzene and carbon monoxide, national emission ceilings for sulphur dioxide, nitrogen oxides, volatile organic compounds and ammonia and large combustion plants.

3.6.1. Wastes

Waste is defined as an unwanted material that is being discarded. Waste includes items being taken for further use, recycling or reclamation. Waste produced at household, commercial and industrial premises are control waste and come under the waste regulations. Waste Incineration Directive (WID) emissions limit values will favour efficient, inherently cleaner technologies that do not rely heavily on abatement. For existing plant, the requirements are likely to lead to improved control of:

- NO_x emissions, by the adoption of infurnace combustion control and abatement techniques.
- Acid gases, by the adoption of abatement techniques and optimisation of their control.
- Particulate control techniques, and their optimisation, e.g., of bag filters and electrostatic precipitators.

The waste and resources action programme has been working hard to reduce demand for virgin aggregates and market uptake of recycled and secondary alternatives. The programme targets are:

- To deliver training and information on the role of recycling and secondary aggregates in sustainable construction for influences in the supply chain, and
- To develop a promotional programme to highlight the new information on websites.

3.6.2. Global Warming

This results in the following requirements:

- Relevant climate variables should be generated (solar radiation: global, diffuse, direct solar direction, temperature, humidity, wind speed and direction) according to the statistics of the real climate.
- The average behaviour should be in accordance with the real climate.
- Extremes should occur in the generated series in the way it will happen in a real warm period. This means that the generated series should be long enough to capture these extremes, and series based on average values from nearby stations.

On some climate change issues (such as global warming), there is no disagreement among the scientists. The greenhouse effect is unquestionably real; it is essential for life on earth. Water vapour is the most important GHG; followed by carbon dioxide (CO_2). Without a natural greenhouse effect, scientists estimate that the earth's average temperature would be $-18°C$ instead of its present $14°C$ (Raphael, 2012).

There is also no scientific debate over the fact that human activity has increased the concentration of the GHGs in the atmosphere (especially CO_2 from combustion of coal, oil and gas). The greenhouse effect is also being amplified by increased concentrations of other gases, such as methane, nitrous oxide, and CFCs as a result of human emissions. Most scientists predict that rising global temperatures will raise the sea level and increase the frequency of intense rain or snowstorms.

Climate change scenarios sources of uncertainty and factors influencing the future climate are:

- The future emission rates of the GHGs (Table 14).
- The effect of this increase in concentration on the energy balance of the atmosphere.
- The effect of these emissions on the GHGs concentrations in the atmosphere, and
- The effect of this change in energy balance on global and regional climate.

It has been known for a long time that urban centres have mean temperatures higher than their less developed surroundings. The urban heat increases the average and peak air temperatures, which in turn affect the demand for heating and cooling.

Higher temperatures can be beneficial in the heating season, lowering fuel use, but they exacerbate the energy demand for cooling in the summer times. Neither heating nor cooling may dominate the fuel use in a building in temperate climates, and the balance of the effect of the heat is less.

Table 14. West European states GHG emissions

Country	1990	1999	Change 1990-99	Reduction target
Austria	76.9	79.2	2.6%	-13%
Belgium	136.7	140.4	2.8%	-7.5%
Denmark	70.0	73.0	4.0%	-21.0%
Finland	77.1	76.2	-1.1%	0.0%
France	545.7	544.5	-0.2%	0.0%
Germany	1206.5	982.4	-18.7%	-21.0%
Greece	105.3	123.2	16.9%	25.0%
Ireland	53.5	65.3	22.1%	13.0%
Italy	518.3	541.1	4.4%	-6.5%
Luxembourg	10.8	6.1	-43.3%	-28.0%
Netherlands	215.8	230.1	6.1%	-6.0%
Portugal	64.6	79.3	22.4%	27.0%
Spain	305.8	380.2	23.2%	15.0%
Sweden	69.5	70.7	1.5%	4.0%
United Kingdom	741.9	637.9	-14.4%	-12.5%
Total EU-15	4199	4030	-4.0%	-8.0%

As the provision of cooling is expensive with higher environmental cost, ways of using innovative alternative systems, like the mop fan will be appreciated. The solar gains would affect energy consumption. Therefore, lower or higher percentages of glazing, or shading devices might affect the balance between annual heating and cooling loads.

In addition to conditioning energy, the fan energy needed to provide mechanical ventilation can make a significant further contribution to energy demand.

Much depends on the efficiency of design, both in relation to the performance of fans themselves and to the resistance to flow arising from the associated ductwork. Figure 6 illustrates the typical fan and thermal conditioning needs for a variety of ventilation rates and climate conditions.

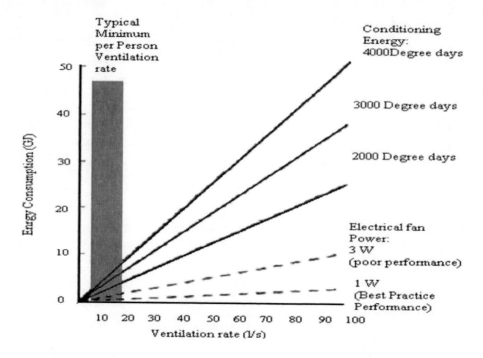

Figure 6. Energy impact of ventilation.

3.7. Ground Source Heat Pumps

The term "ground source heat pump" has become an all-inclusive term to describe a heat pump system that uses the earth, ground water, or surface water as a heat source and/or sink. Some of the most common types of ground source ground-loop heat exchangers configurations are classified in Figure 7.

The GSHP systems consist of three loops or cycles as shown in Figure 8. The first loop is on the load side and is either an air/water loop or a water/water loop, depending on the application. The second loop is the refrigerant loop inside a water source heat pump. Thermodynamically, there is no difference between the well-known vapour-compression refrigeration cycle and the heat pump cycle; both systems absorb heat at a low temperature level and reject it to a higher temperature level. However, the difference between the two systems is that a refrigeration application is only concerned with the low temperature effect produced at the evaporator, while a heat pump may be concerned with both the cooling effect produced at the evaporator and the heating effect produced at the condenser. In these dual-mode GSHP systems, a reversing valve is used to switch between heating and cooling modes by reversing the refrigerant flow direction. The third loop in the system is the ground loop in which water or an antifreeze solution exchanges heat with the refrigerant and the earth.

The GSHPs utilise the thermal energy stored in the earth through either vertical or horizontal closed loop heat exchange systems buried in the ground. Many geological factors impact directly on site characterisation and subsequently the design and cost of the system. The solid geology of the United Kingdom varies significantly. Furthermore there is an extensive and variable rock head cover.

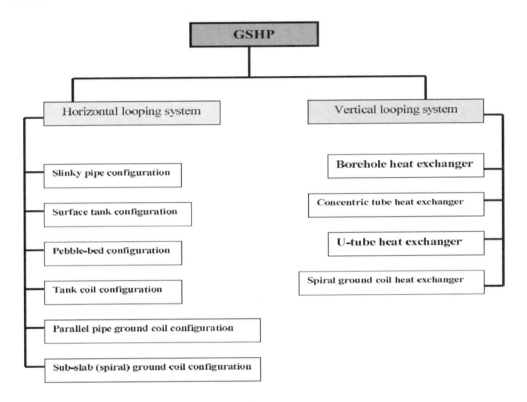

Figure 7. Common types of ground-loop heat exchangers.

The geological prognosis for a site and its anticipated rock properties influence the drilling methods and therefore system costs. Other factors important to system design include predicted subsurface temperatures and the thermal and hydrological properties of strata.

The GSHP technology is well established in Sweden, Germany and North America, but has had minimal impact in the United Kingdom space heating and cooling market. Perceived barriers to uptake include geological uncertainty, concerns regarding performance and reliability, high capital costs and lack of infrastructure. System performance concerns relate mostly to uncertainty in design input parameters, especially the temperature and thermal properties of the source. These in turn can impact on the capital cost, much of which is associated with the installation of the external loop in horizontal trenches or vertical boreholes. The climate in the United Kingdom makes the potential for heating in winter and cooling in summer from a ground source less certain owing to the temperature ranges being narrower than those encountered in continental climates. This project will develop an impartial GSHP function on the site to make available information and data on site-specific temperatures and key geotechnical characteristics. The GSHPs are receiving increasing interest because of their potential to reduce primary energy consumption and thus reduce emissions of greenhouse gases. The technology is well established in the North Americas and parts of Europe, but is at the demonstration stage in the United Kingdom. The information will be delivered from digital geoscience's themes that have been developed from observed data held in corporate records. This data will be available to the GSHP installers and designers to assist the design process, therefore reducing uncertainties. The research will also be used to help inform the public as to the potential benefits of this technology.

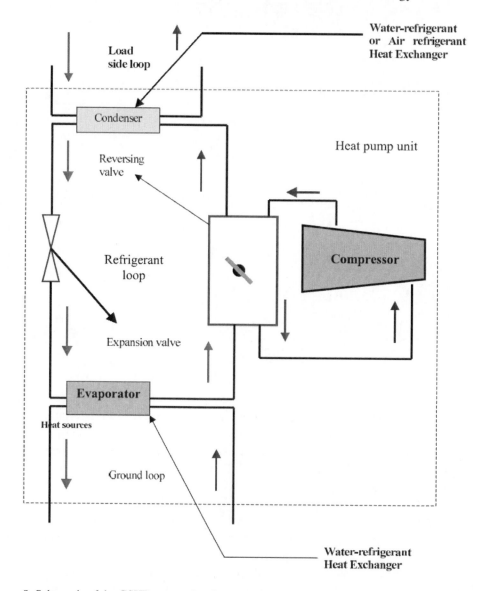

Figure 8. Schematic of the GSHP system (heating mode operation).

The GSHPs play a key role in geothermal development in Central and Northern Europe. With borehole heat exchangers as heat source, they offer de-central geothermal heating with great flexibility to meet given demands at virtually any location. No space cooling is included in the vast majority of systems, leaving ground-source heat pumps with some economic constraints. Nevertheless, a promising market development first occurred in Switzerland and Sweden, and now also in Austria and Germany. Approximately 20 years of research and development (R& D) focusing on borehole heat exchangers resulted in a well-established concept of sustainability for this technology, as well as in sound design and installation criteria. The market success brought Switzerland to the third rank worldwide in geothermal direct use. The future prospects are good, with an increasing range of applications including large systems with thermal energy storage for heating and cooling, ground-source heat pumps in densely populated development areas, borehole heat exchangers for cooling of telecommunication equipment, etc.

Loops can be installed in three ways: horizontally, vertically or in a pond or lake (Fig. 9). The type chosen depends on the available land area, soil and rock type at the installation site. These factors help to determine the most economical choice for installation of the ground loop. The GSHP delivers (3-4 times) as much energy as it consumes when heating, and cools and dehumidifies for a lower cost than conventional air conditioning. It can cut homes or business heating and cooling costs by 50% and provide hot water free or with substantial savings. The GSHPs can reduce the energy required for space heating, cooling and service water heating in commercial/institutional buildings by as much as 50%.

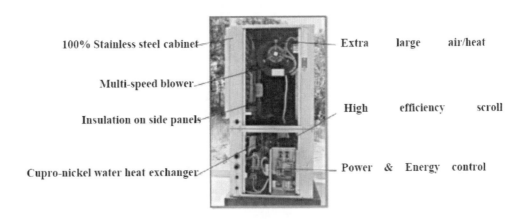

Figure 9. The GSHPs extract solar heat stored in the upper layers of the earth.

Efficiencies of the GSHP systems are much greater than conventional air-source heat pump systems. A higher COP (coefficient of performance) can be achieved by a GSHP because the source/sink earth temperature is relatively constant compared to air temperatures. Additionally, heat is absorbed and rejected through water, which is a more desirable heat transfer medium because of its relatively high heat capacity. The GSHP systems rely on the fact that, under normal geothermal gradients of about $0.5°F/100$ ft ($30°C/km$), the earth temperature is roughly constant in a zone extending from about 20 ft (6.1 m) deep to about 150 ft (45.7 m) deep. This constant temperature interval within the earth is the result of a complex interaction of heat fluxes from above (the sun and the atmosphere) and from below (the earth interior). As a result, the temperature of this interval within the earth is approximately equal to the average annual air temperature (Roriz, 2001). Above this zone (less than about 20 feet (6.1 m) deep), the earth temperature is a damped version of the air temperature at the earth's surface. Below this zone (greater than about 150 ft (45.7 m) deep), the earth temperature begins to rise according to the natural geothermal gradient. The storage concept is based on a modular design that will facilitate active control and optimisation of thermal input/output, and it can be adapted for simultaneous heating and cooling often needed in large service and institutional buildings (Strauss, 2013). Loading of the core is done by diverting warm and cold air from the heat pump through the core during periods with excess capacity compared to the current need of the building (Tukahirwa, 2013; Valkila and Saari, 2012). The cool section of the core can also be loaded directly with air during the night, especially in spring and fall when nights are cold and days may be warm (Vargas-Parra, 2013).

CONCLUSION

There is strong scientific evidence that the average temperature of the earth's surface is rising. This is a result of the increased concentration of carbon dioxide and other GHGs in the atmosphere as released by burning fossil fuels. This global warming will eventually lead to substantial changes in the world's climate, which will, in turn, have a major impact on human life and the built environment. Therefore, effort has to be made to reduce fossil energy use and to promote green energy, particularly in the building sector. Energy use reductions can be achieved by minimising the energy demand, rational energy use, recovering heat and the use of more green energy. This study was a step towards achieving this goal.

The adoption of green or sustainable approaches to the way in which society is run is seen as an important strategy in finding a solution to the energy problem. The key factors to reducing and controlling CO_2, which is the major contributor to global warming, are the use of alternative approaches to energy generation and the exploration of how these alternatives are used today and may be used in the future as green energy sources. Even with modest assumptions about the availability of land, comprehensive fuel-wood farming programmes offer significant energy, economic and environmental benefits. These benefits would be dispersed in rural areas where they are greatly needed and can serve as linkages for further rural economic development.

However, by adopting coherent strategy for alternative clean sustainable energy sources, the world as a whole would benefit from savings in foreign exchange, improved energy security, and socio-economic improvements. With a nine-fold increase in forest – plantation cover, every nation's resource base would be greatly improved while the international community would benefit from pollution reduction, climate mitigation, and the increased trading opportunities that arise from new income sources.

The non-technical issues related to clean energy, which have recently gained attention, include:

(1) Environmental and ecological factors, e.g., carbon sequestration, reforestation and revegetation.
(2) Renewables as a CO_2 neutral replacement for fossil fuels.
(3) Greater recognition of the importance of renewable energy, particularly modern biomass energy carriers, at the policy and planning levels.
(4) Greater recognition of the difficulties of gathering good and reliable renewable energy data, and efforts to improve it.

(5) Studies on the detrimental health efforts of biomass energy particularly from traditional energy users.

The present study is one effort in touching all these aspects.

RECOMMENDATIONS

The following are recommended:

- Launching of public awareness campaigns among local investors particularly small-scale entrepreneurs and end users of RET to highlight the importance and benefits of renewable, particularly solar, wind, and biomass energies.
- Amendment of the encouragement of investment act, to include furthers concessions, facilities, tax holidays, and preferential treatment to attract national and foreign capital investment.
- Allocation of a specific percentage of soft loans and grants obtained by governments to augment budgets of (R&D) related to manufacturing and commercialisation of the RET.
- Governments should give incentives to encourage the household sector to use renewable energy instead of conventional energy. Execute joint investments between the private sector and the financing entities to disseminate the renewable information and literature with technical support from the research and development entities.
- Availing of training opportunities to personnel at different levels in donor countries and other developing countries to make use of their wide experience in application and commercialisation of the RET particularly renewable energy.
- The governments should play a leading role in adopting renewable energy devices in public institutions, e.g., schools, hospitals, government departments, police stations, etc., for lighting, water pumping, water heating, communication and refrigeration.
- Encouraging the private sector to assemble, install, repair and manufacture renewable energy devices via investment encouragement and more flexible licensing procedures.

ARTICLE SUMMARY

The use of renewable energy sources is a fundamental factor for a possible energy policy in the future. Taking into account the sustainable character of the majority of renewable energy technologies, they are able to preserve resources and to provide security, diversity of energy supply and services, virtually without environmental impact. Sustainability has acquired great importance due to the negative impact of various developments on environment. The rapid growth during the last decade has been accompanied by active construction, which in some instances neglected the impact on the environment and human activities. Policies to promote the rational use of electric energy and to preserve natural non-renewable resources are of paramount importance. Low energy design of urban environment and buildings in densely populated areas requires consideration of wide range of factors, including urban setting, transport planning, energy system design and architectural and engineering details. The focus of the world's attention on environmental issues in recent years has stimulated response in many countries, which have led to a closer examination of energy conservation strategies for conventional fossil fuels. One way of reducing building energy consumption is to design buildings, which are more economical in their use of energy for heating, lighting, cooling, ventilation and hot water supply. Passive measures, particularly natural or hybrid ventilation rather than air-conditioning, can dramatically reduce primary energy consumption. However, exploitation of renewable energy in buildings and agricultural greenhouses can, also, significantly contribute towards reducing dependency on fossil fuels. Therefore, promoting innovative renewable applications and reinforcing the renewable energy market will contribute to preservation of the ecosystem by reducing emissions at local and global levels. This will also contribute to the amelioration of environmental conditions by replacing conventional fuels with renewable energies that produce no air pollution or greenhouse gases. This study describes various designs of low energy buildings. It also, outlines the effect of dense urban building nature on energy consumption, and its contribution to climate change. Measures, which would help to save energy in buildings, are also presented.

ACKNOWLEDGMENT

The financial support for this research work is from the Energy Research Institute (ERI). Thanks to my wife Kawthar Abdelhai Ali for her warmth and love. Her unwavering faith in me, her intelligence, humour, spontaneity, curiosity and wisdom added to this article and to my life.

REFERENCES

Abdeen, M. O. (2008). Green energies and environment. *Renewable and Sustainable Energy Reviews,* 12: pp. 1789-1821.

Abdeen, M. O. (2009). Principle of low energy building design: Heating, ventilation and air conditioning, Cooling India: India's Premier Magazine on the Cooling Industry, Mumbai, India. 5 (4): pp. 26-46.

Abdeen, M. O. (2012). Clean and green energy technologies: Sustainable development and environment. *Sky J. Agric. Res.* 1 (2): pp. 28–50.

ASHRAE. (2003). Energy efficient design of new building except new low-rise residential buildings. BSRIASHRAE proposed standards 90-2P-1993, alternative GA. American Society of Heating, Refrigerating, and Air Conditioning Engineers Inc., USA. 2003.

ASHRAE, (2005). Commercial/Institutional Ground Source Heat Pump Engineering Manual. American Society of heating, Refrigeration and Air-conditioning Engineers, Inc. Atlanta, GA: USA.

Barabaro, S., Coppolino, S., Leone, C., Sinagra, E. (1978). Global solar radiation in Italy. *Solar Energy* 1978; 20: pp. 431-38.

Baruah, D. (1995). Utilisation pattern of human and fuel energy in the plantation. *Journal of Agriculture and Soil Science* 1995; 8(2): pp. 189-92.

Bendewald, M., Zhai, Z. J. (2013). Using carrying capacity as a base line for building sustainability assessment. *Habitat International*, 37: pp. 1-30.

Bos, E., My, T., Vu, E., Bulatao, R. (1994). World population projection: 1994-95. Baltimore and London: *World Bank by the John Hopkins University Press;* 1994.

CAEEDAC. (2000). A descriptive analysis of energy consumption in agriculture and food sector in Canada. *Final Report,* February 2000.

Cantrell, J., Wepfer, W. (1984). Shallow Ponds for Dissipation of Building Heat: A case Study. *ASHRAE Transactions* 90 (1): pp. 239-246.

D'Apote, S. L. (1998). IEA biomass energy analysis and projections. In: *Proceedings of Biomass Energy Conference: Data, analysis and Trends, Paris: OECD;* pp. 23-24 March 1998.

David, J. M. (2000). Developing hydrogen and fuel cell products. *Energy World* 2002; 303: pp. 16-17.

Duchin, F. (1995). Global scenarios about lifestyle and technology, the sustainable future of the global system. Tokyo: United Nations University; 1995.

Duffie, J. A., Beckman, W. A. (1980). Solar Engineering of Thermal Processes. New York: *J. Wiley and Sons;* 1980.

Dutt, B. (1982). Comparative efficiency of energy use in rice production. *Energy* 1982; 6: pp. 25.

EWEA. (2003). Wind force 12. Brussels, 2003.

Gerald, S. (2012).Why we disagree about climate change: A different viewpoint. *Energy and Environment,* 23 (8): pp. 76-95.

Givoni, B. (1998). Climate consideration in building and urban design. New York: Van Nostrand Reinhold; 1998.

Glaas, E., Juhola, S. (2013). New levels of climate adaptation policy: Analysing the institutional interplay in the Baltic Sea Region. *Sustainability,* 1: 56-122.

Hall, O., Scrase, J. (1998). Will biomass be the environmentally friendly fuel of the future? *Biomass and Bioenergy* 1998: 15: pp. 357-67.

IHA. (2003). World Atlas & Industry Guide. *The International Journal Hydropower & Dams,* United Kingdom, 2003.

John, A., James, S. (1989). The power of place: bringing together geographical and sociological imaginations, 1989.

Kammerud, R., Ceballos, E., Curtis, B., Place, W., Anderson, B. (1984). Ventilation cooling of residential buildings. *ASHRAE Trans:* 90 Part 1B, 1984.

Kavanaugh, S., Rafferty, K. (1997). Ground source heat pumps. Design of Geothermal Systems for Commercial and Institutional Buildings. American Society of heating, Refrigeration and Air-conditioning Engineers, Inc. Atlanta, GA: USA.

Lin, K., .Chang, T. C. (2013). Everyday crises: Marginal society livelihood vulnerability and adaptability to hazards. *Progress in Development Studies,* 13 (1): pp. 45-78.

Morrow, K. (2012). Rio+20, the green economy and re-orienting sustainable development. *Environmental Law Review,* 14 (4): pp. 15-23.

Njeru, J. (2013). 'Donor-driven' neoliberal reform processes and urban environmental change in Kenya: The case of Karura Forest in Nairobi. *Progress in Development Studies* 13, (1): pp. 17-25.

Odeku, K. O., Maveneka, A., Konanani, R. H. (2013). Consequences of Country's Withdrawal from Climate Change Agreements: Implications for Carbon Emissions Reduction. *J. Human. Ecol.* 41(1): pp. 17-24.

Okkan, P. (1993). Reducing CO_2 emissions - How do heat pumps compete with other options? *IEA Heat Pump Centre Newsletter* 1993; 11 (3): pp. 24-26.

OECD/IEA. (2004). Renewables for power generation: status and prospect. UK, 2004.

Pernille, M. (2004). Feature: Danish lessons on district heating. Energy Resource Sustainable Management and Environmental March/April 2004: pp. 16-17.

Raphael, A. T. (2012). Energy and Climate Change: Critical reflection on the African Continent; Is There an Ideal REDD+ Program? *An Analysis of Journal Sustainable Development in Africa,* 14 (6): pp. 25-36.

Rees, W. E. (1999). The built environment and the ecosphere: a global perspective. *Building Research and information* 1999; 27(4): pp. 206-220.

Roriz, L. (2001). Determining the potential energy and environmental effects reduction of air conditioning systems. *Commission of the European Communities DG TREN.*45-80.

Singh, J. (2000). On farm energy use pattern in different cropping systems in Haryana, India. Germany: International Institute of Management-University of Flensburg, Sustainable Energy Systems and Management, Master of Science; 2000.

Shaviv, E. (1989). The influence of the thermal mass on the thermal performance of buildings in summer and winter. In: Steemers TC, Palz W., editors. Science and Technology at the service of architecture. *Dordrecht: Kluwer Academic Publishers*, 1989: pp. 470-472.

Sitarz, D. (1992). Agenda 21: The Earth Summit Strategy to save our planet. Boulder, CO: *Earth Press*; 1992.

Sivkov, S. I. (1964). To the methods of computing possible radiation in Italy. *Trans. Main Geophys. Obs.* 1964; 160.

Sivkov, S. I. (1964). On the computation of the possible and relative duration of sunshine. Trans. Main Geophys Obs, 160. 1964.

Steele, J. (1997). Sustainable architecture: principles, paradigms, and case studies. New York: *McGraw-Hill Inc;* 1997.

Strauss, J. (2013). Does housing drive state-level job growth? Building permits and consumer expectations forecast a state's economic activity. *J. Urban. Econ.* 73: pp. 65-87.

Thakur, C., Mistra, B. (1993). Energy requirements and energy gaps for production of major crops in India. *Agricultural Situation of India* 1993; 48: 665-89.

Tukahirwa, J. T. (2013). Comparing urban sanitation and solid waste management in East African metropolises: The role of civil society organisations. *Cities,* 30: 45-76.

UNFCCC (The United Nations Framework Convention on Climate Change). (2009). The draft of the Copenhagen Climate Change Treaty: pp. 3-181.

UN (United Nations). (2003). World urbanisation project: the 2002 revision. New York: The United Nations Population Division.

Valkila, N., Saari, A. (2012). Perceptions Held by Finnish Energy Sector Experts Regarding Public Attitudes to Energy Issues. *J. Sustainable. Dev.,* 5(11): pp. 23-45.

Vargas-Parra, M. V. (2013). Applying exergy analysis to rainwater harvesting systems to assess resource efficiency. *Resources, Conservation and Recycling,* 72: pp. 35-65.

Wu, J., Boggess, W. (1999). The optimal allocation of conservation funds. *Journal Environmental Economic Management.* 1999: pp. 38.

Yaldiz, O., Ozturk, H., Zeren, Y. (1993). Energy usage in production of field crops in Turkey. In: 5th International Congress on Mechanisation and Energy Use in Agriculture. Turkey: Kusadasi; 11-14 October 1993.

Chapter 2

TOWARDS SUSTAINABILITY BY USING GROUND SOURCE FOR COOLING, HEATING AND HOT WATER SYSTEMS IN BUILDINGS

ABSTRACT

Geothermal energy is the natural heat that exists within the earth and that can be absorbed by fluids occurring within, or introduced into, the crystal rocks. Although, geographically, this energy has local concentrations, its distribution globally is widespread. The amount of heat that is, theoretically, available between the earth's surface and a depth of 5 km is around 140×10^{24} joules. Of this, only a fraction (5×10^{21} joules) can be regarded as having economic prospects within the next five decades, and only about 500×10^{18} joules is likely to be exploited by the year 2020. Three main techniques used to exploit the heat available are: geothermal aquifers, hot dry rocks and ground source heat pumps (GSHPs). The GSHPs play a key role in geothermal development in Central and Northern Europe. With borehole heat exchangers as heat source, they offer de-central geothermal heating at virtually any location, with great flexibility to meet given demands. In the vast majority of systems, no space cooling is included, leaving the GSHPs with some economic constraints. Nevertheless, a promising market development first occurred in Switzerland and Sweden, and now also is obvious in Austria and Germany. Approximately 20 years of R&D focusing on borehole heat exchangers resulted in a well-established concept of sustainability for this technology, as well as in sound design and installation criteria. The market success brought Switzerland to the third rank worldwide in geothermal direct use. The future prospects are good, with an increasing range of applications including large systems with thermal energy storage for heating and cooling, The GSHPs in densely populated development areas, borehole heat exchangers for cooling of telecommunication equipment, etc. This communication reviews some interactions between buildings and environment. The correct assessment of climate helps to create buildings, which are successful in their external environment, while knowledge of sick buildings helps to avoid unsuccessful internal environments. The sections on energy conservation and green buildings suggest how the correct design and use of buildings can help to improve total environment.

Keywords: green buildings, ground source heat pump, environment

1. INTRODUCTION

Today, the challenge before many cities is to support large numbers of people while limiting their impact on the natural environment. Buildings are significant users of energy and materials in a modern society and, hence, energy conservation in buildings plays an important role in urban environmental sustainability (ASHRAE, 1993). A challenging task of architects and other building professionals, therefore, is to design and promote low energy buildings in a cost effective and environmentally responsive way (Jones and Cheshire, 1996). Passive and low energy architecture has been proposed and investigated in different locations around the world (Yuichiro, Cook and Simos, 1991) and design guides and handbooks have been produced for promoting energy efficient buildings (Givoni, 1994). However, at present, little information is available for studying low energy building design in densely populated areas (Abdeen, 2008b). Designing low energy buildings in high-density areas requires special treatment of the planning of urban structure, co-ordination of energy systems, integration of architectural elements, and utilisation of space. At the same time, the study of low energy buildings will lead to a better understanding of the environmental conditions and improved design practices. This may help people study and improve the quality of the built environment and living conditions (CIBSE, 1998).

Industry's use of fossil fuels has been blamed for warming the climate. When coal, gas and oil are burnt, they release harmful gases, which trap heat in the atmosphere and cause global warming. However, there has been an ongoing debate on this subject, as scientists have struggled to distinguish between changes, which are human induced, and those, which could be put down to natural climate variability. Nevertheless, industrialised countries have the highest emission levels, and must shoulder the greatest responsibility for global warming. However, action must also be taken by developing countries to avoid future increases in emission levels as their economies develop and populations grows, as clearly captured by the Kyoto Protocol (FSEC, 1998). Notably, human activities that emit carbon dioxide (CO_2), the most significant contributor to potential climate change, occur primarily from fossil fuel production. Consequently, efforts to control CO_2 emissions could have serious, negative consequences for economic growth, employment, investment, trade and the standard of living of individuals everywhere (Abdeen, 2008a).

Scientifically, it is difficult to predict the relationship between global temperature and greenhouse gas (GHG) concentrations. The climate system contains many processes that will change if warming occurs. Critical processes include heat transfer by winds and tides, the hydrological cycle involving evaporation, precipitation, runoff and groundwater and the formation of clouds, snow, and ice, all of which display enormous natural variability.

The equipment and infrastructure for energy supply and use are designed with long lifetimes, and the premature turnover of capital stock involves significant costs. Economic benefits occur if capital stock is replaced with more efficient equipment in step with its normal replacement cycle. Likewise, if opportunities to reduce future emissions are taken in a timely manner, they should be less costly. Such a flexible approach would allow society to take account of evolving scientific and technological knowledge, while gaining experience in designing policies to address climate change (SP, 1993).

However, the RETs have the benefit of being environmentally benign when developed in a sensitive and appropriate way with the full involvement of local communities. In addition, they are diverse, secure, locally based and abundant. In spite of the enormous potential and the multiple benefits, the contribution from renewable energy still lags behind the ambitious claims for it due to the initially high development costs, concerns about local impacts, lack of research funding and poor institutional and economic arrangements (Watson, 1993).

Hence, an approach is needed to integrate renewable energies in a way that meets high building performance requirements. However, because renewable energy sources are stochastic and geographically diffuse, their ability to match demand is determined by adoption of one of the following two approaches (Abdel, 1994): the utilisation of a capture area greater than that occupied by the community to be supplied, or the reduction of the community's energy demands to a level commensurate with the locally available renewable resources.

The term low energy is often not uniquely defined in many demonstration projects and studies (Todesco, 1996). It may mean achieving zero energy requirements for a house or reduced energy consumption in an office building. A major goal of low energy building projects and studies is usually to minimise the amount of external purchased energy such as electricity and fuel gas. Yet, sometimes the target may focus on the energy costs or a particular form of energy input to the building. As building design needs to consider requirements and constraints, such as architectural functions, indoor environmental conditions, and economic effectiveness, a pragmatic goal of low energy building is also to achieve the highest energy efficiency, which requires the lowest possible need for energy within the economic limits of reason. Since many complicated factors and phenomena influence energy consumption in buildings, it is not easy to define low energy building precisely or to measure and compare the levels of building energy performance. The loose fit between form and performance in architectural design also makes quantitative analysis of building energy use more difficult. Nevertheless, it is believed that super-efficient buildings, which have significantly lower energy consumption, can be achieved through good design practices and effective use of energy efficient technology (Owens, 1986).

In an ideal case, buildings can even act as producers rather than consumers of energy. Besides the operational energy requirements of buildings, it is important to consider two related energy issues. The first one is the transport energy requirements as a result of the building and urban design patterns and the second one is the embodied energy or energy content of the building materials, equipment or systems being used. Transport energy is affected by the spatial planning of the built environment, transport policies and systems, and other social and economic factors. It is not always possible to study the effect of urban and building design on transport energy without considering the context of other influencing factors. The general efficiency rules are to promote spatial planning and development, which reduce the need to travel, and to devise and enforce land-use patterns that are conducive to public transport (Treloar, Fay, and Trucker, 1998). Embodied energy, on the other hand, is the energy input required to quarry, transport and manufacture building materials, plus the energy used in the construction process. It represents the total life-cycle energy use of the building materials or systems and can be used to help determine design decisions on system or materials selection (Omer, 2009). At present, the field of embodied energy analysis is generally still only of academic interest and it is difficult to obtain reliable data for embodied energy. Research findings in some countries indicate that the operating energy often represents the largest component of life-cycle energy use. Therefore, most people, when studying low energy buildings, would prefer to focus on operating energy, and perhaps carry out only a general assessment of embodied energy (Abdeen, 2008c).

This approach comprises a comprehensive review of energy sources, the environment and sustainable development. It includes the renewable energy technologies, energy efficiency systems, energy conservation scenarios, energy savings and other mitigation measures necessary to reduce climate change.

2. CLIMATE AND ENERGY PERFORMANCE

A building site may have natural microclimates caused by the presence of hills, valleys, slopes, streams and other features. Buildings themselves create further microclimates by shading the ground, by drying the ground, and by disrupting the flow of wind. Further microclimates occur in different parts of the same building, such as parapets and corners, which receive unequal exposures to the sun, wind and rain. An improved microclimate around a building brings the following types of benefits:

- Lower heating costs in winter.
- Reduction of overheating in summertime.
- Longer life for building materials.
- Pleasant outdoor recreation areas.
- Better growth for plants and trees.
- Increased user satisfaction and value.

These factors can vary by the hour, by the day and by the season. Some of the variations will cycle in a predictable manner like the sun, but others such as wind and cloud cover will be less predictable in the short term. Information about aspects of climatic factors is collected over time and made available in a variety of data forms including the followings:

- Maximum or minimum values.
- Average values.
- Probabilities or frequencies.

The type of climatic data that is chosen depends upon design requirements. Peak values of maximum or minimum are needed for some purposes, such as sizing heating plant or designing wind loads. Longer terms averages, such as seasonal information, are needed for prediction of energy consumption (Abdeen, 2008d).

A fundamental reason for the existence of a building is to provide shelter from the climate, such as the cold and the heat, the wind and the rain. The climate for a building is the set of environmental conditions, which surround that building and links to the inside of the building by means of heat transfer. Climate has important effects on the energy performance of buildings, in both winter and summer, and on the durability of the building fabric. Although the overall features of the climate are beyond our control, the design of a building can have a significant influence on the climatic behaviour of the building. The following measures can be used to enhance the interaction between buildings and climate:

- Selection of site to avoid heights and hollows.
- Orientation of buildings to maximise or minimise solar gains.
- Spacing of buildings to avoid unwanted wind and shade effects.
- Design of windows to allow maximum daylight in buildings.
- Design of shade and windows to prevent solar overheating.
- Selection of trees and wall surfaces to shelter buildings from driving rain and snow.
- Selection of ground surfaces for dryness.

The large-scale climate of the earth consists of inter-linked physical systems powered by the energy of the sun. The built environment generally involves the study of small systems for which the following terms are used:

- Macroclimate: the climate of a larger area, such as a region or a country.
- Microclimate: the climate around a building and upon its surfaces.

A building site may have natural microclimates caused by the presence of hills, valleys, slopes, streams and other features. Buildings themselves create further microclimates by shading the ground, by drying the ground, and by disrupting the flow of wind. Further microclimates occur in different parts of the same building, such as parapets and corners, which receive unequal exposures to the sun, wind and rain.

An improved microclimate around a building brings the following types of benefits:

- Lower heating costs in winter.
- Reduction of overheating in summertime.
- Longer life for building materials.
- Pleasant outdoor recreation areas.
- Better growth for plants and trees.
- Increased user satisfaction and value.

2.1. Temperature

In a qualitative manner, the temperature of an object determines the sensation of warmth or coldness felt from contact with it. A thermometer is an instrument that measures the temperature of a system in a quantitative way. The air in the bulb is referred to as the thermometric medium, i.e., the medium whose property changes with temperature. Fahrenheit measured the boiling point of water to be 212 and the freezing point of water to be 32, so that the interval between the boiling and freezing points of water could be represented by the more rational number 180. To convert from Celsius to Fahrenheit, multiply by 1.8 and add 32.

$$^{\circ}F = 1.8^{\circ}C + 32 \tag{1}$$

To convert from Celsius to Kelvin, add 273.

$$K = {}^{\circ}C + 273 \tag{2}$$

The method of degree-days or accumulated temperature difference (ATD) is based on the fact that the indoor temperature of an unheated building is, on average, higher than the outdoor. In order to maintain an internal design temperature of 18.5°C, the building needs heating when the outdoor temperature falls below 15.5°C. This base temperature is used as a reference for counting the degrees of outside temperature drop and the number of days for which such a drop occurs. The accumulated temperature difference total for a locality is a measure of climatic severity during a particular season and typical values are given in Table 1. This data, averaged over the years, can be used in the calculation of heat loss and energy consumption. Accumulated temperature differences were used base temperature of 15.5°C, September to May.

Table 1. Climatic severities (ASHRAE, 1993)

Area	Degree-days
England	
South west	1800-2000
South east	2000-2100
Midlands	2200-2400
North	2300-2500
Wales	2000-2200
Scotland	2400-2600

2.2. Wind

The main effects of wind on a building are those of force, heat loss and rain penetration. These factors needed to be considered in the structural design and in the choice of building materials. Wind chill factor relates wind to the rate of heat loss from the human body rather than the loss from buildings. The unfavourable working conditions caused by wind chill have particular relevance to operations on exposed construction sites and tall buildings.

The force of a wind increases with the square of the velocity, so that a relatively small increase in wind speed produces a larger than expected force on a surface such as a building. The cooling effect of wind, measured by wind chill, also greatly increases with the speed of the wind. Typical wind speeds range between 0 m/s and to 25 m/s, as described below:

- 5 m/s wind disturbs hair and clothing.
- 10 m/s wind force felt on body.
- 15 m/s wind causes difficulty walking.
- 20 m/s wind blows people over.

The airflow around some parts of a building, especially over a pitched roof, may increase sufficiently to provide an aerodynamic lifting force by using the principle of Bernoulli. The force can be strong enough to lift roofs and also to pull out windows on the downwind side of buildings. The direction of the wind on a building affects both the structural design and the thermal design. The directional data of wind can be diagrammatically shown by a 'rose' of arms around a point which represent the frequency that the wind blows from each direction as shown in Figure 1. Directions with longer arms will indicate colder winds, which affects energy construction. Other systems of roses may indicate the direction of wind chill factors, which affect human comfort and operations on a building site.

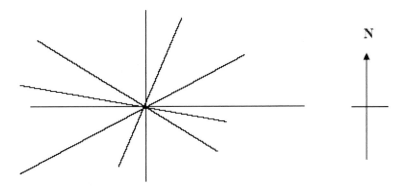

Figure 1. Directional wind rose of mean wind speeds.

The existence of buildings can produce unpleasantly high winds at ground level. It is possible to estimate the ratio of these artificial wind speeds to the wind speed that would exist without the building being present. A typical value of wind speed ratio around low buildings is 0.5, while around tall buildings the ratio might be as high as 2. Wind speed ratios of 2 will double normal wind speed (Jones and Cheshire, 1996). A maximum wind speed of 5 m/s is a suitable design figure for wind around buildings at pedestrian level.

General rules for the reduction of wind effects are given below:

- Reduce the dimensions, especially the height and the dimensions facing the prevailing wind.
- Avoid large cubical shapes.
- Use pitched roofs rather than flat roofs; use hips rather than gable ends.
- Avoid parallel rows of buildings.
- Avoid funnel-like gaps between buildings.
- Use trees, mounds and other landscape features to provide shelter.

2.3. Solar

The effects of the sun on buildings requires the following categories of knowledge about the sun, position in the sky and the angle made with building surfaces, quantity of radiant energy received upon the ground or other surface, and obstructions and reflections caused by clouds, landscape features and buildings. The path that the sun makes across the sky changes each day but repeats in a predictable manner, which has been recorded for centuries. For any position of the sun, the angle that the solar radiation makes with the wall or roof of a building can be predicted by geometry. The angle of incidence has a large effect as the energy received as solar radiation obeys the Cosine Law of Illumination. The intensity of solar radiation falling on a surface, such as the ground, can be measured in Watts per square metre (W/m^2) of that surface. The Watt is defined as a joule per second so this is an instantaneous measurement of the energy received per second on each square metre. When the solar energy is measured over a period of time, such as a day or year, the units will be joules or megajoules per square metre (MJ/m^2). The local metrological conditions the measured radiation (global radiation, diffuse radiation), and the ambient temperatures for Nottingham as summarised in Figures 2-3.

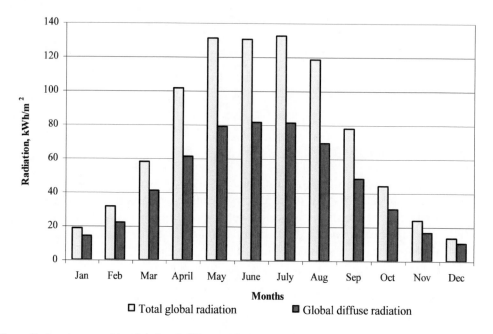

Figure 2. Average monthly global and diffuse radiations over Nottingham.

All buildings gain some casual heat from the sun during winter but more use can be made of solar energy by the design of the building and its services. Despite the high latitude and variable weather of countries in the North Western Europe, like the United Kingdom, there is considerable scope for using solar energy to reduce the energy demands of buildings. The utilisation of solar energy need not depend upon the use of special 'active' equipment such as heat pumps. Passive solar design is a general technique, which makes use of the conventional elements of a building to perform the collection, storage and distribution of solar energy. For example, the afternoon heat in a glass conservatory attached to a house can be stored by the thermal capacity of concrete or brick walls and floors. When this heat is given off in the cool of the evening it can be circulated into the house by natural convection of the air. Heat is transferred from the sun to the earth through space where conduction and convection is not possible. The process of radiation is responsible for the heat transfer through space and for many important effects on buildings as shown in Figure 4. Heat radiation occurs when the thermal energy of surface atoms in a material generates electromagnetic waves in the infrared range of wavelengths. These waves belong to the large family of electromagnetic radiations, including light and radio waves. The rate at which a body emits or absorbs radiant heat depends upon the nature and temperature of its surface. The wavelengths of the radiation emitted by a body depend upon the temperature of the body. High temperature bodies emit a larger proportion of short wavelengths, which have a better penetration than longer wavelengths.

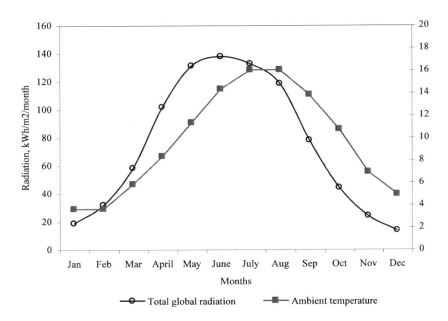

Figure 3. Average monthly solar radiation and ambient temperature over Nottingham.

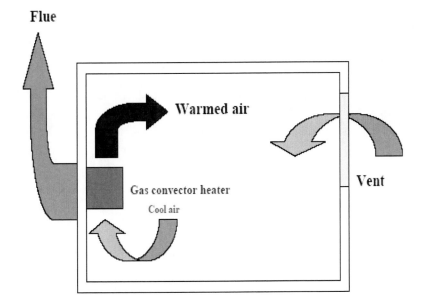

Figure 4. Convection currents in room.

2.4. Rain

The annual driving rain index (DRI) is a combined measure of rainfall and wind speed. The DRI takes account of the fact that rain does not always fall vertically upon a building and that rain can therefore penetrate walls. The DRI is also associated with the moisture content of exposed masonry walls whose thermal properties, such as insulation, vary with moisture content. Driving rain is usually caused by storms but intense driving rain can also occur in heavy showers, which last for minutes rather than hours. These conditions are more likely in exposed areas such as coasts, where high rainfall is accompanied by high winds. Table 2 gives typical values of driving rain index for different types of area. High buildings, or buildings of any height on a hill, usually have an exposure one degree more than indicated.

Table 2. Driving-rain indices for British Isles

Exposure grading	Driving-rain index	Example
Sheltered	3 or less	Within towns
Moderate	3 to 7	Countryside
Severe	7 or more	West coastal areas

3. ENERGY CONSERVATION

At present, most of the energy used to heat buildings, including electrical energy, comes from fossil fuels such as oil and coal. This energy originally came from the sun and was used in the growth of plants such as trees. Then, because of changes in the earth's geology, those ancient forests eventually became a coal seam, an oil field or a natural gas field. The existing stocks of fossil fuels on earth cannot be replaced and unless conserved, they will eventually run out. Primary energy is used for building services such as heating, lighting and electricity. Most of the energy consumed in the domestic sector is used for space heating. Reducing the use of energy in buildings will therefore be of great help in conserving energy resources and in saving money for the occupants of buildings.

Methods of conserving energy in buildings are influenced by the costs involved and in turn, these costs vary with the types of building and the current economic conditions. There are alternative energy sources available for buildings and the energy consumption of buildings can also be greatly reduced by changes of their design and use (Abdeen, 2009a).

3.1. Energy Efficiency

Large amounts of energy are contained in the world's weather system, which is driven by the sun, in the oceans, and in heat from the earth's interior, caused by radioactivity in rocks. This energy is widely available at no cost except for the installation and running of conversion equipment. Devices in use include electricity generators driven by wind machines, wave motion and geothermal steam (Abdeen, 2009b). With improvement of the people's living standards and development of the economies, heat pumps have become widely used for air conditioning.

The total energy of the universe always remains constant but when energy converted from one form to another some of the energy is effectively lost to use by the conversion process. For example, hot gases must be allowed to go up the chimney flue when a boiler converts the chemical energy stored in a fuel into heat energy. Around 90 percent of the electrical energy used by a traditional light bulb is wasted as heat rather than light.

New techniques are being used to improve the conversion efficiency of devices used for services within buildings. Condensing boilers, for example recover much of the latent heat from flue gases before they are released. Heat pumps can make use of low temperature heat sources, such as waste air, which have been ignored in the past. Although electrical appliances have a high-energy efficiency at the point of use, the overall efficiency of the electrical system is greatly reduced by the energy inefficiency of large power stations built at remote locations. The 'cooling towers' of these stations are actually designed to waste large amounts of heat energy (Abdeen, 2009c). It is possible to make use of this waste heat from power stations both in industry and for the heating of buildings. These techniques of combined heat and power (CHP) can raise the energy efficiency of electricity generation. The CHP techniques can also be applied on a small-scale to meet the energy needs of one building or a series of buildings. Electrical energy will still be required for devices such as lights, motors and electronics but need not be used for heating.

3.2. Thermal Insulation

External walls, windows, roof and floors are the largest areas of heat loss from a building. The upgrading of insulation in existing buildings can be achieved by techniques of roof insulation, cavity fill, double-glazing, internal wall lining, and exterior wall cladding.

3.3. Renewable Energy

Renewable energy is energy obtained from sources that will not run out. It does not depend on the continued extraction of fossil fuels and does not contribute to CO_2 emissions. The sources of renewable energy that appear most promising are:

- The sun - used directly to heat water or buildings or indirectly to generate electricity.
- Hydroelectric - using turbines to convert the energy of rivers and streams into electricity.
- Wind - again using turbines sited either on or offshore.
- Biomass - burning energy crops or agricultural waste to produce heat for electricity generation.
- Energy from waste - burning solid waste collected by local authorities or landfill gas.

The term sustainable development is generally defined as development that leads to economic growth and social improvement without harming the environment and without depleting the earth's reserves of resources. Buildings would be designed to be energy efficient utilising low power input for lighting and environment comfort (Abdeen, 2009c).

3.4. Ventilation

The warm air released from a building contains valuable heat energy, even if the air is considered 'state' for ventilation purposes. The heat lost during the opening of doors or windows becomes a significant area of energy conservation, especially when the cladding of buildings is insulated to high standards. These ventilation loses are reduced by better seals in the construction of the buildings, by air-sealed door lobbies, and the use of controlled ventilation. Some of the heat contained in exhausted air can be recovered by heat exchange techniques such as heat pumps. Larger windows provide better daylighting but also cause greater heat losses in winter and larger heat gains in summer. The accompanying Table 3 of environmental factors indicates some of the major interactions between different design decisions.

Natural and international bodies, societies and environmentalists are aware that air-conditioning systems constitute about 55% of our ozone layer depletion through the use of refrigerants such as chlorofluorocarbons (CFCs) and hydro fluorocarbons (HFCs). In most homes air-conditioning systems are used predominantly for internal environmental comfort. The truth of the fact is that, we endanger our environment by the use of these air-conditioning systems for internal environmental comfort. Due to high levels of carbon dioxide (CO_2) and GHG emissions, which occur daily through these systems, the environment now relies heavily on green issues for sustenance. There are currently about 325 parts per million (ppm) of CO_2 in the atmosphere (Fordham, 2000). When the earth's ozone layer, which acts as protective shield from the sun's ultra-violet rays, is broken or torn, we experience global climatic changes. This leads to an increase in global temperature, which tends to cause rising sea levels that lead to flooding in many coastal areas. It is predicted that by 2080 global temperatures would rise by $3°C$ as a result of which 80 million people could be flooded and displaced each year in the coastal areas (Fordham, 2000). The GHG emissions are introduced into the environment by two main sources: fossil fuels usage and burning, and refrigerants leakages in air-conditioning system. For positive reduction of the GHG emissions the following should be addressed:

- Societal awareness to renewable energy utilisation and benefits.
- Efficient design and installation of air-conditioning systems.
- Environmental impact and services.

There are four key features of innovative building design, which are:

- Natural ventilation.
- External day lighting.
- Heat reclamation if necessary.
- Use of borehole cooling where affordable.

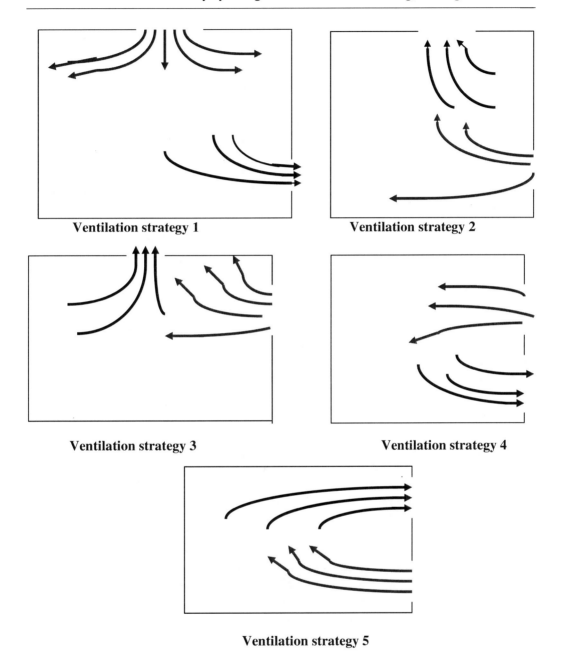

Figure 5. Different ventilation strategies.

Architects, consultants and contractors have to introduce new initiatives, ideas and concepts to the outlook of air-conditioning and ventilation systems operations. Such innovations and ideas would exist in the areas of:

- Design of energy efficient systems, which would make use of sources of renewable energies as a medium of heat transfer.

- Introduction of solar architectural designs in buildings reducing power consumption through electric bulbs during the day.
- Educating clients and users of buildings to understand the role of their buildings with respect to ozone layer depletion and environmental awareness.

Particulate pollutants in buildings can have damaging effects on the health of occupants. Studies have shown that indoor aerosol particles influence the incidence of sick building syndrome (Fordham, 2000). Some airborne particles are associated with allergies because they transport viruses and bacteria. The concentration of indoor aerosol particles can be reduced by using different ventilation strategies (Figure 5) such as displacement and perfect mixing. However, there are insufficient data to quantify the effectiveness of these methods, as removal of particles is influenced by particle deposition rate, particle type, size, source and concentrations.

3.4.1. Natural Ventilation

Generally, buildings should be designed with controllable natural ventilation. A very high range of natural ventilation rates is necessary so that the heat transfer rate between inside and outside can be selected to suit conditions (Fordham, 2000). The ventilation rates required to control summertime temperatures are very much higher than these required to control pollution or odour. Any natural ventilation system that can control summer temperatures can readily provide adequate ventilation to control levels of odour and carbon dioxide production in a building. Theoretically, it is not possible to achieve heat transfer without momentum transfer and loss of pressure.

3.4.2. Mechanical Ventilation

Most medium and large size buildings are ventilated by mechanical systems designed to bring in outside air, filter it, supply it to the occupants and then exhaust an approximately equal amount of stale air. Ideally, these systems should be based on criteria that can be established at the design stage. To return afterwards in attempts to mitigate problems may lead to considerable expense and energy waste, and may not be entirely successful (Fordham, 2000).

The key factors that must be included in the design of ventilation systems are: code requirement and other regulations or standards (e.g., fire), ventilation strategy and systems sizing, climate and weather variations, air distribution, diffuser location and local ventilation, ease of operation and maintenance and impact of system on occupants (e.g., acoustically). These factors differ for various building types and occupancy patterns. For example, in office buildings, pollutants tend to come from sources such as occupancy, office equipment, and automobile fumes.

Occupant pollutants typically include metabolic carbon dioxide emission, odours and sometimes smoking. The occupants (and not smoking) are the prime source of pollution. Carbon dioxide acts as a surrogate and can be used to cost-effectively modulate the ventilation, forming what is known as a demand controlled ventilation system. Generally, contaminant sources are varied but, often, well-defined and limiting values are often determined by occupational standards (Table 3).

Table 3. Interactions of environmental decisions

Some design options	Heating	Ventilation	Lighting	Sound
Sheltered site loss and gain	Less heat	-	Less daylight	Less noise intrusion
Deep building shape	Less heat loss and gain	Reduced natural ventilation	Less daylight	-
Narrow building plan	More heat loss and gain	More natural ventilation	More daylight intrusion	More noise
Heavy building materials	Slower heating and cooling	-	-	Better sound insulation
Increased window area	More heat loss and gain	-	More daylight	More noise intrusion
Smaller, sealed windows	Less heat loss and gain	Reduced natural ventilation	Less daylight	Less noise intrusion

There are increasing challenges facing people throughout the world to secure a reliable, safe and sustainable energy supply to meet their needs. In developing countries the demand for commercial energy is growing quickly. These countries are faced with substantial financial, environmental and energy security problems. In both developed and less developed countries pressure is growing to find workable alternatives to traditional energy supplies and to improve the efficiency of energy use in an attempt to limit emissions of gases that cause global climate change.

4. GROUND SOURCE HEAT PUMPS: A TECHNOLOGY REVIEW

This section provides a detailed literature based review of ground source heat pump (GSHP) technology and looks more briefly at applications of the technology, applicable standards and regulations, financial and other benefits and the current market status.

The first documented suggestion of using the ground as a heat source appears to be in 1912 in Switzerland in a patent filed by H. Zölly (Wirth, 1955) but at that time the efficiency of heat pumps was poor and energy prices were low so the idea was not followed up. In the forties investigation into the GSHPs started up again both in the USA and the UK. In the UK, Sumner first used the ground as a source for a heat pump for space heating in a single house in the mid 1940s (Sumner, 1976). A horizontal collector at a depth of about 1 m was used to supply heat via copper pipes buried in a concrete floor. A coefficient of performance (COP) of 2.8 was achieved. In 1948 he installed 12 prototype heat pump systems using ground collectors each with a 9 kW output. The average COP of these installations was 3. However, this study was stopped after two years (Wirth, 1955). The first ground source heat pump in the North America was installed in a house in Indianapolis in October 1945 (Crandall, 1946). This consisted of copper tubes buried at a depth of about 1.5 m in the ground with the refrigerant circulating directly through them. In the next few years virtually all the methods of exploiting the ground as a heat source/sink, which are used today, were investigated in the USA (Kemler, 1947) and a study in 1953 listed 28 experimental installations. Studies were also carried out in Canada, where the emphasis was on understanding the theoretical basis of using heat from the ground (Kemler, 1947). The first Canadian system was in an experimental house at Toronto University (Kusuda and Achenbach, 1965). Commercial use of the ground as a heat source/sink did not begin until after the first oil shock in 1973 but was well established by the end of the seventies by which time there were over 1000 ground source heat pumps installed in Sweden (Granryd, 1979). The vertical earth heat exchanger was introduced into Europe in the late 70s (Rosenblad, 1979), and from that time on has been used in various types mainly in Sweden, Germany, Switzerland and Austria (Drafz, 1982). Since 1980 there have not been any major technological advances in the heat pump itself except for improved reliability. However, considerable progress has been made in other areas such as system integration, reducing costs for the ground heat exchanger, improving collector configuration and control systems and strategies (Sanner, Hopkirk, Kabus, Ritter and Rybach, 1996).

Today the GSHPs are an established technology with over 400,000 units installed worldwide (around 62% of which are in the USA) and about 45,000 new units installed annually (Drafz, 1982). They are receiving increasing interest in the North America and Europe because of their potential to reduce primary energy consumption and thus reduce the emission of the GHGs and other pollutants (Drafz, 1982).

The GSHPs as the technology for space heating and cooling had the highest potential energy efficiency. The Geothermal Heat Pump Consortium was thus set up in 1994 with the aim of stimulating uptake of the technology and increasing the number of installations from approximately (40,000 units/year) to (400,000 units/year) by the year 2000. It was estimated that this could save over 300×10^9 MJ/year and reduce GHG emissions by 1.5 million metric tons/year of carbon equivalent. Although this target was very ambitious, and will not be met, there has been sustained interest in the technology (Drafz, 1982).

Overall efficiencies for the GSHPs are high because the ground maintains a relatively stable source/sink temperature, allowing the heat pump to operate close to its optimal design point. Efficiencies are inherently higher than for air source heat pumps because the air temperature varies both daily and seasonally and air temperatures are lowest at times of peak heating demand and highest at times of peak cooling demand.

Heat recovery using waste ventilation air to heat water is another possibility; this is more worthwhile on larger flats and offices. There are also many commercial and industrial processes where heat is wasted: a heat pump can recover this energy to provide heat at a useful temperature. Often referred to as "Geothermal", the GSHP is becoming the most common system to be installed in the Northern Europe. The efficiency of any system will be greatly improved if the heated water is kept as low as possible. For this reason, underfloor heating is preferred to radiators. It is vital to ensure that the underfloor layout is designed to use low water temperatures, i.e., plenty of pipe and high flow-rates. Heat pumps have a different design emphasis to boiler systems (ACRI, 1991).

Most underfloor systems use zone valves that reduce the flow-rate. The heat pump can maintain the correct flow-rate in buildings. A buffer tank is suggested. If radiators are to be used, they must be large enough. Double the normal sizing (as used with a boiler) is a good starting point. Whilst this type of heat pump installation could provide all the heating needs, it is common practice, and often-economic sense to have a back-up boiler linked to the system to cope with the very cold periods. Electric back up is not ideal. This is putting a high load on the main supply at a time of peak demand. At this time the power station's net fuel efficiency is lower.

The ground pipe system must be planned carefully, especially as it will be there for well over 50 years. Any mistakes may be too difficult or costly to rectify later. The highest energy efficiency will result from systems that do not go below freezing point, therefore, the bigger the pipe system/ground area, the better; however, this is costly and gives diminishing returns. The pressure drop in the pipes should be compatible with standard low-head pumps.

Weather compensation will greatly improve the annual energy efficiency, by reducing the heated temperature to the minimum required, depending on outside temperature. Most heat pumps incorporate this in the controller; however, this facility can be retrofitted as an extra. To keep energy efficiency high, keep the heated water temperature as low as possible. Then, keep some zone valves fully open and control the temperature down by carefully adjusting the weather compensation controller. If there is not any weather compensation, simply adjust the water temperature as low as possible such that adequate heating is attained. If domestic hot water is provided by the heat pump, have a big enough cylinder such that the water can be stored at a slightly lower temperature. Avoid "thermal store" type systems. They require temperatures higher than heat pumps can efficiently provide. Heat pump compressors like to run for long periods. Stop-starts should be minimised. The use of buffer tanks, correctly set thermostat differentials and correctly positioned cylinder sensors will all help to maximise run periods. Noise could be a problem if not considered properly, at the design stage and this problem should be eliminated.

5. HYDROPOWER SYSTEMS

The mechanical power from a water turbine or wheel is usually used to generate electricity. A heat pump can extract energy from the water to produce a heat energy output of three or more times that of a conventional hydroelectric system. Since space heating is by far the biggest single energy load, it is sometimes better to put all the hydro energy into a heat pump, while remaining grid-connected for general electrical needs. The heat pump can be driven directly by mechanical belt-drive, etc. However, this system requires a lot of maintenance. An electric drive heat pump driven from a hydroelectric source is probably the most practical.

6. AIR-CONDITIONING

Air-to-air systems are used throughout the world for air-cooling. The reversible version of these is the most common type of heat pump. In its heating mode, the efficiency is not as good as the previously mentioned water systems. In cooling mode, they consume large amounts of energy. In the UK, the suggestion that a large ventilation-rate combined with sun shading is more appropriate for cooling, as fans use far less power. If mechanical air-cooling is necessary, then a water - or ground-coupled heat pump system will be the most energy efficient.

If air-conditioning is used, then good housekeeping to reduce energy consumption should not be overlooked. This should include good shading from sunlight by use of automatic or manually controlled blinds. The room temperature should not be set too low, as is often the case. The term coefficient of performance's (COP) usually used to describe a heat pump's efficiency. A COP of 3 is typical, i.e., 1 kilowatt (kWatt) of power input will provide 3 kWatts of useful output. This is equivalent to 300% efficiency. (The extraction of heat from outside makes this possible). The COP depends on the type of application. In general, the closer the difference in temperature between the source and the sink, the higher the efficiency, e.g., a COP of 5 can be attained with a good heat pump with a spring source feeding well designed underfloor heating, whereas a COP as low as 2 may result in heating bath water from an air source system in winter.

1) The word "Efficiency" is defined as the ratio of useful heat output to energy input, e.g., if an open fireplace loses half its energy up the chimney it is said to be 50% efficient.
2) The COP or "Coefficient of performance" is found by dividing the useful heat output by the energy input, e.g., a heat pump that produces 3 kWatts of heat for 1 kWatt of input power has a COP of 3. The open fireplace example with 50% efficiency would have a COP of 0.5 (1/2).
3) The heat "Source" is the outside air, river or ground, wherever the heat is being extracted from. Sometimes is referred to as an ambient source.
4) The "Sink" is the name given to the part where the heat is usefully dissipated, such as radiators in the room, underfloor heating, hot water cylinder, etc.

6.1. Technical Definitions

Slinky:
The name Slinky is given to the way that ground collector pipes can be coiled before buying in a trench.

Horizontal collector:
This can be either coiled 'Slinky' or straight pipes that are buried 1.5 m to 2 m deep in open ground (in gardens). The pipe is usually plastic and contains a Glycol antifreeze solution.

Antifreeze:
This is simply an additive to water that makes its freezing point lower. Common salt does the same thing, but Ethylene or Propylene Glycol is more practical for heat pump systems.

Refrigerant:
This is the working fluid within the heat pump. It evaporates in one part and condenses in another. By doing so, heat is transferred from cold to hot. This fluid is sealed in and will not degrade within the heat pumps life.

Heat exchanger:

This is a simple component that transfers heat from one fluid to another. It could be liquid-to-liquid, or liquid-to-air, or air-to-air. Two heat exchangers are housed within the heat pump, one for the hot side (the condenser), and one for the cold side (the evaporator).

Passive heat exchange:

When waste hot water preheats cold input water, it is said to be 'passive'. This costs nothing to run. A heat pump is said to be 'active' it can extract heat from cold waste water but requires a relatively small power input.

7. COMMERCIAL BUILDING APPLICATIONS

Ground-coupled heat pumps (GCHPs) are often confused with a much more widely used commercial system, the water loop heat pump (WLHP) or water source heat pump. Although the piping loop inside the building is similar, there are several important differences. Water-to-air heat pumps are located throughout the building. Local zone temperature is achieved with conventional on-off thermostats. Ductwork is minimised because the units are in the zone they serve. A central piping loop is connected to all the units. The temperature of this loop is typically maintained between 60°F (16°C) and 90°F (32°C). A cooling tower is used to remove heat when the loop temperature exceeds 90°F and heat is added with a boiler if the temperature falls below 60°F (ACRI, 1991).

The WLHPs are most successful when internal building loads are sufficient to balance the heat loss through the external surfaces and ventilation. If heat losses exceed internal loads, the energy requirements of the WLHPs can become significant. Energy must be added in both the boiler and heat pumps. This is not true in cooling because the heat is dissipated through the cooling tower, which only has pump and fan motor requirements.

The WLHPs are designed to operate in the narrow range of 60 to 90°F. This will not perform adequately in a GCHP system. The units used in the GCHP systems must be extended range water-to-air heat pumps. Some manufacturers create extended range heat pumps by replacing the fixed expansion device of a WLHP with a thermostatic expansion valve (TEV). Others make this modification and add improved compressors, air and water coils, fans, and controls. This has resulted in units that operate with higher efficiencies than conventional WLHPs even when operating with water temperatures outside the 60 to 90°F range. It is obvious that the ground coil can add to the cost of the system. Also many high-rise commercial applications may not have sufficient land area to accommodate a full size ground coil. A hybrid ground-coupled water-loop heat pump (GCWLHP) would be a viable option to reduce the size of the ground coil. The coil would be sized to meet the heating requirement of the building. This is typically one-half the size required for meeting the cooling load. There are several reasons for the smaller size:

- In the heating mode only about 70% of the heat requirement of the building must come from the ground coil. The remaining 30% comes from the power input to the compressor and fan motors. So the coil transfers about 8,400 Btuh/ton. In cooling the coil must transfer the building load and the added heat of the motors. This means 130% or 15,600 Btuh/ton must be moved through the ground coil (MGA, 1992).

- The cooling requirement of commercial buildings with high lighting and internal loads usually exceeds the heating requirement.
- The heating requirement of commercial buildings is often in the form of a morning "spike" followed by a reduced load. Ground coils are well suited to handling spikes because of the large thermal mass of the earth. Therefore, lengths can be reduced compared to systems designed for continuous loads. Since the ground coil for a GCWLHP would not be able to meet the cooling load in most climates, a downsized cooling tower would be added to the loop.

The GCHPs can also be integrated into "free cooling" or thermal storage schemes. For example, hydronic coils could be added to core heat pumps of a GCWLHP system. When the outdoor temperature was cold enough the cooling tower could be started. This would bring the loop temperatures below 50°F (10°C) to cool the core zones without activating the compressors. The heat pumps in perimeter zones could operate simultaneously in the heating mode if required. A variety of other systems are possible because of the simplicity and flexibility of ground-coupled heat pumps.

7.1. Examples of Commercial Installations

Very few building owners, engineers, and architects consider GCHPs because in the past implementation was difficult. There were very few qualified loop installers, design guides were hard to find, and the traditional HVAC&R network balked at the thought of linking equipment to plastic pipe buried in the ground. However, the experiences of those who tried this "new" concept have led to a sound methodology for the design and installation of highly reliable and efficient systems. One such firm operates in Pennsylvania (MGA, 1992). This firm designs, installs and operates GCHP systems. The ground coils are typically 200 to 500 ft deep with 1½ inch (4 cm) polyethylene U-bends. Drilling in the area is very difficult compared to the rest of the USA (MGA, 1992). However, several successful systems have been and are continuing to be installed and operated. A listing is given in Table 4.

Similar firms are operating profitably in areas all over the USA. Texas has several new schools and other commercial buildings that have GCHPs. Activities in Canada are very high compared to the USA (MGA, 1992) with utilities promoting the technology with rebates and technical assistance. Oklahoma, a state that derives much of its income from oil and gas, is in the process of installing a GCHP system to heat and cool its state capital complex. The common thread in successful GCHP programmes appears to be an individual or set of individuals in a particular location who recognise the advantages of the GCHPs. These individuals have the initiative to push forward in spite of the many skeptics who contend that GCHPs will not work (MGA, 1992).

In heating applications, heat pumps save energy by extracting heat from a natural or waste source, using a mechanism similar to that found in a refrigerator. They can be used for any normal heating needs. However, this technology is not new. Several heat pumps were installed in the

1950's in a bid to save energy and fuel costs. One of the most famous of these was used to heat the Royal Festival Hall in London by extracting heat from the River Thames (MGA, 1992).

Table 4. Listing of systems installed and operated by Pennsylvania GCHP Firm (MGA, 1992)

Building Type	Area (sq. ft)	Capacity (tons)	Units	Bores
Bank	5,500	13	3	3
Retired Community	420,000	840	316	187
Elementary School	24,000	59	21	20
Doctor's Office	11,800	35	7	7
Condominiums	88,000	194	74	40
Middle School	110,000	412	96	106
Restaurant	6,500	36	6	7
Office/Lab	104,00	252	43	62
Elderly Apts.	25,000	89	76	12
Life Care Comm.	390,000	1,100	527	263
Ron. McDonald House	2,000	5	4	1

7.2.1. Efficiency

Heat pump efficiency is primarily dependent upon the temperature difference between the building interior and the environment. If this difference can be minimised, heat pumps efficiency (and capacity) will improve. Ground temperatures are almost always closer to room temperature than air temperatures. Therefore, the GCHPs are inherently more efficient than units that use outdoor air as a heat source or sink if the ground coil is correctly designed. This principle is one of Mother Nature's rules and is referred to as Carnot's Law. Secondly, it is important to have large coils for high efficiency. Water is far superior to air with regard to "convection" heat through coil surfaces. Therefore, the water coils in the GCHP are smaller and much more "efficient" in transferring heat. The part-load efficiency of a GCHP is actually improved compared to full load efficiency. When the ground coil is partially loaded, the water loop temperature more closely approaches the earth temperature. This temperature is cooler in the cooling mode and warmer in the heating mode. Therefore, system efficiency is improved. Auxiliary power requirement can be reduced significantly compared to conventional systems.

In addition to being a better heat transfer fluid compared to air, water requires much less energy to be circulated. Although it is heavier than air, a given volume of water contains 3500 times the thermal capacity of atmospheric air. Therefore, the pump motors circulating water through a GCHP system are much smaller than outdoor air or cooling tower fan motors of conventional systems. The indoor fan power is also reduced because the units are in the zone and duct runs are very short and non-existent. Therefore, low pressure (and power) fans can be used. Unfortunately, the efficiency ratings used for the GCHPs do not match with the ratings of conventional equipment. For comparison consider a central chiller and GCHP. Table 5 is a comparison of the full load efficiency of a 100 tons central chiller to a GCHP (HPA, 1992).

7.2. Advantages of the GCHPS

Major advantages of the GCHPS are summarised as follows:

7.2.2. Space

The GCHP systems require a small amount of space if properly designed. A 5 tons (18 kW) water-to-air heat pump in a horizontal package can be as small as 20x25x45 inches (50x64x115cm) and easily located above the ceiling in a typical office. Vertical packages can be placed in small closets. Some units used in the WLHP systems may have excessive noise levels for these locations. However, the improved units recommended for the GCHP systems have quieter compressors and large, low velocity fan wheels that reduce noise levels.

Table 5. Efficiency comparison of high efficiency central chiller with the GCHP (HPA, 1992)

1. Chiller @ 85°F Cond. water and 45°F chilled water (0.6 kW/ton)	60 kW
2. Indoor fans (40,000 cfm, 3.0 in. ESP, 90% motors)	25 kW
3. Cooling tower fan (5 hp)	4 kW
4. Cooling tower pump (5 hp)	4 kW
5. Indoor circulation pump (5 hp)	4 kW
Total	97 kW

The central distribution system requires relatively little space, and the relative requirements for a high velocity air system, a low velocity air system, and a GCHP. It also indicates the required power of the GCHP pump is much smaller than the fans of either air systems.

7.2.3. Aesthetics

One pleasant advantage of a GCHP system is the absence of unsightly outdoor equipment. The ground above outdoor coil can become a greenspace or a parking lot. This is especially suited to schools where outdoor equipment may pose a safety hazard to small children or a vandalism target for not so small children.

7.2.4. Simplicity

The conventional GCHP system is extremely simple. The water-to-air unit consists of a compressor, a small water coil, a conventional indoor air coil, one bi-flow expansion device, and a few electrical controls. The flow control can be either a single circulation pump on each unit (that is turned on with the compressor relay) or a normally closed two-way valve for systems with a central circulation pump. If the designer chooses an extended range heat pump as recommended, no water regulating control valves are necessary.

7.2.5. Control

Control is also very simple. A conventional residential thermostat is sufficient. Since units are located in every zone, a single thermostat serves each unit. Zones can be as small as ½ ton (1.8 kW). However, each unit can be linked to a central energy management system if desired. Air volume control is not required. Larger water-to-air heat pumps are available if multi-zone systems are required. This would complicate one of the most attractive benefits of the GCHPs, local zone control. The simple system can be installed and serviced by technicians with moderate training and skills. The building owner would no longer be dependent on the controls vendor or outside maintenance personnel. The simple control scheme would interface with any manufacturers' thermostats.

7.2.6. Comfort

The GCHPs eliminate the Achilles' heel of conventional heat pumps in terms of comfort, "cold blow". Commercial systems can be designed to deliver air temperatures in the 100° to 105°F (38° to 41°C) range without compromising efficiency (Bose, 1988). Moisture removal capability is also very good in humid climates. The previously discussed advantage of local zone control is also critical to occupant comfort.

7.2.7. Maintenance

One of the most attractive benefits of the GCHPs is the low level of maintenance. The heat pumps are closed packaged units that are located indoors. The most critical period for a heat pump compressor is start-up after defrost. The GCHPs do not have a defrost cycle. The simple system requires fewer components. Logic dictates that the fewer components, the lower the maintenance. Because of the limited amount of data for the GCHPs, not a great deal of data is available to support the claim of low maintenance in commercial buildings. However, a detailed study of a WLHP system was conducted and the median service life of compressors in perimeter heat pumps was projected to be 47 years (Ross, 1990).

7.3. Disadvantages

Various disadvantages are as follows:

7.3.1. New Technology

The GCHPs face the typical barriers of any new technology in the heating and cooling industry. Air source heat pump and natural gas heating technology has been successful. The technical personnel have been needed to install and service this equipment. A great deal of research and development has been successfully devoted to improving this technology. The GCHP technology faces an additional barrier in the lack of an infrastructure to bridge two unrelated networks: the HVAC industry and the drilling/trenching industry. There is little motivation on either side to unite. The HVAC industry prefers to continue marketing a proven technology and well drillers continue to profit on existing water well, environmental monitoring well, and core sampling work. It is the task of the two sectors that benefit the most from the GCHPs, customers and electric utilities, to force a merger of the two networks.

7.3.2. Limited Profit for HVAC Equipment Manufacturers

Some equipment manufacturers are resistant to the GCHPs because of the reduced need of their products. Water-to-air heat pumps are relatively simple and potentially inexpensive devices. The control network is especially simple and inexpensive. There will be no need for manufacturer's technicians to trouble-shoot and service control systems. There will be no need to lock into one manufacturer's equipment because of incompatibility. The GCHPs are not inexpensive. However, approximately 50% of the system's first cost must be shared with a driller/trencher. Therefore, some HVAC manufacturers may be reluctant to support the implementation of the GCHPs.

7.3.3. Installation Cost

The most formidable barrier to the GCHP systems is currently high installation costs. While this is especially true in the residential sector, it also applies to commercial applications. Residential premiums compared to a standard electric cooling/natural gas heating system (9.0 SEER, 65% AFUE) are typically $600 to $800 per ton for horizontal systems and $800 to $1000 per ton for vertical systems. Simple payback is typically five to eight years (ASHRAE, 2000). The percent increase is somewhat less for commercial GCHPs as shown in the following section.

7.3.4. Projected Cost of Commercial GCHPS

The cost of vertical ground coil ranges from ($2.00 to $5.00 per ft) of bore. Required bore lengths range between 125 ft per ton for cold climate, high internal load, and commercial buildings to 250 ft per ton for warm climate installations. Pipe cost can be as low as $0.20 per ft of bore ($.10/ft of pipe) for 3/4 inch (2.0 cm) and as high as $1.00 per ft of bore ($0.50/ft of pipe) for 1½ inch (4.0 cm) polyethylene pipe. Drilling costs range from less than $1.00 per ft to as high as $12.00 per ft. However, $5.00 per ft is typically the upper limit for a drilling rig designed for the small diameter holes required for the GCHP bores even in the most difficult conditions. It should be noted that larger diameter pipes result in shorter required bore lengths. Table 6 gives typical costs for low and high drilling cost conditions for 3/4 and 1½ inch U-bends for a 10 ton system (Pietsch, 1988).

The total cost will be in the range of $400 to $850 per ton. If the cost of the WLHP boiler, drains, and cooling tower is deducted from the total and the cost of the ground coil and current cost of improved heat pumps ($100/ton) is added, a cost range for the GCHPs results (Pietsch, 1988). For low cost drilling sites the cost of the GCHPs is actually lower than conventional 2-pipe VAV systems (Cavanaugh, 1992).

Table 6. Cost of vertical ground coils (Pietsch, 1988)

Systems	$1.50/ft Drilling cost	$4.00/ft. Drilling cost
	¾" (2000')	1.5" (1700')
Drilling	$3000	$2550
Pipe	$600	$1360
Fittings	$300	$300
Total	$3900	$4210
Cost/ton	$390	$421

7.3.5. Operating Costs

Limited data is available documenting the operating cost of the GCHPs in commercial applications. The steady state and part load cooling efficiencies of vertical GCHPs appear to be superior to high efficiency central systems. The heating efficiencies are very good, especially when the ground coil is sized to meet the cooling requirement. However, these high efficiencies will not be realised if ground coils are undersized or low and moderate efficiency water-to-air heat pumps are used.

A comprehensive study of the GCHP operating costs in commercial buildings must be conducted. This study is needed to expand the limited design guidelines currently available for the GCHPs.

7.3.6. Facing the Competition with the GCHPS

Electric utilities may argue that these figures may be exaggerated. There is a strong element of truth in each. Air source heat pumps do not heat well in cold weather (Mother Nature will not allow it). Air heat pumps do blow large amounts of air at temperatures below body temperature. In the heating mode, the air heat pump consumes more net energy than high efficiency gas furnaces.

The electrical utility has focused the bulk of its response on aggressive marketing that in some cases is equally as misleading document. This marketing includes advertising and reduced rates for customers who choose electric heat. To a lesser degree, the response has included development of advanced heat pumps. In the commercial sector, this has generally excluded the GCHPs. Most of the developmental activity for the GCHPs has been confined to the residential market.

The electric utility would be the benefactor. Even when the air is 0°F and below, the ground is 45°F and above, the GCHPs can operate very well and can deliver air at a comfortable temperature. For every 10 Btus used at the power plant at least 10 Btus are delivered to the home. They offer all of the many advantages of conventional electric heat pumps plus higher efficiency, simplicity, reliability, reduced demand, removal of unsightly outdoor equipment, comfort, and long life. They offer a system whose performance cannot be matched by natural gas equipment.

8. DISCUSSIONS

At present, the field of embodied energy analysis is generally still only of academic interest and it is difficult to obtain reliable data for embodied energy. However, research findings in some countries indicate that the operating energy often represents the largest component of life-cycle energy use. Accordingly, most people, when studying low energy buildings, would prefer to focus on operating energy, and perhaps carry out only a general assessment of embodied energy.

The increased availability of reliable and efficient energy services stimulates new development alternatives (Omer, 2009). This study discusses the potential for such integrated systems in the stationary and portable power market in response to the critical need for a cleaner energy technology. Anticipated patterns of future energy use and consequent environmental impacts (acid precipitation, ozone depletion and the greenhouse effect or global warming) are comprehensively discussed in this approach. Throughout the theme several issues relating to renewable energies, environment and sustainable development are examined from both current and future perspectives. It is concluded that renewable environmentally friendly energy must be encouraged, promoted, implemented and demonstrated by full-scale plant (device) especially for use in remote rural areas.

In many countries, global warming considerations have led to efforts to reduce fossil energy use and to promote renewable energies in the building sector. Energy use reductions can be achieved by minimising the energy demand, by rational energy use, by recovering heat and cold and by using energy from the ambient air and from the ground. To keep the environmental impact of a building at sustainable levels (e.g., by greenhouse gas (GHG) neutral emissions), the residual energy demand must be covered with renewable energy. In this theme integral concepts for buildings with both excellent indoor environment control and sustainable environmental impact are presented. Special emphasis is put on ventilation concepts utilising ambient energy from the air, the ground and other renewable energy sources, and on the interaction with heating and cooling. It is essential to avoid the need for mechanical cooling, e.g., by peak load cutting, load shifting and the use of ambient heat or cold from the air or the ground. Techniques considered are hybrid (controlled natural and mechanical) ventilation including night ventilation, thermo-active building mass systems with free cooling in a cooling tower, and air intake via ground heat exchangers. For both residential and office buildings, the electricity demand remains one of the crucial elements to meet sustainability requirements. The electricity demand of ventilation systems is related to the overall demand of the building and the potential of photovoltaic systems and advanced co-generation units.

In climate-sensitive architecture, strategies are adopted to meet occupants' needs, taking into account local solar radiation, temperature, wind and other climatic conditions. Different strategies are required for the various seasons. These strategies can themselves be subdivided into a certain number of concepts, which represent actions.

The heating strategy includes four concepts:

- Solar collection: collection of the sun's heat through the building envelope.
- Heat storage: storage of the heat in the mass of the walls and floors.
- Heat distribution: distribution of collected heat to the different spaces, which require heating.
- Heat conservation: retention of heat within the building.

The cooling strategy includes five concepts:

- Solar control: protection of the building from direct solar radiation.
- Ventilation: expelling and replacing unwanted hot air.
- Internal gains minimisation: reducing heat from occupants, equipments and artificial lighting.
- External gains avoidance: protection from unwanted heat by infiltration or conduction through the envelope (hot climates).
- Natural cooling: improving natural ventilation by acting on the external air (hot climates).

The daylighting strategy includes four concepts:

- Penetration: collection of natural light inside the building.
- Distribution: homogeneous spreading of light into the spaces or focusing.
- Protect: reducing by external shading devices the sun's rays penetration into the building.
- Control: control light penetration by movable screens to avoid discomfort.

The admission of daylight into buildings alone does not guarantee that the design will be energy efficient in terms of lighting. In fact, the design for increased daylight can often raise concerns relating to visual comfort (glare) and thermal comfort (increased solar gain in the summer and heat losses in the winter from larger apertures). Such issues will clearly need to address in the design of the window openings, blinds, shading devices, heating systems, etc. Simple techniques can be implemented to increase the probability that lights are switched off (Omer, 2009). These include: (1) making switches conspicuous (2) locating switches appropriately in relation to the lights (3) switching banks of lights independently, and (4) switching banks of lights parallel to the main window wall.

Large energy savings cover a wide range of issues including:

- Guidelines on low energy design.
- Natural and artificial lighting.
- Solar gain and solar shading.
- Fenestration design.
- Energy efficient plant and controls.
- Examining the need for air conditioning.

The strategy:

- Integration of shading and daylighting: an integral strategy is essential and feasible where daylighting and shading can be improved simultaneously.
- Effect of shading on summer comfort conditions: solar shading plays a central role in reducing overheating risks and gives the potential for individual control, but should be complimented with other passive design strategies.
- Effect of devices on daylighting conditions: devices can be designed to provide shading whist improving the daylight conditions, notably glare and the distribution of light in a space, thus improving the visual quality.
- Energy savings: energy savings from the avoidance of air conditioning can be very substantial, whilst daylighting strategies need to integrated with artificial lighting systems to be beneficial in terms of energy use.

The energy potential of daylighting is thus inextricably linked with the energy use of the associated artificial lighting systems and their controls. The economics of daylighting are not only related to energy use but also to productivity. Good daylighting of workspaces helps to promote efficient productive work, and simultaneously increases the sense of well-being.

However, energy and economics should not become the sole concern of daylighting design to the exclusion of perceptual considerations.

The comfort in a greenhouse depends on many environmental parameters. These include temperature, relative humidity, air quality and lighting. Although greenhouse and conservatory originally both meant a place to house or conserve greens (variegated hollies, cirrus, myrtles and oleanders), a greenhouse today implies a place in which plants are raised while conservatory usually describes a glazed room where plants may or may not play a significant role. Indeed, a greenhouse can be used for so many different purposes. It is, therefore, difficult to decide how to group the information about the plants that can be grown inside it. Whereas heat loss in winter a problem, it can be a positive advantage when greenhouse temperatures soar considerably above outside temperatures in summer. Indoor relative humidity control is one of the most effective long-term mite control measures. There are many ways in which the internal relative humidity can be controlled including the use of appropriate ventilation, the reduction of internal moisture production and maintenance of adequate internal temperatures through the use of efficient heating and insulation.

The introduction of a reflecting wall at the back of a greenhouse considerably enhances the solar radiation that reaches the ground level at any particular time of the day. The energy yield of the greenhouse with any type of reflecting wall was also significantly increased. The increase in energy efficiency was obtained by calculating the ratio between the total energy received during the day in greenhouse with a reflecting wall, compared to that in a classical greenhouse. Hence, the energy balance was significantly shifted towards conservation of classical energy for heating or lighting. The four-fold greater amount of energy that can be captured by virtue of using a reflecting wall with an adjustable inclination and louvers during winter attracts special attention. When sky (diffuse) radiation that was received by the ground in amounts, were taken into account, the values of the enhancement coefficients were reduced to some extent: this was due to the fact that they added up to the direct radiation from the sun in both new and classical greenhouses. However, this is a useful effect as further increases overall energy gain. There is also an ironing out effect expressed in terms of the ratios between peak and average insolations.

Finally, the presented theory can be used to calculate the expected effects of the reflecting wall at any particular latitude, under different weather conditions, and when the average numbers of clear days are taken into account. Thereby an assessment of the cost of a particular setup can be obtained. Under circumstances of a few clear days, it may still be worthwhile from a financial point of view to turn a classical greenhouse into one with a reflecting wall by simply covering the glass wall on the north-facing side with aluminum foil with virtually negligible expenditure.

CONCLUSION

With environmental protection posing as the number one global problem, man has no choice but to reduce his energy consumption. One way to accomplish this is to resort to passive and low-energy systems to maintain thermal comfort in buildings. The conventional and modern designs of wind towers can successfully be used in hot arid regions to maintain thermal comfort (with or without the use of ceiling fans) during all hours of the cooling season, or a fraction of it. Climatic design is one of the best approaches to reduce the energy cost in buildings. Proper design is the first step of defence against the stress of the climate. Buildings should be designed according to the climate of the site, reducing the need for mechanical heating or cooling. Hence maximum natural energy can be used for creating a pleasant environment inside the built envelope. Technology and industry progress in the last decade diffused electronic and informatics' devices in many human activities, and also in building construction. The utilisation and operating opportunities components, increase the reduction of heat losses by varying the thermal insulation, optimising the lighting distribution with louver screens and operating mechanical ventilation for coolness in indoor spaces. In addition to these parameters the intelligent envelope can act for security control and became an important part of the building revolution. Application of simple passive cooling measures is effective in reducing the cooling load of buildings in hot and humid climates. 43% reductions can be achieved using a combination of well-established technologies such as glazing, shading, insulation, and natural ventilation. More advanced passive cooling techniques such as roof pond, dynamic insulation, and evaporative water jacket need to be considered more closely. The building sector is a major consumer of both energy and materials worldwide, and that consumption is increasing. Most industrialised countries are in addition becoming more and more dependent on external supplies of conventional energy carriers, i.e., fossil fuels. Energy for heating and cooling can be replaced by new renewable energy sources. New renewable energy sources, however, are usually not economically feasible compared with the traditional carriers. In order to achieve the major changes needed to alleviate the environmental impacts of the building sector, it is necessary to change and develop both the processes in the industry itself, and to build a favourable framework to overcome the present economic, regulatory and institutional barriers. Ground source heat pumps are receiving increasing interest because of their potential to reduce primary energy consumption and reduce emissions of the GHGs. The technology is well established in the North America and parts of Europe, but is at the demonstration stage in the United Kingdom. Benefits to the community at large will result from the reduction in fossil consumption and the resulting environment benefits. By reducing primary energy consumption, the use of the GSHPs has the potential to reduce the quantity of CO_2 produced by the combustion of fossil fuels and thus to reduce global warming.

REFERENCES

Abdeen M. Omer. (2008a). People, power and pollution, *Renewable and Sustainable Energy Reviews,* Vol.12 No.7, pp.1864-1889, United Kingdom, September 2008.

Abdeen M. Omer. (2008b). Energy, environment and sustainable development, *Renewable and Sustainable Energy Reviews*, Vol.12, No.9, pp.2265-2300, United Kingdom, December 2008.

Abdeen M. Omer. (2008c). Focus on low carbon technologies: the positive solution, Renewable and Sustainable Energy Reviews, Vol.12, No.9, pp.2331-2357, United Kingdom, December 2008.

Abdeen M. Omer. (2008d). Chapter 10: Development of integrated bioenergy for improvement of quality of life of poor people in developing countries, In: Energy *in*

Europe: Economics, Policy and Strategy- IB, Editors: Flip L. Magnusson and Oscar W.
    Bengtsson, 2008 NOVA Science Publishers, Inc., pp.341-373, New York, USA.

Abdeen M. Omer. (2009a). Environmental and socio-economic aspect of possible
    development in renewable energy use, In: *Proceedings of the 4th International
    Symposium on Environment*, Athens, Greece, 21-24 May 2009.

Abdeen M. Omer. (2009b). Energy use, environment and sustainable development, In:
    *Proceedings of the 3rd International Conference on Sustainable Energy and
    Environmental Protection (SEEP 2009)*, Paper No.1011, Dublin, Republic of Ireland, 12-
    15 August 2009.

Abdeen M. Omer. (2009c). Energy use and environmental: impacts: a general review, *Journal
    of Renewable and Sustainable Energy*, Vol.1, No.053101, pp.1-29, United State of
    America, September 2009.

Abdeen M. Omer. (2009d). Chapter 3: Energy use, environment and sustainable development,
    In: *Environmental Cost Management*, Editors: Randi Taylor Mancuso, 2009 NOVA
    Science Publishers, Inc., pp.129-166, New York, USA.

Abdel E. (1994). Low-energy buildings. *Energy and Buildings*. 1994: 21(3): 169-74.

Air-Conditioning and Refrigeration Institute (ACRI). (1991). Directory of Certified Applied
    Air-Conditioning Products. Arlington, VA.

American Gas Association (MGA). (1992). Electric Heat Pumps: How Good Are They?

ASHRAE. (1993). Principle of the heating, ventilation and air conditioning. *Handbook-
    Fundamentals. American Society of Heat and Refrigeration and Air Conditioning
    Engineers*. USA.

ASHRAE Transactions. (2000). V.96, Pt.1, Atlanta, GA.

Bose, J.E. (1988). Closed-loop/ground-source Heat Pump Systems. International Association
    Ground-Source Heat Pumps.

Cavanaugh, S.P. (1992). Ground and Water Source Heat Pumps. IGSHPA. Stillwater, UK.

CIBSE. (1998). Energy efficiency in buildings: CIBSE guide. London: Chattered Institution
    of Building Services Engineers (CIBSE).

Crandall, A.C. (1946). House heating with earth heat pump. Electrical World, 126(19): 94-95.
    New York.

Drafz, H.J. (1982). *Ground source heat pumps*. ETA, 40(A): 222-226.

Fordham, M. (2000). Natural ventilation. *Renewable Energy*, 19: 17-37.

FSEC. (1998). Principles of low energy building design in warm, humid climates. Cape
    Canaveral (FL): *Florida Solar Energy Centre* (FSEC).

Givoni B. (1994). *Passive and low energy cooling of buildings*. New York: Van Nostrand
    Reinhold.

Granryd, E.G. (1979). *Ground source heat pump systems in a northern climate*. 15th IIR
    Congress, Venice Paper No. E: 1-82.

Heat Pump Association (HPA). (1992). *Ground source heat pumps*. Stillwater, OK.

Jones, P.G., and Cheshire, D. (1996). Bulk data for marking non-domestic building energy
    consumption. CIBSE/ASHRAE Joint National Conference, pp.203-213. Harrogate: 29
    September – 1 October 1996.

Kemler, E.N. (1947). Methods of earth heat recovery for the heat pump. Housing and
    Ventilating. New York.

Kusuda, T., and Achenbach, P. (1965). Earth temperature and thermal diffusivity at selected
    stations in the United States. *ASHRAE Trans.*, 71 (1): 61-75.

Omer, A. M. (2009). Energy use, environment and sustainable development, In: *Proceedings of the 3rd International Conference on Sustainable Energy and Environmental Protection (SEEP 2009),* Paper No.1011, Dublin, Republic of Ireland, 12-15 August 2009.

Owens S. (1986). *Energy, planning and urban form.* London: Pion.

Pietsch, J.A. (1988). *Water-Loop Heat Pump Systems: Assessment Study.* Electric Power.

Rosenblad, G. (1979). Earth heat pumps system with vertical pipes for heat extraction and storage. *Proc. Nordic Symposium of Earth Heat Pump Systems*, Goteborg, pp. 102-110.

Ross, D.P. (1990). Service Life of Water-Loop Heat Pump Compressors in Commercial Buildings.

Sanner, B., Hopkirk R., Kabus, F., Ritter W., and Rybach, L. (1996). Practical experiences in Europe of the combination of geothermal energy and heat pumps. *Proc. 5th IEA Conference on Heat Pumping Technologies,* pp. 111-125.

State Projects (SP). (1993). *Building Energy Manual.* Australia: NSW Public Works.

Sumner, J.A. (1976). *Domestic heat pumps.* Prism Press.

Todesco G. (1996). Super-efficient buildings: how low can you go? *ASHRAE Journal,* 1996; 38(12): 35-40.

Treloar G., Fay R, and Trucker S. (1998). Proceedings of the Embodied Energy: the current state of play, seminar held at Deakin University. Woodstores Campus. Geelong. Australia, 28-29 November 1996. Australia: School of Architecture and Building, Deakin University.

Watson D. (1993). *The energy design handbook.* Washington (DC): American Institute of Architects Press.

Wirth, P.E. (1955). *History of the development of heat pumps.* Schweiz Bouzeits, 73(42): 47-51.

Yuichiro K., Cook J., and Simos Y. (1991). Passive and low energy architecture. In: *Special Issue of Process Architecture,* Tokyo: Japan.

SOME USEFUL FIGURES AND CONVERSIONS

1 kW (kilowatt) is a unit of power, or a rate of energy (A 1 bar fire consumes 1 kW)

There are 3,411 Btus in 1 kWatt, i.e., 10 kWatts = 31,400 Btu/hr

There are 860 kcal/h in 1 kW

1 kWh (Kilowatt hour) is a quantity of energy

(A 1 kW heater would use 24 kWhr per day)

I kWatt hr = 1 unit of electricity = 1 bar fire used for one hour

Gas bills now use kWhr instead of the old confusing units Thermo, etc.

1 kJoule x 3,600 = 1 kWhr

If 10 kWatts were extracted from water having a flow rate of 0.8 lit/sec, then the temperature would drop by $3°C$ (3K).

$0°C = 32°F$ (freezing point of water)

$20°C = 68°F$ (room temperature)

$100°C = 212°F$ (boiling point of water)

$(°F-32)/(9x5)=°C$, or $°Cx9/5+32=°F$

1 lit/sec = 13.19 Galls (UK)/min

Chapter 3

SUSTAINABLE DEVELOPMENT OF INTEGRATED BIOMASS POTENTIAL, WASTER MANAGEMENT AND PROPER UTILISATION IN AFRICA

ABSTRACT

This communication discusses a comprehensive review of biomass energy sources, environment and sustainable development. This includes all the biomass energy technologies, energy efficiency systems, energy conservation scenarios, energy savings and other mitigation measures necessary to reduce emissions globally. The current literature is reviewed regarding the ecological, social, cultural and economic impacts of biomass technology. This study gives an overview of present and future use of biomass as an industrial feedstock for production of fuels, chemicals and other materials. However, to be truly competitive in an open market situation, higher value products are required. Results suggest that biomass technology must be encouraged, promoted, invested, implemented, and demonstrated, but especially in remote rural areas.

Keywords: biomass resources, wastes, energy, environment, sustainable development

1. INTRODUCTION

This study highlights the energy problem and the possible saving that can be achieved through the use of biomass sources energy. Also, this study clarifies the background of the study, highlights the potential energy saving that could be achieved through use of biomass energy source and describes the objectives, approach and scope of the theme.

The aim of any modern biomass energy systems must be:

- To maximise yields with minimum inputs.
- Utilisation and selection of adequate plant materials and processes.
- Optimum use of land, water, and fertiliser.
- Create an adequate infrastructure and strong R and D base.

There is strong scientific evidence that the average temperature of the earth's surface is rising. This was a result of the increased concentration of carbon dioxide (CO_2), and other greenhouse gases (GHGs) in the atmosphere as released by burning fossil fuels (Robinson, 2007; Omer, 2008). This global warming will eventually lead to substantial changes in the world's climate, which will, in turn, have a major impact on human life and the environment. Energy use can be achieved by minimising the energy demand, by rational energy use, by recovering heat and the use of more green energies. This will lead to fossil fuels emission reduction. This study was a step towards achieving this goal. The adoption of green or sustainable approaches to the way in which society is run is seen as an important strategy in finding a solution to the energy problem. The key factors to reducing and controlling CO_2, which is the major contributor to global warming, are the use of alternative approaches to energy generation and the exploration of how these alternatives are used today and may be used in the future as green energy sources. Even with modest assumptions about the availability of land, comprehensive fuel-wood farming programmes offer significant energy, economic and environmental benefits. These benefits would be dispersed in rural areas where they are greatly needed and can serve as linkages for further rural economic development. The nations as a whole would benefit from savings in foreign exchange, improved energy security, and socio-economic improvements. With a nine-fold increase in forest – plantation cover, the nation's resource base would be greatly improved. The non-technical issues, which have recently gained attention, include: (1) Environmental and ecological factors, e.g., carbon sequestration, reforestation and revegetation. (2) Renewables as a CO_2 neutral replacement for fossil fuels. (3) Greater recognition of the importance of renewable energy, particularly modern biomass energy carriers, at the policy and planning levels. (4) Greater recognition of the difficulties of gathering good and reliable biomass energy data, and efforts to improve it. (5) Studies on the detrimental health efforts of biomass energy particularly from traditional energy users. There is a need for some further development to suit local conditions, to minimise spares holdings, to maximise interchangeability both of engine parts and of the engine application. Emphasis should be placed on full local manufacture (Abdeen, 2008a).

Energy is an essential factor in development since it stimulates, and supports economic growth and development. Fossil fuels, especially oil and natural gas, are finite in extent, and should be regarded as depleting assets, and efforts are oriented to search for new sources of energy. The clamour all over the world for the need to conserve energy and the environment has intensified as traditional energy resources continue to dwindle whilst the environment becomes increasingly degraded. Alternatively energy sources can potentially help fulfill the acute energy demand and sustain economic growth in many regions of the world. Bioenergy is beginning to gain importance in the global fight to prevent climate change. The scope for exploiting organic waste as a source of energy is not limited to direct incineration or burning refuse-derived fuels. Biogas, biofuels and woody biomass are other forms of energy sources that can be derived from organic waste materials. These biomass energy sources have significant potential in the fight against climate change (Abdeen, 2008b).

Conservation of energy and rationing in some form will however have to be practised by most countries, to reduce oil imports and redress balance of payments positions. Meanwhile development and application of nuclear power and some of the traditional solar, wind, biomass and water energy alternatives must be set in hand to supplement what remains of the fossil fuels. The encouragement of greater energy use is an essential component of development. In the short-term it requires mechanisms to enable the rapid increase in energy/capita, and in the long-term we should be working towards a way of life, which makes use of energy efficiency and without the impairment of the environment or of causing safety problems. Such a programme should as far as possible be based on renewable energy resources (Abdeen, 2008c).

Large-scale, conventional, power plant such as hydropower has an important part to play in development. It does not, however, provide a complete solution. There is an important complementary role for the greater use of small-scale, rural based-power plants. Such plant can be used to assist development since it can be made locally using local resources, enabling a rapid built-up in total equipment to be made without a corresponding and unacceptably large demand on central funds. Renewable resources are particularly suitable for providing the energy for such equipment and its use is also compatible with the long-term aims. In compiling energy consumption data one can categorise usage according to a number of different schemes:

- Traditional sector - industrial, transportation, etc.
- End-use- space heating, process steam, etc.
- Final demand - total energy consumption related to automobiles, to food, etc.
- Energy source - oil, coal, etc.
- Energy form at point of use - electric drive, low temperature heat, etc.

2. BIOENERGY DEVELOPMENT

Bioenergy is energy from the sun stored in materials of biological origin. This includes plant matter and animal waste, known as biomass. Plants store solar energy through photosynthesis in cellulose and lignin, whereas animals store energy as fats. When burned, these sugars break down and release energy exothermically, releasing carbon dioxide (CO_2), heat and steam. The by-products of this reaction can be captured and manipulated to create power, commonly called bioenergy. Biomass is considered renewable because the carbon (C) is taken out of the atmosphere and replenished more quickly than the millions of years required for fossil fuels to form. The use of biofuels to replace fossil fuels contributes to a reduction in the overall release of carbon dioxide into the atmosphere and hence helps to tackle global warming (Abdeen, 2008d).

The biomass energy resources are particularly suited for the provision of rural power supplies and a major advantage is that equipment such as flat plate solar driers, wind machines, etc., can be constructed using local resources and without the high capital cost of more conventional equipment. Further advantage results from the feasibility of local maintenance and the general encouragement such local manufacture gives to the build up of small scale rural based industry.

Table 1 lists the energy sources available. Currently the 'non-commercial' fuels wood, crop residues and animal dung are used in large amounts in the rural areas of developing countries, principally for heating and cooking; the method of use is highly inefficient. Table 2 presented some renewable applications. Table 3 lists the most important of energy needs. Table 4 listed methods of energy conversion.

Table 1. Sources of energy (Omer, 2008)

Energy source	Energy carrier	Energy end-use
Vegetation	Fuel-wood	Cooking Water heating Building materials Animal fodder preparation
Oil	Kerosene	Lighting Ignition fires
Dry cells	Dry cell batteries	Lighting Small appliances
Muscle power	Animal power	Transport Land preparation for farming Food preparation (threshing)
Muscle power	Human power	Transport Land preparation for farming Food preparation (threshing)

Table 2. Renewable applications (Omer, 2008)

Systems	Applications
Water supply	Rain collection, purification, storage and recycling
Wastes disposal	Anaerobic digestion (CH_4)
Cooking	Methane
Food	Cultivate the 1 hectare plot and greenhouse for four people
Electrical demands	Wind generator
Space heating	Solar collectors
Water heating	Solar collectors and excess wind energy
Control system	Ultimately hardware
Building fabric	Integration of subsystems to cut costs

Table 3. Energy needs in rural areas (Omer, 2008)

Transport, e.g., small vehicles and boats
Agricultural machinery, e.g., two-wheeled tractors
Crop processing, e.g., milling
Water pumping
Small industries, e.g., workshop equipment
Electricity generation, e.g., hospitals and schools
Domestic, e.g., cooking, heating, and lighting

Considerations when selecting power plant include the following:

- Power level - whether continuous or discontinuous.
- Cost - initial cost, total running cost including fuel, maintenance and capital amortised over life.
- Complexity of operation.
- Maintenance and availability of spares.
- Life and suitability for local manufacture.

The internal combustion engine is a major contributor to rising CO_2 emissions worldwide and some pretty dramatic new thinking is needed if our planet is to counter the effects. With its use increasing in developing world economies, there is something to be said for the argument that the vehicles we use to help keep our inner-city environments free from waste, litter and grime should be at the forefront of developments in low-emissions technology. Materials handled by waste management companies are becoming increasingly valuable. Those responsible for the security of facilities that treat waste or manage scrap will testify to the precautions needed to fight an ongoing battle against unauthorised access by criminals and crucially, to prevent the damage they can cause through theft, vandalism or even arson. Of particular concern is the escalating level of metal theft, driven by various factors including the demand for metal in rapidly developing economies such as India and China (Abdeen, 2008e).

Table 4. Methods of energy conversion (Omer, 2007)

Muscle power	Man, animals
Internal combustion engines	
Reciprocating	Petrol- spark ignition
	Diesel- compression ignition
	Humphrey water piston
Rotating	Gas turbines
Heat engines	
Vapour (Rankine)	
Reciprocating	Steam engine
Rotating	Steam turbine
Gas Stirling (Reciprocating)	Steam engine
Gas Brayton (Rotating)	Steam turbine
Electron gas	Thermionic, thermoelectric
Electromagnetic radiation	Photo devices
Hydraulic engines	Wheels, screws, buckets, turbines
Wind engines (wind machines)	Vertical axis, horizontal axis
Electrical/mechanical	Dynamo/alternator, motor

There is a need for greater attention to be devoted to this field in the development of new designs, the dissemination of information and the encouragement of its use. International and government bodies and independent organisations all have a role to play in biomass energy technologies. Environment has no precise limits because it is in fact a part of everything. Indeed, environment is, as anyone probably already knows, not only flowers blossoming or birds singing in the spring, or a lake surrounded by beautiful mountains. It is also human settlements, the places where people live, work, rest, the quality of the food we eat, the noise or silence of the street they live in. Environment is not only the fact that our cars consume a good deal of energy and pollute the air, but also, that we often need them to go to work and for holidays. Obviously man uses energy just as plants, bacteria, mushrooms, bees, fish and rats do (Figure 1).

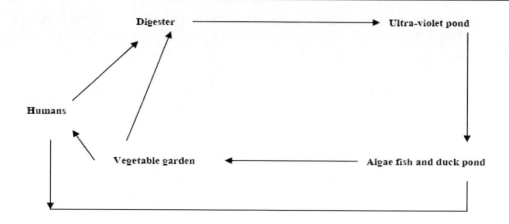

Figure 1. Biomass utilisation cycle concepts (Omer, 2006).

Man largely uses solar energy- food, hydropower, wood- and thus participates harmoniously in the natural flow of energy through the environment. But man also uses oil, gas, coal and nuclear power. We always modify our environment with or without this source of energy (Brain, and Mark, 2007). Economic importance of environmental issue is increasing, and new technologies are expected to reduce pollution derived both from productive processes and products, with costs that are still unknown. This is due to market uncertainty, weak appropriability regime, lack of a dominant design, and difficulties in reconfiguring organisational routines. The degradation of the global environment is one of the most serious energy issues (Abdeen, 2009a).

3. Energy Use and the Environment

The range of waste treatment technologies that are tailored to produce bioenergy is growing. There are a number of key areas of bioenergy from wastes including (but not limited to) biogas, biofuels and bioheat. When considering using bioenergy, it is important to take into account the overall emission of carbon in the process of electricity production. Energy use is one of several essential components for every country:

- The overall situation and the implications of increased energy use in the future.
- The problem of the provision of power in rural areas, including the consideration of energy resources and energy conversion.

Table 5. Annual GHG emissions from different types of power plants (Robinson, 2007)

Primary source of energy	Emissions (x 10^3 metric tons CO_2)		Waste (x 10^3 metric tons CO_2)	Area (km^2)
	Atmosphere	Water		
Coal	380	7-41	60-3000	120
Oil	70-160	3-6	Negligible	70-84
Gas	24	1	-	84
Nuclear	6	21	2600	77

Table 6. Energy consumption per person (Robinson, 2007)

Region	Population (millions)	Energy per person (Watt/m^2)
Africa	820	0.54
Asia	3780	2.74
Central America	180	1.44
North America	335	0.34
South America	475	0.52
Western Europe	445	2.24
Eastern Europe	130	2.57
Oceania	35	0.08
Russia	330	0.29

In addition to the drain on resources, such an increase in consumption consequences, together with the increased hazards of pollution and the safety problems associated with a large nuclear fission programmes. It would be equally unacceptable to suggest that the difference in energy between the developed and developing countries and prudent for the developed countries to move towards a way of life which, whilst maintaining or even increasing quality of life, reduce significantly the energy consumption per capita. Such savings can be achieved in a number of ways:

- Improved efficiency of energy use, for example environmental cost of thermal insulation must be taken into account, energy recovery, and total energy.
- Conservation of energy resources by design for long life and recycling rather than the short life throwaway product and systematic replanning of our way of life, for example in the field of transport.

Energy ratio (Er) is defined as the ratio of Energy content (Ec) of the food product / Energy input (Ei) to produce the food.

$$Er = Ec/Ei \qquad (1)$$

4. COMBINED HEAT AND POWER (CHP)

The atmospheric emissions of fossil fuelled installations are mostly aldehydes ($CH_3CH_2CH_2CHO$), carbon monoxide (CO), nitrogen oxides (NO_X), sulpher oxides (SOx) and particles (i.e., ash) as well as carbon dioxide. Table 5 shows estimates include not only the releases occurring at the power plant itself but also cover fuel extraction and treatment, as well as the storage of wastes and the area of land required for operation (Table 6). A review of the potential range of recyclables is presented in Table 7.

Combined heat and power (CHP) installations are quite common in greenhouses, which grow high-energy, input crops (e.g., salad vegetables, pot plants, etc.). Scientific assumptions for a short-term energy strategy suggest that the most economically efficient way to replace the thermal plants is to modernise existing power plants to increase their energy efficiency and to improve their environmental performance (Pernille, 2004).

**Table 7. Summary of material recycling practices in the construction sector
(Robinson, 2007)**

Construction and demolition material	Recycling technology options	Recycling product
Asphalt	Cold recycling: heat generation; Minnesota process; parallel drum process; elongated drum; microwave asphalt recycling system; finfalt; surface regeneration	Recycling asphalt; asphalt aggregate
Brick	Burn to ash, crush into aggregate	Slime burn ash; filling material; hardcore
Concrete	Crush into aggregate	Recycling aggregate; cement replacement; protection of levee; backfilling; filter
Ferrous metal	Melt; reuse directly	Recycled steel scrap
Glass	Reuse directly; grind to powder; polishing; crush into aggregate; burn to ash	Recycled window unit; glass fibre; filling material; tile; paving block; asphalt; recycled aggregate; cement replacement; manmade soil
Masonry	Crush into aggregate; heat to 900°C to ash	Thermal insulating concrete; traditional clay
Non-ferrous metal	Melt	Recycled metal
Paper and cardboard	Purification	Recycled paper
Plastic	Convert to powder by cryogenic milling; clopping; crush into aggregate; burn to ash	Panel; recycled plastic; plastic lumber; recycled aggregate; landfill drainage; asphalt; manmade soil
Timber	Reuse directly; cut into aggregate; blast furnace deoxidisation; gasification or pyrolysis; chipping; moulding by pressurising timber chip under steam and water	Whole timber; furniture and kitchen utensils; lightweight recycled aggregate; source of energy; chemical production; wood-based panel; plastic lumber; geofibre; insulation board

However, utilisation of wind power and the conversion of gas-fired CHP plants to biomass would significantly reduce the dependence on imported fossil fuels. Although a lack of generating capacity is forecasted in the long-term, utilisation of the existing renewable energy potential and the huge possibilities for increasing energy efficiency are sufficient to meet future energy demands in the short-term (Pernille, 2004).

A total shift towards a sustainable energy system is a complex and long process, but is one that can be achieved within a period of about 20 years. Implementation will require initial investment, long-term national strategies and action plans. However, the changes will have a number of benefits including a more stable energy supply than at present and major improvement in the environmental performance of the energy sector, and certain social benefits (Figure 2).

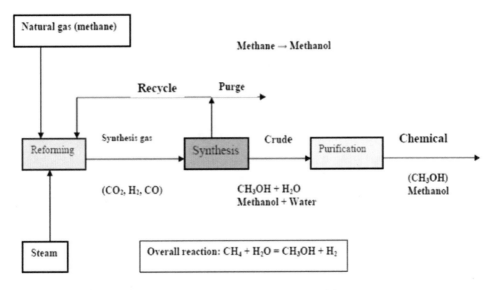

Figure 2. Schematic diagram shows methanol production (Omer, 2006).

A vision that used methodologies and calculations based on computer modelling can utilised:

- Data from existing governmental programmes.
- Potential renewable energy sources and energy efficiency improvements.
- Assumptions for future economy growth.
- Information from studies and surveys on the recent situation in the energy sector.

The main advantages are related to energy, agriculture and environment problems, are foreseeable both at national level and at worldwide level and can be summarised as follows:

- Reduction of dependence on import of energy and related products.
- Reduction of environmental impact of energy production (greenhouse effect, air pollution, and waste degradation).
- Substitution of food crops and reduction of food surpluses and of related economic burdens, and development of new know-how and production of technological innovation.
- Utilisation of marginal lands and of set aside lands and reduction of related socio-economic and environmental problems (soil erosion, urbanisation, landscape deterioration, etc.).

In some countries, a wide range of economic incentives and other measures are already helping to protect the environment. These include:

- Taxes and user charges that reflect the costs of using the environment, e.g., pollution taxes and waste disposal charges.
- Subsidies, credits and grants that encourage environmental protection.

- Deposit-refund systems that prevent pollution on resource misuse and promote product reuse or recycling.
- Financial enforcement incentives, e.g., fines for non-compliance with environmental regulations.
- Tradable permits for activities that harm the environment.

District Heating (DH), also known as community heating can be a key factor to achieve energy savings, reduce CO2 emissions and at the same time provide consumers with a high quality heat supply at a competitive price. The DH should generally only be considered for areas where the heat density is sufficiently high to make DH economical. In countries like Denmark DH may today be economical even to new developments with lower density areas due to the high level of taxation on oil and gas fuels combined with the efficient production of the DH. To improve the opportunity for the DH local councils can adapt the following plan:

- Analyse the options for heat supply during local planning stage.
- In areas where DH is the least cost solution it should be made part of the infrastructure just like for instance water and sewage connecting all existing and new buildings.
- Where possible all public buildings should be connected to the DH.
- The government provides low interest loans or funding to minimise conversion costs for its citizens.
- Use other powers, for instance national legislation to ensure the most economical development of the heat supply and enable an obligation to connect buildings to a DH scheme.

Denmark has broadly seen three scales of the CHP which where largely implemented in the following chronological order (Pernille, 2004):

- Large-scale CHP in cities (>50 MWe).
- Small (5 kWe – 5 MWe) and medium-scale (5-50 MWe).
- Industrial and small-scale CHP.

Combined heat and power (CHP) installations are quite common in greenhouses, which grow high-energy, input crops (e.g., salad vegetables, pot plants, etc.). Most of the heat is produced by large CHP plants (gas-fired combined cycle plants using natural gas, biomass, waste or biogas). DH is energy efficient because of the way the heat is produced and the required temperature level is an important factor. Buildings can be heated to temperature of $21^{o}C$ and domestic hot water (DHW) can be supplied with a temperature of $55^{o}C$ using energy sources that are most efficient when producing low temperature levels ($<95^{o}C$) for the DH water. Most of these heat sources are CO_2 neutral or emit low levels. Only a few of these sources are available to small individual systems at a reasonably cost, whereas DH schemes because of the plant's size and location can have access to most of the heat sources and at a low cost. Low temperature DH, with return temperatures of around $30-40^{o}C$ can utilise the following heat sources:

- Efficient use of the CHP by extracting heat at low calorific value (CV).
- Efficient use of biomass or gas boilers by condensing heat in economisers (Table 8).
- Efficient utilisation of geothermal energy.
- Direct utilisation of excess low temperature heat from industrial processes.
- Efficient use of large-scale solar heating plants.

Table 8. Final energy projections including biomass (Mtoe) (D'Apote, 1998)

Region 2011				
	Biomass	Conventional Energy	Total	Share of Biomass (%)
Africa	205	136	341	60
China	206	649	855	24
East Asia	106	316	422	25
Latin America	73	342	415	18
South Asia	235	188	423	56
Total developing countries	825	1632	2457	34
Other non-OECD * countries	24	1037	1061	2
Total non-OECD* countries	849	2669	3518	24
OECD countries	81	3044	3125	3
World	930	5713	6643	14
Region 2020				
	Biomass	Conventional Energy	Total	Share of Biomass (%)
Africa	371	266	637	59
China	224	1524	1748	13
East Asia	118	813	931	13
Latin America	81	706	787	10
South Asia	276	523	799	35
Total developing countries	1071	3825	4896	22
Other non-OECD * countries	26	1669	1695	2
Total non-OECD* countries	1097	5494	6591	17
OECD countries	96	3872	3968	2
World	1193	9365	10558	11

* Organisation for Economic Co-operation and Development.

Heat tariffs may include a number of components such as a connection charge, a fixed charge and a variable energy charge. Also, consumers may be incentivised to lower the return temperature. Hence, it is difficult to generalise but the heat practice for any DH company no matter what the ownership structure can be highlighted as follows:

- To develop and maintain a development plan for the connection of new consumers.
- To evaluate the options for least cost production of heat.
- To implement the most competitive solutions by signing agreements with other companies or by implementing own investment projects.
- To monitor all internal costs and with the help of benchmarking, and improve the efficiency of the company.

- To maintain a good relationship with the consumer and deliver heat supply services at a sufficient quality.

Installing DH should be pursued to meet the objectives for improving the environment through the improvement of energy efficiency in the heating sector. At the same time DH can serve the consumer with a reasonable quality of heat at the lowest possible cost. The variety of possible solutions combined with the collaboration between individual companies, the district heating association, the suppliers and consultants can, as it has been in Denmark, be the way forward for developing DH in the United Kingdom. Implementation will require initial investment, long-term national strategies and action plans. However, the changes will have a number of benefits including a more stable energy supply than at present and major improvement in the environmental performance of the energy sector, and certain social benefits (Pernille, 2004).

5. BIOMASS UTILISATION AND DEVELOPMENT OF CONVERSION TECHNOLOGIES

Sustainable energy is energy that, in its production or consumption, has minimal negative impacts on human health and the healthy functioning of vital ecological systems, including the global environment. It is an accepted fact that renewable energy is a sustainable form of energy, which has attracted more attention during recent years. A great amount of renewable energy potential, environmental interest, as well as economic consideration of fossil fuel consumption and high emphasis of sustainable development for the future will be needed. Explanations for the use of inefficient agricultural-environmental polices include: the high cost of information required to measure benefits on a site-specific basis, information asymmetries between government agencies and farm decision makers that result in high implementation costs, distribution effects and political considerations (Wu and Boggess, 1999). Achieving the aim of agric-environment schemes through:

- Sustain the beauty and diversity of the landscape.
- Improve and extend wildlife habitats.
- Conserve archaeological sites and historic features.
- Improve opportunities for countryside enjoyment.
- Restore neglected land or features, and
- Create new habitats and landscapes.

The data required to perform the trade-off analysis simulation can be classified according to the divisions given in Table 9: the overall system or individual plants, and the existing situation or future development. The effective economic utilisations of these resources are shown in Table 10, but their use is hindered by many problems such as those related to harvesting, collection, and transportation, besides the sanitary control regulations. Biomass energy is experiencing a surge in an interest stemming from a combination of factors, e.g., greater recognition of its current role and future potential contribution as a modern fuel, global environmental benefits, its development and entrepreneurial opportunities, etc. Possible routes of biomass energy development are shown in Table 11.

The key to successful future appears to lie with successful marketing of the treatment by products. There is also potential for using solid residue in the construction industry as a filling agent for concrete. Research suggests that the composition of the residue locks metals within the material, thus preventing their escape and any subsequent negative effect on the environment (Abdeen, 2009b).

The use of biomass through direct combustion has long been, and still is, the most common mode of biomass utilisation as shown in Tables (9-11). Examples for dry (thermo-chemical) conversion processes are charcoal making from wood (slow pyrolysis), gasification of forest and agricultural residues (fast pyrolysis – this is still in demonstration phase), and of course, direct combustion in stoves, furnaces, etc. Wet processes require substantial amount of water to be mixed with the biomass. Biomass technologies include:

- Briquetting.
- Improved stoves.
- Biogas.
- Improved charcoal.
- Carbonisation.
- Gasification.

Table 9. Classifications of data requirements (Omer, 2008)

	Plant data	System data
Existing data	Size	Peak load
	Life	Load shape
	Cost (fixed and var. O and M)	Capital costs
	Forced outage	Fuel costs
	Maintenance	Depreciation
	Efficiency	Rate of return
	Fuel	Taxes
	Emissions	
Future data	All of above, plus	System lead growth
	Capital costs	Fuel price growth
	Construction trajectory	Fuel import limits
	Date in service	Inflation

The increased demand for gas and petroleum, food crops, fish and large sources of vegetative matter mean that the global harvesting of carbon has in turn intensified. It could be said that mankind is mining nearly everything except its waste piles. It is simply a matter of time until the significant carbon stream present in municipal solid waste is fully captured. In the meantime, the waste industry needs to continue on the pathway to increased awareness and better-optimised biowaste resources. Optimisation of waste carbon may require widespread regulatory drivers (including strict limits on the landfilling of organic materials), public acceptance of the benefits of waste carbon products for soil improvements/crop enhancements and more investment in capital facilities (Abdeen, 2009c). In short, a significant effort will be required in order to capture a greater portion of the carbon stream and put it to beneficial use. From the standpoint of waste practitioners, further research and pilot programmes are necessary before the available carbon in the waste stream can be extracted in sufficient quality and quantities to create the desired end products. Other details need to be ironed out too, including measurement methods, diversion calculations, sequestration values and determination of acceptance contamination thresholds (Abdeen, 2009d).

Table 10. Agricultural residues routes for development (Omer, 2006)

Source	Process	Product	End use
Agricultural residues	Direct	Combustion	Rural poor Urban household Industrial use
	Processing	Briquettes	Industrial use Limited household use
	Processing	Carbonisation (small-scale) Briquettes	Rural household (self sufficiency)
	Carbonisation	Carbonised	Urban fuel
	Fermentation	Biogas	Energy services Household Industry
Agricultural, and animal residues	Direct	Combustion	(Save or less efficiency as wood)
	Briquettes	Direct combustion	(Similar end use devices or improved)
	Carbonisation	Carbonised	Use
	Carbonisation	Briquettes	Briquettes use
	Fermentation	Biogas	Use

Table 11. Effective biomass resource utilisation (Omer, 2007)

Subject	Tools	Constraints
Utilisation and land clearance for agriculture expansion	Stumpage fees Control Extension Conversion Technology	Policy Fuel-wood planning Lack of extension Institutional
Utilisation of agricultural residues	Briquetting Carbonisation Carbonisation and briquetting Fermentation Gasification	Capital Pricing Policy and legislation Social acceptability

5.1.1. Briquette Production

Charcoal stoves are very familiar to African society. As for the stove technology, the present charcoal stove can be used, and can be improved upon for better efficiency. This energy term will be of particular interest to both urban and rural households and all the income groups due to the simplicity, convenience, and lower air polluting characteristics. However, the market price of the fuel together with that of its end-use technology may not enhance its early high market penetration especially in the urban low income and rural households.

Briquetting is the formation of a charcoal (an energy-dense solid fuel source) from otherwise wasted agricultural and forestry residues. One of the disadvantages of wood fuel is that it is bulky with a low energy density and is therefore enquire to transport. Briquette formation allows for a more energy-dense fuel to be delivered, thus reducing the transportation cost and making the resource more competitive. It also adds some uniformity, which makes the fuel more compatible with systems that are sensitive to the specific fuel input (Jeremy, 2005).

5.1.2. Improved Cook Stoves

Traditional wood stoves can be classified into four types: three stone, metal cylindrical shaped, metal tripod and clay type. Another area in which rural energy availability could be secured where woody fuels have become scarce, are the improvements of traditional cookers and ovens to raise the efficiency of fuel saving. Also, to provide a constant fuel supply by planting fast growing trees. The rural development is essential and economically important since it will eventually lead to better standards of living, people's settlement, and self sufficient in the following:

- Food and water supplies.
- Better services in education and health care.
- Good communication modes.

5.1.3. Biogas Technology

Biogas technology cannot only provide fuel, but is also important for comprehensive utilisation of biomass forestry, animal husbandry, fishery, agricultural economy, protecting the environment, realising agricultural recycling, as well as improving the sanitary conditions, in rural areas. The introduction of biogas technology on wide scale has implications for macro planning such as the allocation of government investment and effects on the balance of payments. Factors that determine the rate of acceptance of biogas plants, such as credit facilities and technical backup services, are likely to have to be planned as part of general macro-policy, as do the allocation of research and development funds (Hall and Scrase, 1998).

Biogas is a generic term for gases generated from the decomposition of organic material. As the material breaks down, methane (CH_4) is produced as shown in Figure 3. Sources that generate biogas are numerous and varied. These include landfill sites, wastewater treatment plants and anaerobic digesters. Landfills and wastewater treatment plants emit biogas from decaying waste. To date, the waste industry has focused on controlling these emissions to our environment and in some cases, tapping this potential source of fuel to power gas turbines, thus generating electricity. The primary components of landfill gas are methane (CH_4), carbon dioxide (CO_2), and nitrogen (N_2). The average concentration of methane is ~45%, CO_2 is ~36% and nitrogen is ~18%. Other components in the gas are oxygen (O_2), water vapour and trace amounts of a wide range of non-methane organic compounds (NMOCs).

For hot water and heating, renewables contributions come from biomass power and heat, geothermal direct heat, ground source heat pumps, and rooftop solar hot water and space heating systems. Solar assisted cooling makes a very small but growing contribution. When it comes to the installation of large amounts of the PV, the cities have several important factors in common. These factors include:

- A strong local political commitment to the environment and sustainability.
- The presence of municipal departments or offices dedicated to the environment, and sustainability or renewable energy.
- Information provision about the possibilities of renewables.
- Obligations that some or all buildings include renewable energy.

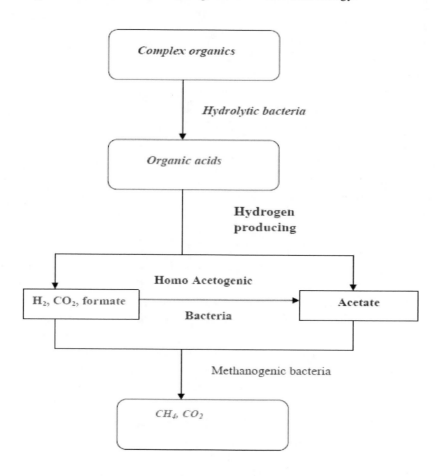

Figure 3. Biogas production process (Omer, 2003).

5.1.4. Improved Forest and Tree Management

Dry cell batteries are a practical but expensive form of mobile fuel that is used by rural people when moving around at night and for powering radios and other small appliances. The high cost of dry cell batteries is financially constraining for rural households, but their popularity gives a good indication of how valuable a versatile fuel like electricity is in rural area. Dry cell batteries can constitute an environmental hazard unless they are recycled in a proper fashion. Direct burning of fuel-wood and crop residues constitute the main usage of biomass, as is the case with many developing countries. However, the direct burning of biomass in an inefficient manner causes economic loss and adversely affects human health. In order to address the problem of inefficiency, research centres around the world have investigated the viability of converting the resource to a more useful form, namely solid briquettes and fuel gas (Figure 4). Biomass resources play a significant role in energy supply in all developing countries. Biomass resources should be divided into residues or dedicated resources, the latter including firewood and charcoal can also be produced from forest residues (Table 12).

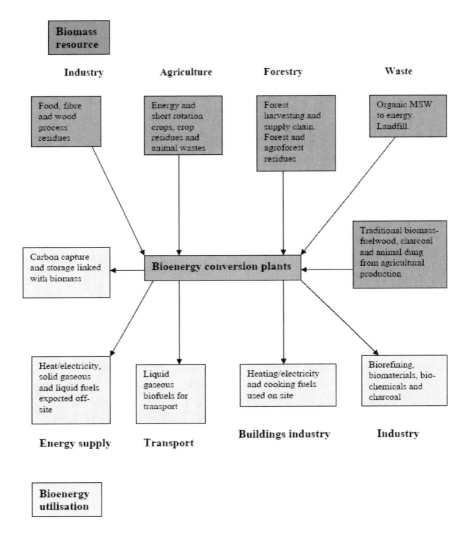

Figure 4. Biomass resources from several sources are converted into a range of products for use by transport, industry and building sectors (Sims, 2007).

Implementing measures for energy efficiency increase at the demand side and in the energy transformation sector. It is common practice to dispose of this waste wood in landfill where it slowly degraded and takes up valuable void space. This wood is a good source of energy and is an alternative to energy crops. Agricultural wastes are abundantly available globally and can be converted to energy and useful chemicals by a number of microorganisms. The success of promoting any technology depends on careful planning, management, implementation, training and monitoring. Main features of gasification project are:

- Networking and institutional development/strengthening.
- Promotion and extension.
- Construction of demonstration projects.
- Research and development, and training and monitoring.

Table 12. Biomass residues and current use (Omer, 2007)

Type of residue	Current use
Wood industry waste	Residues available
Vegetable crop residues	Animal feed
Food processing residue	Energy needs
Sorghum, millet, and wheat residues	Fodder, and building materials
Groundnut shells	Fodder, brick making, and direct fining oil mills
Cotton stalks	Domestic fuel considerable amounts available for short period
Sugar, bagasse, and molasses	Fodder, energy need, and ethanol production (surplus available)
Manure	Fertiliser, brick making, and plastering

5.1.5. Gasification Processes

Gasification is based on the formation of a fuel gas (mostly CO and H_2) by partially oxidising raw solid fuel at high temperatures in the presence of steam or air. The technology can use wood chips, groundnut shells, sugar cane bagasse, and other similar fuels to generate capacities from 3 kW to 100 kW. Three types of gasifier designs have been developed to make use of the diversity of fuel inputs and to meet the requirements of the product gas output (degree of cleanliness, composition, heating value, etc.). The requirements of gas for various purposes, and a comparison between biogas and various commercial fuels in terms of calorific value, and thermal efficiency are presented in Table 13.

Table 13. Comparison of various fuels (Omer, 2003)

Fuel	Calorific value (kcal)	Burning mode	Thermal efficiency (%)
Electricity, kWh	880	Hot plate	70
Coal gas, kg	4004	Standard burner	60
Biogas, m^3	5373	Standard burner	60
Kerosene, l	9122	Pressure stove	50
Charcoal, kg	6930	Open stove	28
Soft coke, kg	6292	Open stove	28
Firewood, kg	3821	Open stove	17
Cow dung, kg	2092	Open stove	11

Sewage sludge is rich in nutrients such as nitrogen and phosphorous. It also contains valuable organic matter, useful for remediation of depleted or eroded soils. This is why untreated sludge has been used for many years as a soil fertiliser and for enhancing the organic matter of soil. A key concern is that treatment of sludge tends to concentrate heavy metals, poorly biodegradable trace organic compounds and potentially pathogenic organisms (viruses, bacteria and the like) present in wastewaters. These materials can pose a serious threat to the environment. When deposited in soils, heavy metals are passed through the food chain, first entering crops, and then animals that feed on the crops and eventually human beings, to whom they appear to be highly toxic. In addition they also leach from soils, getting into groundwater and further spreading contamination in an uncontrolled manner (Levine, and Hirose, 2005).

European and American markets aiming to transform various organic wastes (animal farm wastes, industrial and municipal wastes) into two main by-products:

- A solution of humic substances (a liquid oxidate).
- A solid residue.

Agricultural wastes are abundantly available globally and can be converted to energy and useful chemicals by a number of microorganisms. The organic matter was biodegradable to produce biogas and the variation show a normal methanogene bacteria activity and good working biological process as shown in Figures 5-7. The success of promoting any technology depends on careful planning, management, implementation, training and monitoring. Main features of gasification project are:

- Networking and institutional development/strengthening.
- Promotion and extension.
- Construction of demonstration projects.
- Research and development, and training and monitoring.

Biomass is a raw material that has been utilised for a wide variety of tasks since the dawn of civilisation. Important as a supply of fuel in the third world, biomass was also the first raw material in the production of textiles. The gasification of the carbon char with steam can make a large difference to the surface area of the carbon. The corresponding stream gasification reactions are endothermic and demonstrate how the steam reacts with the carbon charcoal (Bacaoui, 1998).

$$H_2O\ (g) + C_x\ (s) \rightarrow H_2\ (g) + CO\ (g) + C_{x-1}\ (s) \tag{2}$$

$$CO\ (g) + H_2O\ (g) \rightarrow CO_2\ (g) + H_2\ (g) \tag{3}$$

$$CO_2\ (g) + C_x\ (s) \rightarrow 2\ CO\ (g) + C_{x-1}\ (s) \tag{4}$$

The sources to alleviate the energy situation in the world are sufficient to supply all foreseeable needs. Conservation of energy and rationing in some form will however have to be practised by most countries, to reduce oil imports and redress balance of payments positions. Meanwhile development and application of nuclear power and some of the traditional solar, wind and water energy alternatives must be set in hand to supplement what remains of the fossil fuels.

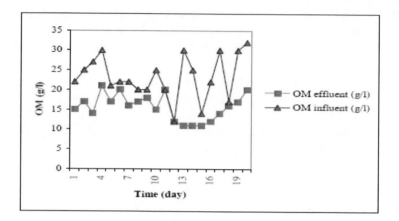

Figure 5. Organic matters before and after treatment in digester (Omer, 2006).

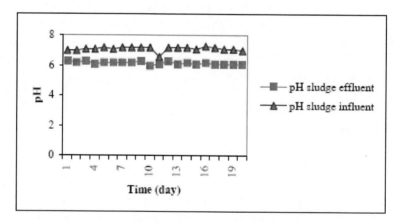

Figure 6. pH sludge before and after treatment in the digester (Omer, 2006).

The encouragement of greater energy use is an essential component of development. In the short-term it requires mechanisms to enable the rapid increase in energy/capita, and in the long-term we should be working towards a way of life, which makes use of energy efficiency and without the impairment of the environment or of causing safety problems. Such a programme should as far as possible be based on renewable energy resources.

6. BIOHEAT

Bioenergy is a growing source of power that is playing an ever-increasing role in the provision of electricity. The potential contribution of the waste industry to bioenergy is huge and has the ability to account for a source of large amount of total bioenergy production.

Woody biomass is usually converted into power through combustion or gasification. Biomass can be specially grown in the case of energy crops. Waste wood makes up a significant proportion of a variety of municipal, commercial and industrial waste streams. It is common practice to dispose of this waste wood in landfill where it slowly degraded and takes up valuable void space. This wood is a good source of energy and is an alternative to energy crops. The biomass directly produced by cultivation can be transformed by different processes into gaseous, liquid or solid fuels (Table 14). The whole process of production of methyl or ethyl esters (biodiesel) is summarised in Figures 8-9.

Table 14. Biomass conversions to energy (Omer, 2006)

Feedstock	Crops	Conversion process	End product
Wood-cellulosic biomass	Short rotation forest (poplar, willow), plant species (sorghum, mischantus, etc.), fibre-crops (cynara, kenaf, etc.)	Direct combustion Gasification Pyrolysis	Heat Methane Hydrogen Oil
Vegetable oils	Oleaginous crops (rapeseed, soybean, sunflower, etc.)	Direct combustion Esterification	Heat Biodiesel
Sugar/starch	Cereals, root and tuber crops, grape, topinambour, etc.	Fermentation	Ethanol

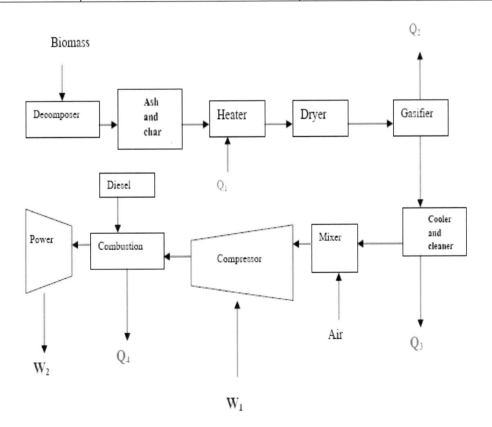

Figure 7. Advanced biomass power with diesel engine (Omer, 2006).

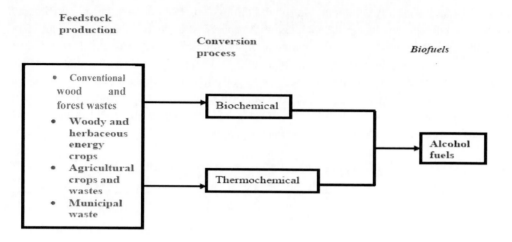

Figure 8. Biofuel pathways for renewable alcohol fuels (Omer, 2006).

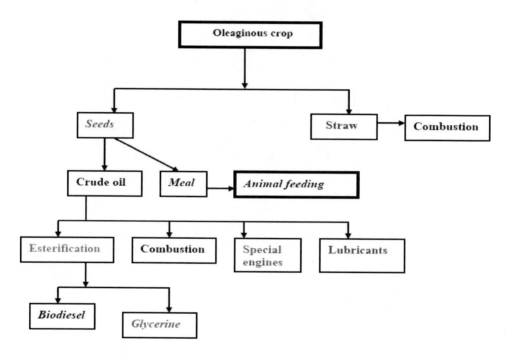

Figure 9. Flow chart of biodiesel production (Omer, 2006).

6.1. Waste Policy in Context

In terms of solid waste management policy, many non-governmental organisations (NGOs) have changed drastically in the past ten years from a mass production and mass consumption society to 'material-cycle society'. In addition to national legislation, municipalities are legally obliged to develop a plan for handling the municipal solid waste (MSW) generated in administrative areas. Such plans contain:

- Estimates of future waste volume.
- Measures to reduce waste.
- Measures to encourage source separation.
- A framework for solid waste disposal and the construction and management of solid waste management facilities.

Landfilling is in the least referred tier of the hierarchy of waste management options: waste minimisation, reuse and recycling, incineration with energy recovery, and optimised final disposal. The key elements are as follows: construction impacts, atmospheric emissions, noise, water quality, landscape, visual impacts, socio economics, ecological impacts, traffic, solid waste disposal and cultural heritage (Barton, 2007).

6.2. Energy from Agricultural Biomass

The main advantages are related to energy, agriculture and environment problems, are foreseeable both at regional level and at worldwide level and can be summarised as follows:

- Reduction of dependence on import of energy and related products.
- Reduction of environmental impact of energy production (greenhouse effect, air pollution, and waste degradation).
- Substitution of food crops and reduction of food surpluses and of related economic burdens.
- Utilisation of marginal lands and of set aside lands and reduction of related socio-economic and environmental problems (soil erosion, urbanisation, landscape deterioration, etc.).
- Development of new know-how and production of technological innovation.

A study (Bacaoui, 1998) individuated on the basis of botanical, genetical, physiological, biochemical, agronomical and technological knowledge reported in literature some 150 species potentially exploitable divided as reported in Table 15.

7. ROLE OF CHEMICAL ENGINEERING

Turning to chemical engineering and the experience of the chemical process industry represents a wakening up but does not lead to an immediate solution to the problems. The traditional techniques are not very kind to biological products, which are controlled by difficulty and unique physico-chemical properties such as low mechanical, thermal and chemical stabilities. There is the question of selectivity. The fermentation broths resulting from microbial growth contain a bewildering mixture of many compounds closely related to the product of interests. By the standards of the process streams in chemical industry, fermenter is highly impure and extremely dilutes aqueous systems (Table 16).

Table 15. Plant species potentially exploitable for production of agricultural biomass for energy or industrial utilisations (Rossi et al., 1990)

Groups of plants	Number of species
Plants cultivated for food purposes that can be reconverted to new uses	9
Plants cultivated in the past, but not in culture any more	46
Plants cultivated in other world areas	46
Wild species, both indigenous and exotic	47
Total	148
Plant product	Number of species
Biomass	8
Sugars and polysaccharides	38
Cellulose	17
Hydrocarbons	3
Polymeric hydrocarbons	5
Gums and resins	12
Tannins and phenolic compounds	3
Waxes	7
Vegetable oils	38
Total	131

Table 16. Typical product concentrations exiting fermenters (Rossi et al., 1990)

Product	Concentration (kg/m^3)
Ethanol	70-120
Organic acids (e.g., citric)	40-100
Vitamin B12	0.02
Interferon	50-70
Single-cell protein	30-50
Antibiotics (e.g., Penicillin G)	10-30
Enzyme protein (e.g., protease)	2-5

The disadvantages of the fermentation media are as the following: mechanically fragile, temperature sensitive, rapidly deteriorating quality, harmful if escaping into the environment, corrosive (acids, chlorides, etc.), and troublesome (solids, theological, etc.), and expensive. Thus, pilot plants for scale-up work must be flexible. In general, they should contain suitably interconnected equipment for: fermentation, primary separation, cell disruption fractionalises and clarifications, purification by means of high-resolution techniques and concentration and dry. The effects of the chlorofluorocarbons (CFCs) molecule can last for over a century.

7.1. Fluidised Bed Drying

An important consideration for operators of wastewater treatment plants is how to handle the disposal of the residual sludge in a reliable, sustainable, legal and economical way. The benefits of drying sludge can be seen in two main treatment options:

- Use of the dewatered sludge as a fertiliser or in fertiliser blends.
- Incineration with energy recovery.

Use as a fertiliser takes advantage of the high organic content 40%-70% of the dewatered sludge and its high levels of phosphorous and other nutrients. However, there are a number of concerns about this route including:

- The chemical composition of the sludge (e.g., heavy metals, hormones and other pharmaceutical residues).
- Pathogen risk (e.g., SALMOELLA, ESCHERICHIA COLI, prionic proteins, etc.).
- Potential accumulation of heavy metals and other chemicals in the soil.

Sludge can be applied as a fertiliser in three forms: liquid sludge, wet cake blended into compost, and dried granules.

The advantages of energy recovery sludge include:

- The use of dewatered sludge is a 'sink' for pollutants such as heavy metals, toxic organic compounds and pharmaceutical residues. Thus, offering a potential disposal route for these substances provided the combustion plant has adequate flue gas cleaning.
- The potential, under certain circumstances, to utilise the inorganic residue from sludge incineration (incinerator ash), such as in cement or gravel.
- The high calorific value (similar to lignite) of dewatered sludge.
- The use of dewatered sludge as a carbon dioxide neutral substitute for primary fuels such as oil, gas and coal.

7.2. Energy Efficiency

Energy efficiency is the most cost-effective way of cutting carbon dioxide emissions and improvements to households and businesses. It can also have many other additional social, economic and health benefits, such as warmer and healthier homes, lower fuel bills and company running costs and, indirectly, jobs. Britain wastes 20 per cent of its fossil fuel and electricity use in transportation. This implies that it would be cost-effective to cut £10 billion a year off the collective fuel bill and reduce CO_2 emissions by some 120 million tones CO_2. Yet, due to lack of good information and advice on energy saving, along with the capital to finance energy efficiency improvements, this huge potential for reducing energy demand is not being realised. Traditionally, energy utilities have been essentially fuel providers and the industry has pursued profits from increased volume of sales. Institutional and market arrangements have favoured energy consumption rather than conservation. However, energy is at the centre of the sustainable development paradigm as few activities affect the environment as much as the continually increasing use of energy. Most of the used energy depends on finite resources, such as coal, oil, gas and uranium. In addition, more than three quarters of the world's consumption of these fuels is used, often inefficiently, by only one quarter of the world's population. Without even addressing these inequities or the precious, finite nature of these resources, the scale of environmental damage will force the reduction of the usage of these fuels long before they run out.

Throughout the energy generation process there are impacts on the environment on local, national and international levels, from opencast mining and oil exploration to emissions of the potent greenhouse gas carbon dioxide in ever increasing concentration. Recently, the world's leading climate scientists reached an agreement that human activities, such as burning fossil fuels for energy and transport, are causing the world's temperature to rise. The Intergovernmental Panel on Climate Change has concluded that "the balance of evidence suggests a discernible human influence on global climate". It predicts a rate of warming greater than any one seen in the last 10,000 years, in other words, throughout human history. The exact impact of climate change is difficult to predict and will vary regionally. It could, however, include sea level rise, disrupted agriculture and food supplies and the possibility of more freak weather events such as hurricanes and droughts. Indeed, people already are waking up to the financial and social, as well as the environmental, risks of unsustainable energy generation methods that represent the costs of the impacts of climate change, acid rain and oil spills. The insurance industry, for example, concerned about the billion dollar costs of hurricanes and floods, has joined sides with environmentalists to lobby for greenhouse gas emissions reduction. Friends of the earth are campaigning for a more sustainable energy policy, guided by the principal of environmental protection and with the objectives of sound natural resource management and long-term energy security. The key priorities of such an energy policy must be to reduce fossil fuel use, move away from nuclear power, improve the efficiency with which energy is used and increase the amount of energy obtainable from sustainable, and renewable sources. Efficient energy use has never been more crucial than it is today, particularly with the prospect of the imminent introduction of the climate change levy (CCL). Establishing an energy use action plan is the essential foundation to the elimination of energy waste. A logical starting point is to carry out an energy audit that enables the assessment of the energy use and determine what actions to take. The actions are best categorised by splitting measures into the following three general groups:

(1) High priority/low cost:
These are normally measures, which require minimal investment and can be implemented quickly. The followings are some examples of such measures:

(1) Good housekeeping, monitoring energy use and targeting waste-fuel practices.
(2) Adjusting controls to match requirements.
(3) Improved greenhouse space utilisation.
(4) Small capital item time switches, thermostats, etc.
(5) Carrying out minor maintenance and repairs.
(6) Staff education and training.
(7) Ensuring that energy is being purchased through the most suitable tariff or contract arrangements.

(2) Medium priority/medium cost:
Measures, which, although involve little or no design, involve greater expenditure and can take longer to implement. Examples of such measures are listed below:

• New or replacement controls.
• Greenhouse component alteration, e.g., insulation, sealing glass joints, etc.

- Alternative equipment components, e.g., energy efficient lamps in light fittings, etc.

(3) Long term/high cost:

These measures require detailed study and design. They can be best represented by the followings:

- Replacing or upgrading of plant and equipment.
- Fundamental redesign of systems, e.g., CHP installations.

This process can often be a complex experience and therefore the most cost-effective approach is to employ an energy specialist to help.

7.3. Policy Recommendations for a Sustainable Energy Future

Sustainability is regarded as a major consideration for both urban and rural development. People have been exploiting the natural resources with no consideration to the effects, both short-term (environmental) and long-term (resources crunch). It is also felt that knowledge and technology have not been used effectively in utilising energy resources. Energy is the vital input for economic and social development of any country. Its sustainability is an important factor to be considered. The urban areas depend, to a large extent, on commercial energy sources. The rural areas use non-commercial sources like firewood and agricultural wastes. With the present day trends for improving the quality of life and sustenance of mankind, environmental issues are considered highly important. In this context, the term energy loss has no significant technical meaning. Instead, the exergy loss has to be considered, as destruction of exergy is possible. Hence, exergy loss minimisation will help in sustainability. In the process of developing, there are two options to manage energy resources: (1) End use matching/demand side management, which focuses on the utilities. The mode of obtaining this is decided based on economic terms. It is, therefore, a quantitative approach. (2) Supply side management, which focuses on the renewable energy resource and methods of utilising it. This is decided based on thermodynamic consideration having the resource-user temperature or exergy destruction as the objective criteria. It is, therefore, a qualitative approach. The two options are explained schematically in Figure 10. The exergy-based energy, developed with supply side perspective is shown in Figure 11. The following policy measures had been identified:

- Clear environmental and social objectives for energy market liberalisation, including a commitment to energy efficiency and renewables.
- Economic, institutional and regulatory frameworks, which encourage the transition to total energy services.
- Economic measures to encourage utility investment in energy efficiency (e.g., levies on fuel bills).
- Incentives for demand side management, including grants for low-income households, expert advice and training, standards for appliances and buildings and tax incentives.

- Research and development funding for renewable energy technologies not yet commercially viable.
- Continued institutional support for new renewables (such as standard cost-reflective payments and obligation on utilities to buy).
- Ecological tax reform to internalise external environmental and social costs within energy prices.
- Planning for sensitive development and public acceptability for renewable energy.

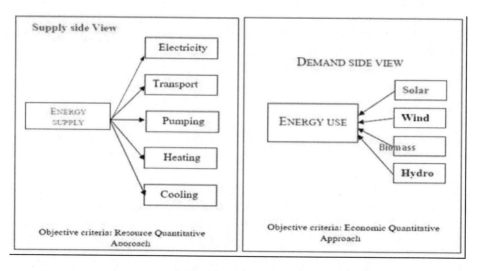

Figure 10. Supply side and demand side management approach for energy (Omer, 2008).

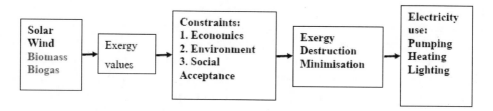

Figure 11. Exergy based optimal energy model (Omer, 2008).

Energy resources are needed for societal development. Their sustainable development requires a supply of energy resources that are sustainably available at a reasonable cost and can cause no negative societal impacts. Energy resources such as fossil fuels are finite and lack sustainability, while renewable energy sources are sustainable over a relatively longer term. Environmental concerns are also a major factor in sustainable development, as activities, which degrade the environment, are not sustainable. Hence, as much as environmental impact is associated with energy, sustainable development requires the use of energy resources, which cause as little environmental impact as possible. One way to reduce the resource depletion associated with cycling is to reduce the losses that accompany the transfer of exergy to consume resources by increasing the efficiency of exergy transfer between resources, i.e., increasing the fraction of exergy removed from one resource that is transferred to another (Erlich, 1991).

As explained above, exergy efficiency may be thought of as a more accurate measure of energy efficiency that accounts for quantity and quality aspects of energy flows. Improved exergy efficiency leads to reduced exergy losses. Most efficiency improvements produce direct environmental benefits in two ways. First, operating energy input requirements are reduced per unit output, and pollutants generated are correspondingly reduced. Second, consideration of the entire life cycle for energy resources and technologies suggests that improved efficiency reduces environmental impact during most stages of the life cycle. Quite often, the main concept of sustainability, which often inspires local and national authorities to incorporate environmental consideration into setting up energy programmes have different meanings in different contexts though it usually embodies a long-term perspective. Future energy systems will largely be shaped by broad and powerful trends that have their roots in basic human needs. Combined with increasing world population, the need will become more apparent for successful implementation of sustainable development (Aroyeun, 2009).

Heat has a lower exergy, or quality of energy, compared with work. Therefore, heat cannot be converted into work by 100% efficiency. Some examples of the difference between energy and exergy are shown in Table 17.

Carnot Quality Factor (CQF) = $(1-T_o/T_s)$ (5)

Exergy = Energy (transferred) x CQF (6)

Where T_o is the environment temperature (K) and T_s is the temperature of the stream (K).

Table 17. Qualities of various energy sources (Omer, 2008)

Source	Energy (J)	Exergy (J)	CQF
Water at 80°C	100	16	0.16
Steam at 120°C	100	24	0.24
Natural gas	100	99	0.99
Electricity/work	100	100	1.00

The terms used in Table 17 have the following meanings:

Various parameters are essential to achieving sustainable development in a society. Some of them are as follows:

- Public awareness.
- Information.
- Environmental education and training.
- Innovative energy strategies.
- Renewable energy sources and cleaner technologies.
- Financing.
- Monitoring and evaluation tools.

Improving access for rural and urban low-income areas in developing countries must be through energy efficiency and renewable energies. Sustainable energy is a prerequisite for development. Energy-based living standards in developing countries, however, are clearly below standards in developed countries. Low levels of access to affordable and environmentally sound energy in both rural and urban low-income areas are therefore a predominant issue in developing countries. In recent years many programmes for development aid or technical assistance have been focusing on improving access to sustainable energy, many of them with impressive results (Omer, 2006).

Apart from success stories, however, experience also shows that positive appraisals of many projects evaporate after completion and vanishing of the implementation expert team. Altogether, the diffusion of sustainable technologies such as energy efficiency and renewable energies for cooking, heating, lighting, electrical appliances and building insulation in developing countries has been slow. Energy efficiency and renewable energy programmes could be more sustainable and pilot studies more effective and pulse releasing if the entire policy and implementation process was considered and redesigned from the outset. New financing and implementation processes are needed which allow reallocating financial resources and thus enabling countries themselves to achieve a sustainable energy infrastructure. The links between the energy policy framework, financing and implementation of renewable energy and energy efficiency projects have to be strengthened and capacity building efforts are required.

8. RESULTS AND DISCUSSION

Alternatively energy sources can potentially help fulfill the acute energy demand and sustain economic growth in many regions of the world. Bioenergy is beginning to gain importance in the global fight to prevent climate change. The scope for exploiting organic waste as a source of energy is not limited to direct incineration or burning refuse-derived fuels. Biogas, biofuels and woody biomass are other forms of energy sources that can be derived from organic waste materials. These biomass energy sources have significant potential in the fight against climate change. Recently, there are many studies on modern biomass energy technology systems published (Bhutto, Bazmi, Zahwdi 2011; Cihan G, Dursun, Bora, Erkan 2009).

Vegetation and in particular forests, can be managed to sequester carbon. Management options have been identified to conserve and sequester up to 90 Pg C in the forest sector in the next century, through global afforestation (Singh, 2008; Duku, 2009). For efficient use of bioenergy resources, it is essential to take account of the intrinsic energy potential. Despite the availability of basic statistics, many differences have been observed between the previous assessments of bioenergy potential (Cheng, 2010; Bessou, 2009).

On some climate change issues (such as global warming), there is no disagreement among the scientists. The greenhouse effect is unquestionably real; it is essential for life on earth. Water vapour is the most important GHG; followed by carbon dioxide (CO_2). Without a natural greenhouse effect, scientists estimate that the earth's average temperature would be $-18^{\circ}C$ instead of its present $14^{\circ}C$ (Kothari, Singal, Rakesh, and Ranjan, 2011). There is also no scientific debate over the fact that human activity has increased the concentration of the GHGs in the atmosphere (especially CO_2 from combustion of coal, oil and gas). The greenhouse effect is also being amplified by increased concentrations of other gases, such as methane, nitrous oxide, and CFCs as a result of human emissions. Most scientists predict that rising global temperatures will raise the sea level and increase the frequency of intense rain or snowstorms (Andrea, and Fernando, 2012).

Globally, buildings are responsible for approximately 40% of the total world annual energy consumption. Most of this energy is for the provision of lighting, heating, cooling, and air conditioning. Increasing awareness of the environmental impact of CO_2, NO_x and CFCs emissions triggered a renewed interest in environmentally friendly cooling, and heating technologies. Under the 1997 Montreal Protocol, governments agreed to phase out chemicals used as refrigerants that have the potential to destroy stratospheric ozone. It was therefore considered desirable to reduce energy consumption and decrease the rate of depletion of world energy reserves and pollution of the environment. One way of reducing building energy consumption is to design buildings, which are more economical in their use of energy for heating, lighting, cooling, ventilation and hot water supply. Passive measures, particularly natural or hybrid ventilation rather than air-conditioning, can dramatically reduce primary energy consumption. However, exploitation of renewable energy in buildings and agricultural greenhouses can, also, significantly contribute towards reducing dependency on fossil fuels. Therefore, promoting innovative renewable applications and reinforcing the renewable energy market will contribute to preservation of the ecosystem by reducing emissions at local and global levels.

CONCLUSION

Even with modest assumptions about the availability of land, comprehensive fuel-wood farming programmes offer significant energy, economic and environmental benefits. These benefits would be dispersed in rural areas where they are greatly needed and can serve as linkages for further rural economic development. The nations, as a whole would benefit from savings in foreign exchange, improved energy security, and socio-economic improvements. With a nine-fold increase in forest – plantation cover, the nation's resource base would be greatly improved. The international community would benefit from pollution reduction, climate mitigation, and the increased trading opportunities that arise from new income sources. Furthermore, investigating the potential is needed to make use of more and more of its waste. Household waste, vegetable market waste, and waste from the cotton stalks, leather, and pulp; and paper industries can be used to produce useful energy either by direct incineration, gasification, digestion (biogas production), fermentation, or cogeneration. Therefore, effort has to be made to reduce fossil energy use and to promote green energies, particularly in the building sector. Energy use reductions can be achieved by minimising the energy demand, by rational energy use, by recovering heat and the use of more green energies. This study was a step towards achieving that goal. The adoption of green or sustainable approaches to the way in which society is run is seen as an important strategy in finding a solution to the energy problem. The key factors to reducing and controlling CO_2, which is the major contributor to global warming, are the use of alternative approaches to energy generation and the exploration of how these alternatives are used today and may be used in the future as green energy sources. Even with modest assumptions about the availability of land, comprehensive fuel-wood farming programmes offer significant energy, economic and environmental benefits. These benefits would be dispersed in rural areas where they are greatly needed and can serve as linkages for further rural economic development. The nations as a whole would benefit from savings in foreign exchange, improved energy security, and socio-economic improvements. With a nine-fold increase in forest – plantation cover, a nation's resource base would be greatly improved. The international community would benefit from pollution reduction, climate mitigation, and the increased trading opportunities that arise from new income sources.

REFERENCES

Abdeen, M. O. (2008a). Renewable building energy systems and passive human comfort solutions. *Renewable and Sustainable Energy Reviews, 12*(6), 1562-1587.

Abdeen, M. O. (2008b). People, power and pollution. *Renewable and Sustainable Energy Reviews, 12*(7), 1864-1889.

Abdeen, M. O. (2008c). Energy, environment and sustainable development. *Renewable and Sustainable Energy Reviews, 12*(9), 2265-2300.

Abdeen, M. O. (2008d). Focus on low carbon technologies: The positive solution. *Renewable and Sustainable Energy Reviews, 12*(9), 2331-2357.

Abdeen, M. O. (2008e). Chapter 10: Development of integrated bioenergy for improvement of quality of life of poor people in developing countries. In F. L. Magnusson & O. W. Bengtsson (Eds.), Energy in Europe: Economics, policy and strategy (pp. 341-373). New York, NY: NOVA Science Publishers.

Abdeen, M. O. (2009a). Environmental and socio-economic aspect of possible development in renewable energy use. In *Proceedings of the 4th International Symposium on Environment*, Athens, Greece, 21-24 May 2009.

Abdeen, M. O. (2009b). Energy use, environment and sustainable development. In *Proceedings of the 3rd International Conference on Sustainable Energy and Environmental Protection* (SEEP 2009), Paper No.1011, Dublin, Republic of Ireland, 12-15 August 2009.

Abdeen, M. O. (2009c). Energy use and environmental: Impacts: A general review. *Journal of Renewable and Sustainable Energy, 1*(5), 1-29.

Abdeen, M. O. (2009d). Chapter 3: Energy use, environment and sustainable development. In R. T. Mancuso (Ed.), Environmental cost management, (pp. 129-166). New York, NY: NOVA Science Publishers.

Andrea, S., and Fernando, R. (2012). Identifying, developing, and moving sustainable communities through renewable energy, World Journal of Science, Technology and Sustainable Development, 9(4): 273-281.

Aroyeun, S. O. (2009). Reduction of aflatoxin B1 and Ochratoxin A in cocoa beans infected with *Aspergillus* via Ergosterol Value. *World Review of Science, Technology and Sustainable Development, 6*(1), 75-90.

Bacaoui, A., Yaacoubi, A., Dahbi, C., Bennouna, J., and Mazet, A. (1998). Activated carbon production from Moroccan olive wastes-influence of some factors. Environmental Technology 19: 1203-1212.

Barton A. L. (2007). *Focus on sustainable development research advances*, (pp. 189-205). New York, NY: NOVA Science Publishers, Inc.

Bessou, S. (2009). Biofuels, greenhouse gases and climate change. Agronomy for Sustainable Development, DOI: 10. 1051/agro/2009039.

Bhutto A, Bazmi A, Zahwdi G (2011) Greener energy: issues and challenges for Pakistan – Biomass energy prospective. Renewable and Sustainable Energy Reviews, 15 (6): 3207-32-19.

Brain, G., & Mark, S. (2007). Garbage in, energy out: Landfill gas opportunities for CHP projects. *Cogeneration and On-Site Power, 8*(5), 37-45.

Cheng, R. (2010). Advanced biofuel technologies: status and barriers. World Bank Report, WPS5411.

Cihan G, Dursun B, Bora A, Erkan S (2009) Importance of biomass energy as alternative to other sources in Turkey. Energy Policy, 37 (2): 424-431.

D'Apote, S.L. (1998). IEA biomass energy analysis and projections. In: *Proceedings of Biomass Energy Conference: Data, analysis and Trends,* Paris: OECD; 23-24 March 1998.

Duku, B. (2009). Comprehensive review of biomass resources and biofuels potential in Ghana. Renewable and Sustainable Energy Review, 15: 404-415.

Erlich, P. (1991). Forward facing up to climate change, in Global Climate Change and Life on Earth. R.C. Wyman (Ed), Chapman and Hall, London.

Hall O. and Scrase J. (1998). Will biomass be the environmentally friendly fuel of the future? Biomass and Bioenergy 15: 357-67.

Jeremy, L. (2005). The energy crisis, global warming and the role of renewables. *Renewable Energy World, 8*(2).

Kothari, D. P., Singal, K. C., Rakesh, Ranjan (2011). Renewable energy sources and emerging technologies, 2nd Edition, Private Ltd, New Delhi, 2011.

Levine, M., & Hirose, M. (2005). *Energy efficiency improvement utilising high technology: An assessment of energy use in industry and buildings*. Report and Case Studies. London, UK: World Energy Council.

Omer, A.M., Yemen, F. (2003). Biogas energy technology in Sudan. Renewable Energy, 28 (3): 499-507.

Omer, A.M. (2006). Review: Organic waste treatment for power production and energy supply. Cells and Animal Biology 1 (2): 34-47.

Omer, A.M. (2008). Green energies and environment. Renewable and Sustainable Energy Reviews 12: 1789-1821.

Omer, A.M. (2007). Renewable energy resources for electricity generation in Sudan. Renewable and Sustainable Energy Reviews 11: 1481-1497.

Pernille, M. (2004). Feature: Danish lessons on district heating. Energy Resource Sustainable Management and Environmental March/April 2004: 16-17.

Robinson, G. (2007). Changes in construction waste management. Waste Management World, pp. 43-49. May-June 2007.

Rossi, S., Arnone, S., Lai, A., Lapenta, E., and Sonnino, A. (1990). ENEA's activities for developing new crops for energy and industry. In: Biomass for Energy and Industry (G. Grassi, G. Gosse, G. dos Santos Eds.). Vol.1, pp.107-113, Elsevier Applied Science, London and New York.

Sims, R.H. (2007). Not too late: IPCC identifies renewable energy as a key measure to limit climate change. Renewable Energy World 10 (4): 31-39.

Singh, A. (2008). Biomass conversion to energy in India: a critique. Renewable and Sustainable Energy Review, 14: 1367-1378.

Wu, J. and Boggess, W. (1999). The optimal allocation of conservation funds. Journal Environmental Economic Management. 1999: 38.

Chapter 4

SUSTAINABLE WATER RESOURCES MANAGEMENT, FUTURE DEMANDS AND ADAPTATION STRATEGIES IN THE NILE VALLEY

ABSTRACT

For the thirty-nine million, who live in Sudan, environmental pollution is a major concern; therefore industry, communities, local authorities and central government, to deal with pollution issues, should adopt an integrated approach. Most polluters pay little or no attention to the control and proper management of polluting effluents. This may be due to a lack of enforceable legislation and/or the fear of spending money on the treatment of their effluent prior to discharge. Furthermore, the imposed fines are generally low and therefore do not deter potential offenders. The present problems that are related to water and sanitation in Sudan are many and varied, and the disparity between water supply and demand is growing with time due to the rapid population growth and aridity. The situation of the sewerage system in the cities is extremely critical, and there are no sewerage systems in the rural areas. There is an urgent need for substantial improvements and extensions to the sewerage systems treatment plants. The further development of water resources for agriculture and domestic use is one of the priorities to improve the agricultural yield of the country, and the domestic and industrial demands for water. This article discusses the overall problem and identifies possible solutions.

Keywords: Sudan, water resources development, community water supply, effective water-supply management, environment

1. INTRODUCTION

In Sudan, with more than ten million people do not have adequate access to water supply, twenty million inhabitants are without access to sanitation, and a very low proportion of domestic sewage being treated. The investment, which is needed to fund the extension and improvement of these services, is substantial (Omer, 1995). Most governments in developing countries are ready to admit that they lack the financial resources for proper water and sanitation schemes. Moreover, historically, bilateral and multilateral funding accounts for less than 10% of total investment needed. Thus the need for private financing is imperative.

Many water utilities in developing countries need to work in earnest to improve the efficiency of operations. These improvements will not only lead to better services but also to enhanced net cash flows that can be re-invested to improve the quality of service. Staff productivity is another area where significant gains can be achieved. Investment and consumption subsidies have been predicated on the need to help the poor to have, access to basic services and to improve the environment. Failure of subsidies to reach intended objectives is due, in part, to lack of transparency in their allocation. A key element to successful private participation is the allocation of risks. How project risks are allocated and mitigated will determine the financial and operational performance and success of the project, under the basic principle that the risk should be allocated to the party, which is best able to bear it. Many developing countries (Sudan is not an exception) are encouraging the participation of the private-sector as a means to improve productivity in the provision of water and wastewaters services. Private-sector involvement is also needed to increase financial flows to expand the coverage and quality of services. Many successful private-sector interventions have been under taken. Private operators are not responsible for the financing of works, nonetheless they can bring significant productivity gains, which would allow the utility to allocate more resources to improve and extend services. Redressing productivity, subsidy and cross-subsidy issues before the private-sector is invited to participate, has proven to be less contentious. I have previously thought to encourage more private-sector involvement (Omer, 1995).

Sudan is geo-politically well located, bridging the Arab world to Africa. Its large size and extension from south to north provides for several agro-ecological zones with a variety of climatic conditions, rainfall, soils and vegetation. Water resources available to Sudan from the Nile system, together with groundwater resources, provide a potential for thirty years increase in the irrigated sub-sector. There are also opportunities for increased hydropower generation. The strategy of Sudan at the national level aims at the multi-purpose use of water resources to ensure water security for attaining food security, drinking-water security, fibre-security, hydro-energy security, industrial security, navigation, waste disposal and the security at the regional levels within an environmentally sustainable development context and in harmony with the promotion of basin-wide integrated development of the shared water resources (Noureddine, 1997). The government has continued to pay for the development and operation of water systems, but attempts are being sought to make the user communities pay water charges. In order to ensure the sustainability of water supplies, an adequate institutional and legal framework is needed. Funds must be generated (a) for production, (b) for environmental protection to ensure water quality, and (c) to ensure that water abstraction from groundwater remains below the annual groundwater recharge. At present, there are private-sector providers who do not have an enabling environment to offer the services adequately. There is a need for the government to have a mechanism to assist in the regulation and harmonisation of the private-sector providers. Privatisation is part of a solution to improve services delivery in water and sanitation sector. At present, there is a transitional situation characterised by: (i) A resistance to water charge; (ii) Insufficient suitable law/law enforcement; (iii) Insufficient capacities; and (iv) Inadequate interaction between actors.

In a country with a relatively sparsely populated, there are extreme pressures on water and waste systems, which can stunt the country's economic growth. However, Sudan has recognised the potential to alleviate some of these problems by promoting renewable water and utilising its vast and diverse climate, landscape, and resources, and by coupling its solutions for waste disposal with its solutions for water production. Thus, Sudan may stand at the forefront of the global renewable water community, and presents an example of how non-conventional water strategies may be implemented. In Sudan, more than ten million people do not have adequate access to water supply, twenty million inhabitants are without access to sanitation, and a very little domestic sewage is being treated. The investment needed to fund the extension and improvement of these services is great. Most governments in developing countries are ready to admit that they lack the financial resources for proper water and sanitation schemes. Moreover, historically, bilateral and multilateral funding accounts for less than 10% of total investment needed. Thus, the need for private financing is imperative. Water utilities in developing countries need to work in earnest to improve the efficiency of operations. These improvements would not only lead to better services but also to enhanced net cash flows that can be re-invested to improve the quality of service. Staff productivity is another area where significant gains can be achieved. Investment and consumption subsidies have been predicated on the need to (a) help the poor, which have not an access to basic services and (b) improve the environment.

Failure of subsidies to reach intended objectives is in part, from lack of transparency in their allocation. Subsidies are often indiscriminately assigned to support investment programmes that benefit more middle and high-income families that already receive acceptable service. Consumption subsidies often benefit upper-income domestic consumers much more than low-income ones. Many developing countries (Sudan is no exception) are encouraging the participation of the private-sector as a means to improve productivity in the provision of water and of wastewaters services. Private sector involvement is also needed to increase financial flows to expand the coverage and quality of services.

A key element to successful private participation is the allocation of risks. How project risks are allocated and mitigated determines the financial and operational performance and success of the project, under the basic principle that the risk should be allocated to the party, which is best able to bear it. Many successful private-sector interventions have been undertaken. Private operators are not responsible for the financing of works, nonetheless they can bring significant gains in productivity, which would allow the utility to allocate more resources to improve and extend services. Redressing productivity, subsidy and cross-subsidy issues before the private-sector is invited to participate, has proven to be less contentious. I have previously sought to encourage more private-sector involvement (Omer, 1995).

This study comprises a comprehensive review of water sources, the environment and sustainable development. It includes the renewable water resources, water conservation scenarios and other mitigation measures necessary to reduce climate change. This is still very much lacking particularly under developing countries conditions.

2. WATER RESOURCES

Sudan is rich in water (from the Nile system, rainfall and groundwater) and lands resources in Table 1. Surface water resources are estimated at 84 billion m^3 and the annual rainfall varies from almost nil in the arid hot north to more than 1600 mm in the tropical zone of the south. The total quantity of groundwater is estimated to be 260 billion m³, but only 1% of this amount is being utilised. Water-resources assessment in Sudan is not an easy task because of uncertainty of parameters, numerous degrees of freedom of variables, lack of information and inaccurate measurements. However, according to seasonal water availability, Sudan could be globally divided into three zones: (a) areas with water availability throughout the year are the rainy regions (equatorial tropical zones); (b) areas with seasonal water availability; and (c) areas with water deficit throughout the year, which occupy more than half the area of Sudan.

Table 1. Land use, land-resource zones and water resources (Omer, 2002)

(a) Land use (millions of ha)

Geographical area (total Sudan area)	250.6
Land area	237.6
Cultivable area	8.4
Pastures	29.9
Forests and woodland	108.3
Uncultivable land	81.0
Area under crop (irrigated, rain-fed, mechanised, and rain-fed traditional)	10.0

(b) Land-resource zones

Zone	Area as % to total area of Sudan	Persons per km²	Mean average rainfall range (mm)
Desert	44	2	0-200
QOS sands (dune)	10	11	200-800
Central clay plains	14	19	200-800
Southern clay plains	12	8	800-900
Ironstone plateau	12	7	800-1400
Hill area and others	8	16	Variable

(c) Water resources

Water resource	Available number	Static water level (m)	Number
Haffirs	824	0-0	824
Slow sand filters	128	0-0	128
Open shallow wells	3000	0-10	3000
Boreholes deep wells	2259	0-25	1248
		26-50	478
		51-75	287
		76-100	246

(d) Geological formations

Basins	Amount of water recharged (10^6 m^3)	Water level below land (m)	Aquifer thickness (m)	Velocity (m/year)	Abstraction (10^6 m^3/year)
Sahara Nile	136	30-100	300-500	1-2.5	7.3
Sahara Nubian	20.6	10-50	300-500	0.8-1.5	1.5
Central Darfur	47.6	25-100	250-550	0.3-6.0	5.5
Nuhui	15.4	75-120	200-400	1.0-2.75	1.6
Sag El Na'am	13.5	50-1000	300-500	1.0-25.0	2.5
River Atbara	150	100-150	250-300	0.3-5.0	2.3
Sudd	341	10-25	200-400	0.1-1.8	1.8
Western Kordofan	15	50-70	300-500	0.1-0.3	1.7
Baggara	155	10-75	300-500	0.1-2.4	11.9
Blue Nile	70.9	10-50	250-500	0.1-2.5	10.2
The Alluvial	N.A	Shallow	N.A	N.A	N.A
Gedaref	41.7	50-75	200-500	0.1-2.0	1.2
Shagara	1.1	25-30	200-300	0.1-2.5	0.7

The most important research and development policies which have been adopted in different fields of water resources are: (i) the water resource; (ii) irrigation development; (iii) the re-use of drainage water and groundwater; (iv) preventive and canal maintenance; (v) aquatic weed control and river channel development, and (vi) protection plans. The physical and human resources base can provide for sustainable agriculture growth and food security for itself and for others in the region. Failure to do so in the past derives from several causes and constraints, which are manageable. These include misguided policies, poor infrastructure, low level of technology use, recurring droughts and political instability. Perhaps the biggest challenge is that of finding resources for capital improvements in the light of changing water-quality regulations and ageing systems (James, 1994).

The desert environment is fragile and highly affected by human activities. Disturbances in the balanced ecosystems are apt to take place causing serious problems to the environment, and consequently, initiating geotechnical hazards. Urbanisation, climatic conditions, and geomorphic and geologic setting are usually the controlling factors influencing the types of these hazards. One of the potential geotechnical hazards that may occur under desert conditions is sand drifting and dune movement. The problem of sand drifting and dune migration is of special interest in Sudan as moving sand covers approximately one-third of the country. Because sand poses natural erosional-depositional hazards on the existing structures, such as roads and urbanised areas, it become necessary to study the behaviour of the sand forms in the different parts of the country.

Although deserts are known to be simply barren areas, they are scientifically defined in terms of water shortage or aridity, soil type, topography and vegetation. Anon, 1979 presented a map showing the distribution of deserts in the world. Accordingly to this map, most of the Middle Eastern countries lie within the semi-arid, arid, and hyper-arid desert zones, with an aridity index (ratio between annual precipitation and mean annual potential evapotranspiration) ranging between 0.03 and 0.02. Most of the geotechnical hazards are associated with desert environments. The desert environment, being a fragile ecosystem, needs to be treated with care. Intercommunications between different national and international agencies and education of the layman should help to keep the system balanced and reduce the resulting environmental hazards. In addition, any suggested remedial measures should be planned with nature and be engineered with natural materials.

3. WATER AND SANITATION MANAGEMENT

Community water supply and sanitation management is a new form of cooperation between support agencies in the water and sanitation sector and communities.

It involves a common search to identify problems with the local water supply and sanitation systems, to establish the possibilities for, and constraints on, management by communities, and to find possible solutions that may be tested. Some fundamental principles of community water and sanitation management are: (i) Increased management capacities are the basis for improved water and sanitation systems, and each community must develop its own specific management systems; and (ii) Communities own the process of water charge; facilitators and local researchers participate in the community's projects, not the other way around.

Through this approach, the support agency is no longer the provider of technical goods or solutions, but the facilitator of process to enhance the capacity of the community to manage its own water and sanitation systems. Constraints include:(i) A lack of funds or substantial delays in allocating funds for essential requirements such as operation and maintenance of irrigation and drainage projects; (ii) Deterioration in data-collection activities; (iii) A lack of appropriate and consistent policies for water development for both large-and small-scale projects; (iv) Serious delays in completing water projects after major investments such as dams and other hydraulic structures, and main secondary canals not being completed; (v) An absence or inadequacy of monitoring, evaluation, and feedback at both national and international levels; (vi) A lack of proper policies on cost recovery, and water pricing or, if policies exist, absence of their implementation; (vii) A shortage of professional and technical manpower, and training facilities; (viii) A lack of beneficiary participation in planning, implementation, and operation of projects; (ix) Inadequacy of knowledge, and absence of appropriate research to develop new technologies and approaches, and an absence of incentives to adopt them; (x) General institutional weaknesses and a lack of coordination between irrigation, agriculture, energy, healthy, environment, and planning; (xi) Inappropriate project development by donor agencies, e.g., irrigation development with drainage, supporting projects which should not have been supported; and (xii) A lack of donor coordination resulting in differing approaches and methodologies, and thus conflicting advice.

As developing nations strive to provide a safe and reliable drinking-water supply to their growing and increasingly urbanised population, is becoming more evident that new approaches to this problem will be needed. To meet this challenge, new methods of reclaiming and re-using water have been developed in cost-effective and environmentally sound ways (ODA, 1987; Seckler, 1992; and Salih, 1992).

Despite the constraints, over the last decade the rate of implementation of rural and peri-urban water supply and sanitation programmes has increased considerably, and many people are now being served more adequately. The following are Sudan experience in water supply and sanitation projects:

At community level:

- Participatory approaches in planning, implementation and monitoring.
- Establishment and training of water tap committees.
- Clear ownership of improved water supply and sanitation systems.
- Technology and service level selection by consumers.
- Sensitive timing of hygiene and sanitation education.
- Establishment and training of reliable financial and maintenance management.

At district and national level:

- Integrated multi-sectoral approach development.
- Training approach and material development for district and extension staff.
- Continuing support from integrated multi-sectoral extension team.
- Establishment of technical support system.
- Multi-sectoral advisory group including training and research institutions.
- Development and dissemination of relevant information for district and extension staff.

4. WATER RESOURCE MANAGEMENT SYSTEMS

Water is a substance of paramount ecological, economical, and social importance. Interrelationships inherent in water use should encourage integrated water management. Water resources are to be better managed to:

1. Ensure more reliable water availability and efficient water use in the agricultural sector.
2. Mitigate flood damage.
3. Control water pollution.
4. Prevent development of soil salinity and water logging.
5. Reduce the spread of water-borne diseases.

The emerging water crisis, in terms of both water quantity and quality, requires new approaches and actions. Priority areas needing concerted action in various sectors are:

(a) Water use efficiency, (b) Flood control, (c) Management of scarce water resources, (d) Water quality management and provision of safe drinking water, and (e) Coordination and integration of various aspects of water management, and water management with other related resources and societal concern. The following are recommended:

- Community must be the focus of benefits accruing from restructures, legislature to protect community interest on the basis of equity and distribution, handover the assets to the community should be examined; and communities shall encourage the transfer the management of water schemes to a professional entity.
- The private-sector should be used to mobilise, and strengthen the technical and financial resources, from within and without the country to implement the services, with particular emphasis on utilisation of local resources.
- The government should provide the necessary financial resources to guide the process of community management of water supplies. The government to divert from provision of services and be a facilitator through setting up standards, specifications and rules to help harmonise the private-sector and establish a legal independent body by an act of parliament to monitor and control the providers. Governments to assist the poor communities who cannot afford service cost, and alleviate social-economic negative aspects of privatisation.
- The sector actors should create awareness to the community of the roles of the private-sector and government in the provision of water and sanitation services.
- Support agencies assist with the financial and technical support, the training facilities, coordination, development and dissemination of water projects, and then evaluation of projects.

The development of new, modern, and complete water-resources-information systems is one of the basic needs for the implementation of the water-resources - management system. The decision process in drought or flood conditions, and also in over-exploitation cases, can only be correct if based on a reliable information system.

A complete and comprehensive database on water availability, users, water quality monitoring, current technologies (like geographical information systems), is certainly the way to produce an efficient framework for decision-making. Lack of information is one of the most critical points regarding the development and implementation of the new management system (FAO, 1999).

The types of data related to flood management include:

- Topographic data (elevations, land use, soils, vegetation, and hydrography).
- Administrative data (political boundaries, and jurisdictional boundaries).
- Infrastructure data (roads, wells, utilities, bridges and culverts, hydraulic structure, properties, facilities) and imagery (satellite images and aerial photographs).
- Environmental data (threatened and endangered species, critical aquatic and wildlife habitat, archaeological sites, and water quality).
- Hydrometeorology data (stream flows, precipitation, temperature, wind, solar radiation, soil water, discharge rating curves, flood frequency, and flood plain delineation).
- Economic data (stage-damage relationships, insured values, and industries), and
- Emergency management data (emergency plans, census data, and organisational charts).

5. THE POLICY REGIME IN WATER QUALITY MANAGEMENT

Apart from effluent regulations, and sometimes, national water quality guidelines, a common observation is that few developing countries (Sudan is not an exception) include a water-quality-policy context. Whereas water supply is seen as a national issue, pollution is mainly felt at, and dealt with at, the local level.

With few exceptions, national governments have little information on the relative importance of various types of pollution (agriculture, municipal, industrial, animal husbandry, aquaculture), and therefore, have no notion of which is of greatest economic or public health significance.

Usually freshwater quality management is completely divorced from coastal management even through these are intimately linked. Consequently, it is difficult to develop a strategic water quality management plan or to efficiently focus domestic and donor funds on priority issues.

A national water-quality-policy should include the following water quality components:

- A policy framework that provides broad strategic and political directions for future water-quality management.
- A strategic action plan for water-quality management based on priorities that reflect an understanding of economic and social costs of impaired water.

This plan should include the following components:

- A mechanism for identifying national priorities for water-quality management that will guide domestic and donor investment.
- A plan for developing a focused and cost-effective data programme for water quality and related uses, as a basis for economic and social planning.
- A consideration of options for financial sustainability including donor support, public/private-sector partnerships, and regional self-support initiatives.
- A regulatory framework that includes a combination of appropriate water-quality objectives (appropriate to that country and not necessarily based on Western standards) and effluent controls. This includes both surface and groundwater.
- A methodology for public input into goals and priorities.
- A process for tasking specific agencies with implementation so that accountability is firmly established and inter-agency competition is eliminated.
- Specific mechanisms for providing drinking water monitoring capabilities, at the community level if necessary.
- National data standards that must realistically reflect national needs and capabilities. Nevertheless, the objective is to ensure reliable data from those organisations that provide information for national water management purposes and at the community level for drinking water monitoring.

The design criteria in any water-quality programme are to determine the management issues which water quality data are required. Generally, there are four categories of data objectives:

- Descriptive data that are typically used for government policy and planning, meeting international obligations, and for public information.
- Data specific to public health.
- Regulatory concerns, and
- Aquatic ecosystem health.

The last category is not normally included in many developing countries for reasons of cost and complexity. In most developing countries, countries with transitional economies, and some developed countries, the technology of monitoring has changed little since 1970s, yet some of the largest advances in monitoring in recent years involve technical innovation that serve to reduce costs and increase efficiency. Admittedly, not all of these are inexpensive; however when deployed appropriately, they may eliminate traditional monitoring, or reduce costs by increasing the efficiency of more traditional approaches to chemical monitoring. Types of innovation include: biological assessment, use of surrogates, use of enzymatic indicators, miniaturisation, automation, and simplification of laboratory analytical methods.

6. SUSTAINABLE DEVELOPMENTS

In the past decade, sustainability has increasingly become a key concept and ultimate global for socio-economic development in the modern world. Without a doubt, the sustainable development and management of natural resources fundamentally control the survival and welfare of human society. Water is an indispensable component and resource for life and essentially all human activities rely on water in a direct or in direct way. Yet supplying water of sufficient quantity and safe quality has seldom been an easy task. Although sustainability is still a loosely defined and evolving concept, researchers and policy-makers have made tremendous efforts to develop a working paradigm and measurement system for applying this concept in the exploitation, utilisation and management of various natural resources. In water resources arena, recent development has been synthesised and presented in two important documents published by (ASCE, 1998) and (UNESCO, 1999), which attempt to give a specific definition and a set of criteria for sustainable water resource systems. When considering the long-term future as well as the present, sustainability is concept and goal that can only be specified and implemented over a range of spatial scales, of which urban water supply is a local problem with great reliance on the characteristics and availability of regional water resources.

Cleaner, leaner production processes - pursuing improvements and savings in waste minimisation, energy and water consumption, transport and distribution, as well as reduced emissions are needed. Tables (2-4) indicate water conservation, sustainable development and environment. With the debate on climate change, the preference for real measured data has been changed. The analyses of climate scenarios need an hourly weather data series that allows for realistic changes in various weather parameters. By adapting parameters in a proper way, data series can be generated for the site. Weather generators should be useful for:

- Calculation of energy consumption (no extreme conditions are required)
- Design purposes (extremes are essential), and

- Predicting the effect of climate change such as increasing annually average of temperature.

This results in the following requirements:

- Relevant climate variables should be generated (solar radiation: global, diffuse, direct solar direction, temperature, humidity, wind speed and direction) according to the statistics of the real climate.
- The average behaviour should be in accordance with the real climate.
- Extremes should occur in the generated series in the way it will happen in a real warm period. This means that the generated series should be long enough to assure these extremes, and series based on average values from nearby stations.

Table 2. Water and sustainable environment

Technological criteria	Water and environment criteria	Social and economic criteria
Primary water saving in regional scale	Sustainability according to greenhouse gas pollutant emissions	Labour impact
Technical maturity, and reliability	Sustainable according to other pollutant emissions	Market maturity
Consistence of installation and maintenance requirements with local technical known-how	Land requirement	Compatibility with political, legislative and administrative situation
Continuity and predictability of performance	Sustainability according to other environmental impacts	Cost of saved primary water

Table 3. Classification of key variables defining facility sustainability

Criteria	Intra-system impacts	Extra-system impacts
Stakeholder satisfaction	Standard expectations met Relative importance of standard expectations	Covered by attending to extra-system resource base and ecosystem impacts
Resource base impacts	Change in intra-system resource bases Significance of change	Resource flow into/out of facility system Unit impact exerted by flow on source/sink system Significance of unit impact
Ecosystem impacts	Change in intra-system ecosystems Significance of change	Resource flows into/out of facility system Unit impact exerted by how on source/sink system Significance of unit impact

Table 4. Positive impact of durability, adaptability and energy conservation on economic, social and environment systems

Economic system	Social system	Environmental system
Durability	Preservation of cultural values	Preservation of resources
Meeting changing needs of economic development	Meeting changing needs of individuals and society	Reuse, recycling and preservation of resources
Energy conservation and saving	Savings directed to meet other social needs	Preservation of resources, reduction of pollution and global warming

Growing concerns about social and environmental sustainability have led to increased interest in planning for the energy utility sector because of its large resource requirements and production of emissions. A number of conflicting trends combine to make the energy sector a major concern, even though a clear definition of how to measure progress toward sustainability is lacking. These trends include imminent competition in the electricity industry, global climate change, expected long-term growth in population and pressure to balance living standards (including per capital energy consumption).

Table 5. The basket of indicators for sustainable consumption and production

Economy-wide decoupling indicators
1. Greenhouse gas emissions
2. Air pollution
3. Water pollution (river water quality)
4. Commercial and industrial waste arisings and household waste not cycled

Resource use indicators
5. Material use
6. Water abstraction
7. Homes built on land not previously developed, and number of households

Decoupling indicators for specific sectors
8. Emissions from electricity generation
9. Motor vehicle kilometres and related emissions
10. Agricultural output, fertiliser use, methane emissions and farmland bird populations
11. Manufacturing output, energy consumption and related emissions
12. Household consumption, expenditure energy, water consumption and waste generated

Designing and implementing a sustainable energy sector will be a key element of defining and creating a sustainable society. In the electricity industry, the question of strategic planning for sustainability seems to conflict with the shorter time horizons associated with market forces as deregulation replaces vertical integration. Sustainable low-carbon energy scenarios for the new century emphasise the untapped potential of renewable resources. Rural areas can benefit from this transition. The increased availability of reliable and efficient energy services stimulates new development alternatives. It is concluded that renewable environmentally friendly energy must be encouraged, promoted, implemented, and demonstrated by full-scale plant especially for use in remote rural areas (Figure 1).

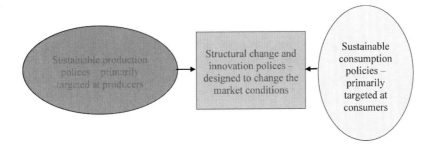

Figure 1. Link between resources and productivity.

This is the step in a long journey to encourage a progressive economy, which continues to provide us with high living standards, but at the same time helps reduce pollution, waste mountains, other environmental degradation, and environmental rationale for future policy-making and intervention to improve market mechanisms. This vision will be accomplished by:

- 'Decoupling' economic growth and environmental degradation. The basket of indicators illustrated shows the progress being made (Table 5). Decoupling air and water pollution from growth, making good headway with CO_2 emissions from energy, and transport. The environmental impact of our own individual behaviour is more closely linked to consumption expenditure than the economy as a whole.
- Focusing policy on the most important environmental impacts associated with the use of particular resources, rather than on the total level of all resource use.
- Increasing the productivity of material and energy use that are economically efficient by encouraging patterns of supply and demand, which are more efficient in the use of natural resources. The aim is to promote innovation and competitiveness. Investment in areas like energy efficiency, water efficiency and waste minimisation.
- Encouraging and enabling active and informed individual and corporate consumers.

7. GOALS AND CHALLENGES

Sudan needs assistance in developing and implementing (a) river-basin management, (b) diffuse source pollution, (c) environmental restoration, and (d) urban storm drainage.

At present the international, bilateral donor agencies, and relevant United Nations bodies provide such assistance. The international associations constitute an additional, but as yet untapped, source of assistance. The solution, which should be seriously explored, is the forging of partnerships with bodies such as the World Bank and the appropriate United Nations agencies.

Advanced research and technology contribute to resolving water shortage and sanitation problems, and non-conventional reliable water supplies cannot be provided unless the environmental impacts are taken into consideration. Looking to the future, Sudan has a set the following priorities for water-resource research and development until the year 2020:

a) Increase overall water-use efficiency to the maximum limit. This could be achieved by (a) improving the irrigation system and assure its flexibility to cope with modern farm irrigation system, (b) developing the farm system, (c) drawing up a proper mechanism for water charges;

b) Modify the cropping pattern; for example (a) planning the different cropping pattern according to water quality, (b) gradually replacing sugar cane by sugar beet, (c) introducing genetic engineering and tissue culture to develop salt tolerance crops, and (d) reducing the area of clover (*Berseem*);

c) Re-use all the possible agricultural drainage water using proper technological means to deal with its quality, especially after implementing the irrigation development programme;

d) Plan properly the re-use of sewage effluent after drawing up guidelines for its use;

e) Research agreements of losses and suggest conservation projects;

f) The conjunctive use and management of reservoirs and groundwater sources in the Nile valley, giving special consideration to drought conditions;

g) Develop non-renewable groundwater resources in the deserts on a sustainable basis;

h) Water harvest rainfall in desert areas and make full use of torrential streams and flash floods;

i) Use new economical technology of seawater desalination;

j) Raise public awareness about water resource scarcity and government management plans;

k) Consider laws to match with the required development and existing scarcity;

l) Establishment of efficient operation, maintenance and repair procedures;

m) Community participation in operation and maintenance;

n) The extent to which initial government investment can or should be recovered from water uses;

o) Domestic potable water supply should reach at least 25 litres per day per person;

p) Water should be available for ten livestock units at 450 l/d;

q) Potable water must be available within two kilometres of individual residences.

The water quantity situation is highly variable in Sudan reflecting different levels of development and different needs for water quality programmes in Table 6.

The conventional paradigm of water quality monitoring is not suitable for the Sudan being too expensive, inefficient, and ineffective. Financial and sustainability issues include cost avoidance and cost reduction, local and accountability frameworks that encourage good business practices by senior programme managers, the use of new cost-effective technologies for monitoring, and a variety of donor/public/private-sector linkages that focus on commercial benefits that permit the transfer of certain parts of water quality programmes to the private-sector.

From a visual investigation of the River Nile in Table 7, the major sources are industrial effluents, crude sewage from blocked, broken or overloaded sewers, sewage effluents, surface runoff, and solid wastes which have been dumped into the river.

Therefore remedial and improvement measures must be taken before the environment becomes further polluted and the natural resources are completely over-exploited (Omer, 2000).

The challenges facing and enhancing the ecology in the twenty-first century are as follows: (a) Drinking-water sources should be treated with chemicals; (b) Suitable toilet facilities should be provided along the main roads to minimise pollution; (c) Proper arrangements should be made for litter dumping and waste disposal; (d) Local people should be fully educated about environment matters and hygiene; (e) Previous damage should not be allowed to continue while planning for a balanced development in the future; (f) The concept of the ecosystem (involving education and interpretation of the natural environment) must be promoted.

Environmental pollution is a major problem facing all nations of the world. People have caused air pollution since they learned to how to use fire, but man-made air pollution (anthropogenic air pollution) has rapidly increased since industrialisation began.

Table 6. Present water management of Sudan

Using of resources	Sources	Institutions	Pricing Principle	Price Details
Urban	Surface and groundwater	National Water Corporation (NWC)	Full cost recovery	Progressive rate with increasing uses. Rates lower in the north
Major rural villages	Mostly groundwater	Rural Water Corporation (RWC)	Stand pipe free, recovery of recurrent costs, charges for yard and house connections	Progressive rates but less comparative to urban cities
Rural villages	Groundwater	District Councils	As above	Not available
Livestock	Surface and groundwater	Rural Water Corporation (RWC)	All investments and recurrent costs	Regressive, no charges on relatively small use
Mines	Surface and groundwater	National Water Corporation (NWC)	Full cost recovery	Progressive rates
Wildlife	Mostly surface	Rural Water Corporation (RWC)	Full cost of boreholes	Regressive

Table 7. Wastes in the River Nile water

Materials	(%)
Paper, and wood	50.0
Ferrous residues	12.5
Glasses	11.0
Organic wastes	10.0
Plastics	5.0
Non-ferrous residues	1.5
Other	10.0

Many volatile organic compounds and trace metals are emitted into the atmosphere by human activities. The pollutants emitted into the atmosphere do not remain confined to the area near the source of emission or to the local environment, and can be transported over long distances, and create regional and global environmental problems. The privatisation and price liberalisation in energy fields has to some secured (but not fully). Availability and adequate energy supplies to the major productive sectors is needed. The result is that, the present situation of energy supplies is for better than ten years ago.

8. THE CHALLENGE OF OVERCOMING THE COUNTRY'S DIVERSITY

Sudan is a federal republic of 2.5 million km^2 located in the eastern Africa. The country is divided into 26 states and a federal district, in which the capital, Khartoum is located. Sudan is known as a country of plentiful water, with highest total renewable fresh water supply in the region. Table 8 shows some of the most significant regional diversities concerning water issues.

Adequate water management is essential to sustain development. Competing needs for this beneficial resource include municipal supply, industry, and agriculture, among others. The National Water Act of 1994 (Law No. 1155) defines the objectives, principles, and instruments of the National Water Resources Policy and the National Water Resources Management system. The law establishes the institutional arrangement under which the country's water policies are to be implemented. The National Water Resources Policy was proposed to achieve:

- Sustainability: to ensure that the present and future generations have an adequate availability of water with suitable quality.
- Integrated management: to ensure the integration among uses in order to guarantee continuing development.
- Security: to prevent and protect against critical events, due either to natural causes or inappropriate uses.

Table 8. Main water resource issue in region

Region	Water resource issues
South	1. Abundant water resources
	2. Localised scarcity of water and untapped water supplies
	3. High hydropower potential
	4. Water conflicts arising from immigration of Bagara Arabs (nomadic) from north to south
	5. Water-borne diseases
	6. International water conflicts (Upstream and downstream countries)
Central	1. Water quality problems from untreated sewage and other pollution
	2. Water-borne diseases
	3. Potential use of rivers for navigation and recreational purposes
	4. Intensive erosion and sedimentation from agriculture
	5. High hydropower potential
	6. Excessive use in large urban and industrialised areas

Region	Water resource issues
North	7. Frequent urban floods 1. Good water quality 2. Scarcity of water resources 3. Intensive erosion and sedimentation from agriculture
Northeast	4. Frequent urban floods 1. Scarcity of water resources 2. Water quality problems from untreated sewage 3. In mining areas, water quality problems from effluent
West	1. Scarcity of water resources 2. Water conflicts between nomadic and non-nomadic tribes 3. Water-borne diseases 4. Soil erosion and degradation caused by agriculture

Table 9. Capacity assessment for flood management: institutional factors

High capacity (plans, etc., in places)	• Basin-wide management plan has been drafted. • Natural mitigation strategy in place. • Basin-wide coordination and communications strategy instituted. • Trained emergency management staff coordinating at the regional level. • Effective regulatory policies that address floodplain occupancy. • Decentralised decision-making with a high degree of local autonomy. • Evidence of an updated national response plan. • Bilateral response agreements.
Medium capacity (evidence of activity on-going)	• Evidence of regional preparedness and response training. • Some trained emergency management staff at the local and/or national level. • Evidence of some regulatory policies designed to address floodplain occupancy. • Attempts to decentralise decision-making, moderate local discretion.
Low capacity (no formalization in place nor apparently evolving)	• No existing flood response plan. • No evidence of mitigation-related activities. • Poor local-and national - level coordination and communications. • Little or no evidence of flood preparedness and response training. • No regulatory policies addressing floodplain occupancy. • Centralised decision-making, no evidence of local autonomy.

To achieve such objectives, water management must be implemented according to the following principles:

- Water is a public good, and it is a finite resource that has economic value.
- The use of water required to meet people's basic needs shall have priority, especially in critical periods.
- Water management shall comprise and induce multiple uses.

- The river basins are the appropriate unit for water management, and water management shall decentralise, with the participation of government, stakeholders and society.

Water resources plans are developed to guide future decisions and are to be developed for each river basin and state, as well as the country. The objective is to coordinate efforts and establish guidelines and priorities for water allocation and water pricing. The priorities established for water allocation will be used in critical drought conditions. Water pricing is the single most controversial instrument of the law. The pricing system is also the most difficult step to implement. The pricing system recognises the economic value of water, as stated in the principles of the policy. The development of a new, modern, and complete water resources information system is one of the basic needs for the implementation of the water resources management system. The decision process in drought or flood conditions, and also in overexploitation cases, can only be correct if based on a reliable information system. A complete and comprehensive database on water availability, users, water quality monitoring, and current technologies (like geographical information systems), is certainly the way to produce an efficient framework for decision-making. Lack of information is one of the most critical points regarding the development and implementation of the new management system. The institutional framework provides the basis by which all actions are taken, and an assessment of its functional character helps determine the collaborative potential. The resulting criteria for measuring a given community's institutional capacity can be found in Table 9.

9. WATER SCARCITY IMPACTS AND POTENTIAL CONFLICTS

The failure of water resources to meet the basic requirements of society has a host of social, economic, environmental, and political impacts. Water scarcity is man-made phenomenon brought about by the increasing demands of the population for water. The imbalance in the population - water resources equation strains society and has an adverse impact on domestic hygiene, public health, and cost of domestic water, and could impart political problems as a serious as bringing down government. On the social side, water scarcity adversely impacts job opportunities, farm incomes, credibility and reliability of agricultural exports, and ability of the vulnerable to meet the cost of domestic water. Economically, the adverse impact is displayed in the loss of production of goods, especially agricultural goods, the loss of working hours because of the hardships society faces as a result of water scarcity. The impacts of water scarcity on regional stability are addressed with reference to water in the Middle East Peace Process, taking into account the serious impacts of conflicts and potential water war.

Conditions of scarcity propel an increase in competition among the different sectors of water use with results, invariably, at the expense of irrigated agriculture. Pure market forces create a gradient under which water flows from the poor to the rich. Tough decisions await politicians, and the consequences are expected to displease one or more parties, and please others. The scene of domestic politics becomes as fluid as water itself, with politicians shifting positions continuously in response to domestic pressures. The political fallout from water resources scarcity on the domestic scene is parallel to the impact the scarcity has on domestic households in terms of basic needs for drinking and food preparation, on domestic hygiene, and on public health. Other important factors have a delayed response to water scarcity, and these pertain to the integrity of the environment, and deterrence it imparts on development investment and economic credibility of the country. The cost of mitigating these problems and of the provision of services to the increased urbanisation could very well be beyond the ability of government to bear. The political consequences resulting from this will not be in favour of domestic stability, and social explosions can be anticipated. A bilateral agreement was reached between Egypt and the Sudan in 1959 by which the two countries share the Nile flow: 55.5 billion cubic meters to Egypt, 18.5 billion to Sudan, and 10 billion were allocated to evaporation. Hopes are high for achieving a more extensive participation by the other riparian parties in what could be a multilateral treaty on the Nile encompassing the other riparian states in addition to Egypt and Sudan. The above agreement is not complete; it lacks the entry of other legitimate riparian states, lacks water quality components, and tends to focus on quantity measures, and miss important management issues. It is to be noted that regional relations, including those among the riparian parties, are connected to the political, economic, and trade network of international relations. Water is not the only determinant factor in shaping the nature of bilateral, regional, or international relations. Water relations can be transformed into a positive sum game by which all parties can be made to win. One common gain to all is the environmental protection of the common watercourse or water body. Lack of cooperation and agreement will most likely lead to environmental neglect and water quality degradation, which is loss to all. International encouragement to attain cooperation can, therefore, be brought to bear on the regional parties, and efforts of international lending agencies can be called upon to pool with the regional and international efforts to achieve this objective. It has been stipulated by many that under conditions of scarcity, water conflicts can lead to hostile actions between riparian parties. Experience in the region indicates that water, in its own right, has not been the cause of any of the wars that have broken out in the region.

Today's advanced societies heavily depend on energy. The principal sources of energy and electricity generation today are solar, wind, biomass, hydropower, and fossil fuel. Energy from hydropower is short of meeting the current or future energy requirements, and the fossil fuel resources, being depleted with time, will eventually run out. For human civilisation to continue at its natural pace, new forms of affordable and clean energy will have to come on line. Failure of human civilisation to introduce new forms of energy will render that civilisation doomed, and the quality of life will deteriorate. If this unlikely scenario actually takes place, the requirements will decrease because the mechanism of making it available for use (pumping) diminishes.

The more likely scenario is more optimistic one, and it is that a new form of energy generation will be introduced in which case water desalination becomes affordable and its pumping from the coastal desalination plants become possible at reasonable cost.

The way out of the looming water crisis rests, therefore, in the invention of new forms of energy generation that will make possible the reliance on desalination and in the recycling of wastewater for reuse in agricultural production and for environmental reasons. Integrated management of the three resources of water, energy, and the environment, will result in better results with a positive sum for society.

10. COMMON LANGUAGE AND CULTURE

A common language and similar culture simplify communication and reduce the potential for misunderstandings. In the Nile basin where several languages are spoken, an international language, English, is used with some success by multi-jurisdictional basin management authorities.

10.1. Primary Factors Promoting Data and Information Exchange

Data and information exchange is more probable when needs are compatible and when there is potential for mutual benefit from cooperation in Table 10. Where countries are working on developments that are beneficial to both countries as well as other riparians, there is little incentive to hide project impacts. This means that since data and information exchange is unlikely to lead to pressure from surrounding countries that might restrict developments, countries have less reason to restrict access to their data and information resources. It is important, therefore to be no perceived clash of interests in development plans and needs. An example of this might be in developing their part of the basin primarily for hydroelectric development, while the lower riparians are more interested in developing the irrigation potential of their portion of the basin. By constructing large storage dams in the upper part of the basin, the river Nile seasonal flow might be evened out, reducing flooding downstream while increasing irrigation water supplies and even making downstream run-of-the-river hydroelectric projects more profitable. Ecosystem effects would have to be considered.

Table 10. Summary of the situation relating to data and information exchange in the Nile basin

River basin	Nile basin
Basin states or territories	Burundi, Democratic Republic of Congo, Egypt, Eritrea, Ethiopia, Kenya, Rwanda, Sudan, Tanzania, Uganda
Cooperative frameworks in place	Nine of the countries of basin are pursuing the development of a cooperative framework
Major languages spoken	More than 6 official languages and numerous unofficial languages
Major water issue facing the basin	Rapid population growth, environmental degradation, under development
External funding of cooperative basin initiatives	Extensive external funding of cooperative initiative
Range of GDP *per capita* of the basin	$550-$3000
Extent of data/information exchange	Information exchange through the cooperative framework being developed is beginning to occur

10.2. Sufficient Levels of Economic Development

Sufficient levels of economic development across a basin are needed to permit joint funding of cooperative processes, particularly data collection and dissemination. Although countries with differing levels and forms of economic development may, at times, have more complementary needs than countries with similarly structured economies, the overall level of economic development is still significant. A wealthier country in a river basin may be able to assist with the funding of data collection activities in the neighbouring country with much needed data and helping to build confidence between the two countries.

10.3. Increasing Water Resources Stress

As *per capita* water resources availability decreases as shown in Table 11, tensions between riparian nations may rise and make cooperation difficult. Stress may, therefore, reduce cooperation and data sharing rather than strife.

The historical background of the basin may have a lasting effect on current negotiations. Past conflicts can have a deleterious effect on the prospects for establishing cooperative practices, such as data sharing. Where there is a history of conflict between two nations, both nations may view the present situation primarily as competitive and focus on conflicting rather than common interests. Democracies may find it easier to negotiate cooperative arrangements with other democracies. Political differences can lead to legacies of mistrust developing between countries.

Table 11. Diverse water challenge (WRI, 2002)

Country	Egypt	Sudan
Per capita annual water resources 2000 (m^3)	34	1187
Per capita annual withdrawal (m^3)	921	666
Per capita annual withdrawal for agriculture (m^3)	86	94

11. DISCUSSIONS

11.1. Water Stress in Sudan

Water stress refers to economic, social, or environmental problems caused by unmet water needs. Lack of supply is often caused by contamination, drought, or a disruption in distribution. In an extreme example, when Sudan split four years ago between the rebel-led west and government-ruled north, the conflict led to unpaid water bills, which precipitated a dangerous health threat in the region, increasing the risk of water-born diseases such as cholera. Some analysts believe the disruption of distribution was a political ploy to put pressure on the rebel-led west.

While water stress occurs throughout the world, no region has been more afflicted than sub-Saharan Africa. The crisis in Darfur stems in part from disputes over water: The conflict that led to the crisis arose from tensions between nomadic farming groups who were competing for water and grazing land-both increasingly scarce due to the expanding Sahara Desert. As Mark Giordano of the International Water Management Institute in Colombo Sri Lanka says, "Most water extracted for development in sub-Saharan Africa-drinking water, livestock watering, and irrigation - is at least in some sense 'transboundary'". Because water sources are often cross-border, conflict emerges.

Improving water and sanitation programmes is crucial to spurring growth and sustaining economic development. Because it takes time to develop these programmes, a paradox emerges: poor economies are unable to develop because of water stress, and economic instability prohibits the development of programmes to abate water stress. Developments in water storage could have prevented that drought from significantly affecting Sudan's economy. Hydropower can also spark economic development. Accordingly, some transboundary water agreements also play a clear role in fostering development, for example, by facilitating investment in hydropower and irrigation.

11.2. The Role of Agriculture in Water Stress

Agricultural development has the potential to improve African economies but requires extensive water supplies. These statics from the Water Systems Analysis Group at the Institute for the Study of Earth, Oceans, and Space at the University of New Hampshire reveal the urgent need for sustainable agricultural development:

- About 64 percent of Africans rely on water that is limited and highly variable;
- Croplands inhabit the driest regions of Africa where some 40 percent of the irrigated land is unsustainable;
- Roughly 25 percent of Africa's population suffers from water stress;
- Nearly 13 percent of the population in Africa experiences drought-related stress once each generation.

Another aspect of water-related stress is the relationship between water, soil, and agriculture. Improved access to quality water is a long-term goal that requires more than humanitarian funds.

- Because sub-Saharan Africa is subject to more extreme climate variability than other regions, it needs improved water storage capacity. Some experts say that large dam projects would create a more sustainable reserve of water resources to combat the burden of climate fluctuations, but other disagrees, stating the harmful environmental impact of large dams.
- Many experts say more water treaties are needed. The transboundary water agreements have cultivated international cooperation and reduced the "probability of conflict and its intensity".
- Better donor emphasis on water development is needed. Small-scale agricultural improvements also offer a solution to water stress, including the harvest of water in shallow wells, drip irrigation for crops, the use of pumps, and other technological innovations.

Farmers can access green water through drip irrigation systems that slowly and consistently deliver water to plant's root system, supplemental irrigation (supplementary to natural rainfall rather than the primary source of moisture during periods of drought) and rainwater harvesting (the collection of rainwater for crops, which reduces reliance on irrigation). Crops can grow poorly even during periods of rainfall, and most farms in Africa suffer from nitrogen and phosphorus depletion in soil.

One way to assuage water stress in terms of food scarcity is to increase water-holding capacity with organic fertilisers that would increase availability and efficacy of green water.

11.3. Water Supply Problems in the Butana Region - Central Sudan with Special Emphasis on Jebel Qeili Area

The Butana region of central Sudan is famous for its animal wealth and extensive pastures. Yet scarcity of water resources in the area especially during the dry seasons handicaps the proper utilisation of these pastures. The area is occupied by non-water-bearing basement rocks and the only source of water is from direct run-off.

Thus large numbers of small-size water reservoirs, "haffirs", were constructed, but these are inadequate to provide enough water for the growing human and animal population. An all-year lake is here proposed to be constructed utilising the ring-structure the Jebel Qeili igneous complex, central Butana. This lake is expected to solve the present water problem and meet the future demand of central Butana at the present rate of human and animal growth (Omer, 2001).

11.4. Southern Sudan

World Vision began its work in Sudan in 1972 through a partnership with the African Committee for Rehabilitation of the Southern Sudan (ACROSS) to provide emergency relief aid to war-effected families. Efforts included the reconstruction of the Rumbek community hospital and surrounding buildings, the provision of medicine and supplies, and education in preventative health care.

Other projects during this period focused on training health and social workers in general medical aid and child welfare and instruction in water development, agriculture, handcrafts, and literacy (Omer, 2004). The 1980s brought constant turmoil to the Sudanese people as the civil war raged on and severe drought parched the country. In 1983, approximately 1,500 refugees entered Sudan daily from violence-torn neighbouring countries, straining the already limited food supply. World Vision, through the ACROSS Refugee Settlement Project, responded by distributing blankets, grain, cooking oil, medical kits, and shelter to more than 50,000 people. Supplemental feeding for children also was provided.

Numerous development projects were initiated during this time that assisted communities in improved crop production, animal husbandry, health care, clean water collection, infrastructure repair, and literacy. In 1989, World Vision became a founding member of Operation Lifeline Sudan (OLS), a partnership of non-governmental organisations (NGOs) and the United Nations (UN) agencies designated to coordinate the southern relief efforts.

During the 1990s World Vision conducted operations in all major regions of southern Sudan. Project objectives included primary health care, water provision, agriculture, local grain purchase, enterprise development, and emergency relief efforts. World Vision focused on an integrated work approach that involved peace and advocacy, gender development, church support, and environment and natural resource initiatives. Some specific projects included:

- The Kapoeta Medical Supplies Project provided health and educational assistance to more than 200,000 people to help reduce incidences of disease and suffering.
- The Agriculture/Livestock Rehabilitation Project assisted the aforementioned families with food, seed, vaccinations, and agricultural consults.
- The South Sudan Relief/Church Support Project coordinated a pastors' conference for 150 pastors and religious leaders from the Western Equatoria province.
- The South Sudan Relief and Rehabilitation Project, a 10-year programme, provided 450,000 Sudanese with agriculture and economic development, food and water, health care, enterprise development opportunities, and emergency relief.

11.5. Beja People's Problems

The Beja, a semi-nomadic group of people, who live in rebel-held areas of eastern Sudan, need a huge amount of humanitarian assistance, a representative from the International Rescue Committee (IRC). Although Beja can be found throughout northeast Africa, tens of thousands are currently trapped in an area of eastern Sudan near the Eritrean border, held by Sudanese rebels since the late 1990s. Only two NGOs, both based in Eritrea, are able to access the 15,000 sq km area at the moment, one of which is the IRC. The organisation estimates the Beja population in the area to be between 45,000 and 186,000 people (Omer, 2008).

Although it did rain in the area in 2004, a shortage of water had also posed serious problems. "Fresh drinking water is incredibly hard to come by. All the settlements have just focused around dry river beds, in which people dig hand-dug wells". Locusts would eat the foliage that usually sustains the Beja's goats and camels—upon which the Beja utterly depend for survival. A few immature locust swarms have formed in northeast Sudan near the Red Sea and the border of Egypt, the UN Food and Agriculture Organisation said in March 2004. Moreover, Beja grazing areas have been severely restricted by a front line between rebel forces and Khartoum government soldiers, the second to be opened by southern rebels during Sudan's 21-year-old civil war.

Sudan is an example that projects the environmental plight of Africa, south of the Sahara – drought and desertification, floods, deforestation, loss of biodiversity, tribal and ethnic conflict and poverty are only too common. As a result, interest and commitment to environmental impact assessment practices have become mandatory by donors when executing new development projects. The ecological zones of Sudan in 1998 as:

- Deserts: cover almost 30% of the northern parts. Annual precipitation is less than 50 mm; soils are sandy. Sparse vegetation grows on seasonal 'Wadis' and the banks of the Nile.

- Semi deserts: cover above 20% south of the desert belt. Rainfall ranges from 50 to 300 mm. It is speckled with few Acacia trees and thorny bushes and zerophytes.
- Low rainfall woodland Savannah: covers about 27% of the area of Sudan with rainfall less than 900 mm, with a nine-month dry period. Annual grasses are dominant. Heavy clay soils lie on the east of the Nile and the west is sandy. Most of the 36 million feddans of rain-fed agriculture and the 4 million irrigated lands fall within this heavily populated belt.
- High rainfall woodland Savannah: 13% of the area with rainfall more than 900 mm and with broad-leafed trees in the southern parts of Sudan.
- Swamps: are probably the largest in the world and cover about l0% and fall in three main areas around the tributaries of the White Nile.
- Highlands: are less than 0.3% of the areas of Sudan and are scattered along the Red Sea coast, the south and the west of the country.
- The Red Sea Coast-Marine ecosystem, mangrove swamps, coral reefs and associated fauna.

Environmental problems include:

- Horizontal expansion in rain-fed and irrigated agriculture;
- The complete absence of the environmental dimensions in policies, strategies, plans and programmes of management of resources;
- Development is random and environmental evaluation does exist before or after execution of projects;
- The economy and society, in spite of the century-long attempts at 'modernisation' are still dominated by subsistence way of living;
- The economy is still affected seriously by the yearly, seasonal and geographical variability of rainfall for crop and livestock production;
- Dependence on imported seeds and agricultural chemicals has increased cost of production;
- Loss of land productivity and marketing policies decreased cash surplus;
- The civil war in the south has grave economic and social costs;
- Population distribution and rural-urban migration due to desertification and civil strife has led to deterioration of natural resources, indigenous knowledge and loss of local culture and dignity;
- Problems of poor sanitation, limited industrial pollution and food hygiene have become more complex;
- The energy crisis is aggravating desertification and affecting climate charge;
- Vast water resources are badly managed;
- Environmental education has only been recently incorporated in school curricula; and
- Laws and legislation concerning the environment are not effective and law enforcement measures are not integrated.

12.6. Western Sudan

El Fasher, Darfur region, Sudan, 24 August 2005 – Torrential rains have caused severe flooding in this city of 400,000 people and in nearby Abu Shook, a camp for people forced to flee their homes as a result of the ongoing Darfur conflict (Figures 2-3). The floods have destroyed hundreds of homes and have made El Fasher's water supply largely unsafe (WHO, 2006).

Figure 2. For some people water collection is a daily need.

Figure 3. A typical donkey-drawn water tank used by water vendors.

UNICEF is mounting a concerted effort to restore basic services to those affected by the flood, and to prevent the outbreak of disease. Since the flood, UNICEF has assisted with the following:

- Reinstalling pipes in Abu Shook and restoring the water supply by linking boreholes with pumps.
- Testing the water quality each day. No bacterial contamination has been found.
- Rebuilding 156 latrines and 88 bath stations.
- Renting five tankers to deliver more water.
- Repairing damaged schools and child-friendly spaces.
- Providing daily door-to-door hygiene-promotion trainings.
- Distributing jerry cans, soap, tarps, and mosquito nets.

CONCLUSION

A booming economy, high population, land-locked location, vast area, remote separated and poorly accessible rural areas, large reserves of oil, excellent sunshine, large mining sector and cattle farming on a large-scale, are factors which are most influential to the total water scene in Sudan. It is expected that the pace of implementation of water infrastructure will increase and the quality of work will improve in addition to building the capacity of the private and district staff in contracting procedures. The financial accountability is also easier and more transparent. The communities should be fully utilised in any attempts to promote the local management of water supply and sanitation systems. There is little notion of 'service, invoice and move on'. As a result, there are major problems looming with sustainability of completed projects. A charge in water and sanitation sector approach from supply-driven approach to demand-responsive approach calls for full community participation. The community should be defined in terms of their primary role as user/clients. Private-sector services are necessary because there are gaps, which exist as a result of the government not being able to provide water services due to limited financial resources and increase in population. The factors affecting the eco-environmental changes are complex, interrelated, and interactive. The deterioration problems of water and sanitation have attracted some attention in recent years. There is an urgent need to study possible rehabilitation measures to ensure a sustainable and excellent water quality and improved sanitation. Water resources plans are developed to guide future decisions and are to be developed for each river basin and state, as well as for the country. The overall objective is to coordinate efforts and establish guidelines and priorities for water allocation and water pricing. The priorities established for water allocation would be used in critical drought conditions. The water quality classification of water bodies by different classes of use is the basis for truly integrating the quality and quality of water management. Water pricing is the single most controversial instrument of the law. The pricing system recognises the economic value of water, as stated in the principles of the policy, but is also the most difficult step to implement. It is expected that the pace of implementation will increase and the quality of work will improve in addition to building the capacity of the private and district staff in contracting procedures. The financial accountability is also easier and more transparent. The communities should be fully utilised in any attempts to promote the local management of water supply and sanitation systems. A charge in water and sanitation sector approach from supply-driven approach to demand-responsive approach calls for full community participation. The community should be defined in terms of their primary role as user/clients. Private-sector services are necessary because there are gaps, which exist as a result of the government not being able to provide water services due to limited financial resources and increase in population. There is little notion of 'service, invoice and move on'. As a result, there are major problems looming with sustainability of completed projects. The factors affecting the eco-environmental changes are complex. There are interrelated and interact. The deterioration problems of water and sanitation have attracted some attention in recent years. There is an urgent need to study possible rehabilitation measures to ensure a sustainable and excellent water quality and improved sanitation.

REFERENCES

Anon. (1979). *'Map of the world distribution regions'*, MAB Tech Note 7.

ASCE. (1998). *'ASCE Task Committee on Sustainability Criteria, Sustainability Criteria for Water Resource Systems'*, Reston, Virginia, USA.

FAO United Nations Food and Agriculture Organisation. (1999). 'The State of Food in Security in the World', Rome: Italy.

James, W. (1994). *'Managing water as economic resources'*, Overseas Development Institute (ODI), UK.

Noureddine, R.M. (1997). 'Conservation planning and management of limited water resources in arid and semi-arid areas', *Proceedings of the 9th Session of the Regional Commission on Land and Water Use in the Near East,* Rabat: Morocco, pp. 15-21.

Omer, A.M. (1995). 'Water resources in Sudan', *NETWAS 2*, Nairobi, pp. 7-8.

Omer, A.M. (2000). 'Water and environment in Sudan: the challenges of the new millennium', *NETWAS 7*(2), pp. 1-3.

Omer, A.M. (2001). Water development in Sudan: Present and future challenges. Arab Organisation for Agriculture Development (AOAD). *Arabic Journal of Irrigation Water Management,* Vo.2, p.48-58, Khartoum: Sudan.

Omer, A.M. (2002). 'Focus on groundwater in Sudan', International Journal of Geosciences Environmental Geology 41(8), pp. 972-976.

Omer, A.M. (2004). Water resources development and management in the Republic of the Sudan. *Water and Energy International,* 61(4): 27-39.

Omer, A.M. (2008). Water resources in the Sudan. *Water International,* 32 (5): 894-903.

Overseas Development Administration (ODA). (1987). 'Sudan profile of agricultural potential', Survey, UK, 1987.

Salih, A.M.A., and Ali, A.A.G. (1992). 'Water scarcity and sustainable development', *Nature and Resources* 28, pp.1.

Seckler, D. (1992). 'Private sector irrigation in Africa-Water resources and irrigation policy studies', Winrock International Institute for Agricultural Development.

UNESCO. (1999). UNESCO Working Group M.IV, 'Sustainability criteria for water resource systems, Cambridge', United Kingdom: Cambridge University Press.

World Health Organisation (WHO). (2006). Water norms and attitudes. Geneva: Switzerland.

World Resources Institute (WRI). (2002). *'World Resources 2000-2001'*, USA.

Chapter 5

Sustainable Food Production at Mediterranean Climate in Greenhouse Environment

Abstract

People are relying upon oil for primary energy and this for a few more decades. Other conventional sources may be more enduring, but are not without serious disadvantages. The renewable energy resources are particularly suited for the provision of rural power supplies and a major advantage is that equipment such as flat plate solar driers, wind machines, etc., can be constructed using local resources and without the advantage results from the feasibility of local maintenance and the general encouragement such local manufacture gives to the buildup of small-scale rural based industry.

This chapter comprises a comprehensive review of energy sources, the environment and sustainable development. It includes the renewable energy technologies, energy efficiency systems, energy conservation scenarios, energy savings in greenhouses environment and other mitigation measures necessary to reduce climate change. This study gives some examples of small-scale energy converters, nevertheless it should be noted that small conventional, i.e., engines are currently the major source of power in rural areas and will continue to be so for a long time to come. There is a need for some further development to suit local conditions, to minimise spares holdings, to maximise interchangeability both of engine parts and of the engine application. Emphasis should be placed on full local manufacture. It is concluded that renewable environmentally friendly energy must be encouraged, promoted, implemented and demonstrated by full-scale plant (device) especially for use in remote rural areas.

Keywords: renewable energy technologies, energy efficiency, sustainable development, emissions, environment

Nomenclature

a	annum
ha	hectares

l   litre
HFU   heat flor unit
MSW   municipal sewage waste

1. INTRODUCTION

Power from natural resources has always had great appeal. Coal is plentiful, though there is concern about despoliation in winning it and pollution in burning it. Nuclear power has been developed with remarkable timeliness, but is not universally welcomed, construction of the plant is energy-intensive and there is concern about the disposal of its long-lived active wastes. Barrels of oil, lumps of coal, even uranium come from nature but the possibilities of almost limitless power from the atmosphere and the oceans seem to have special attraction. The wind machine provided an early way of developing motive power. The massive increases in fuel prices over the last years have however, made any scheme not requiring fuel appear to be more attractive and to be worth reinvestigation. In considering the atmosphere and the oceans as energy sources the four main contenders are wind power, wave power, tidal and power from ocean thermal gradients. The sources to alleviate the energy situation in the world are sufficient to supply all foreseeable needs. Conservation of energy and rationing in some form will however have to be practised by most countries, to reduce oil imports and redress balance of payments positions. Meanwhile development and application of nuclear power and some of the traditional solar, wind and water energy alternatives must be set in hand to supplement what remains of the fossil fuels.

The encouragement of greater energy use is an essential component of development. In the short-term, it requires mechanisms to enable the rapid increase in energy/capita, while in the long-term it may require the use of energy efficiency without environmental and safety concerns. Such programmes should as far as possible be based on renewable energy resources.

Large-scale, conventional, power plant such as hydropower has an important part to play in development although it does not provide a complete solution. There is however an important complementary role for the greater use of small-scale, rural based, and power plants. Such plants can be employed to assist development since they can be made locally. Renewable resources are particularly suitable for providing the energy for such equipment and its use is also compatible with the long-term aims.

In compiling energy consumption data one can categorise usage according to a number of different schemes:

- Traditional sector - industrial, transportation, etc.
- End-use - space heating, process steam, etc.
- Final demand - total energy consumption related to automobiles, to food, etc.
- Energy source - oil, coal, etc.
- Energy form at point of use - electric drive, low temperature heat, etc.

2. RENEWABLE ENERGY POTENTIAL

The increased availability of reliable and efficient energy services stimulates new development alternatives (Omer, 2009a). This communication discusses the potential for such integrated systems in the stationary and portable power market in response to the critical need for a cleaner energy technology. Anticipated patterns of future energy use and consequent environmental impacts (acid precipitation, ozone depletion and the greenhouse effect or global warming) are comprehensively discussed in this approach. Throughout the theme several issues relating to renewable energies, environment and sustainable development are examined from both current and future perspectives. It is concluded that renewable environmentally friendly energy must be encouraged, promoted, implemented and demonstrated by full-scale plant (device) especially for use in remote rural areas. Globally, buildings are responsible for approximately 40% of the total world annual energy consumption. Most of this energy is for the provision of lighting, heating, cooling, and air conditioning. Increasing awareness of the environmental impact of CO_2 and NO_x and CFCs emissions triggered a renewed interest in environmentally friendly cooling, and heating technologies. Under the 1997 Montreal Protocol, governments agreed to phase out chemicals used as refrigerants that have the potential to destroy stratospheric ozone. It was therefore considered desirable to reduce energy consumption and decrease the rate of depletion of world energy reserves and pollution of the environment. One way of reducing building energy consumption is to design buildings, which are more economical in their use of energy for heating, lighting, cooling, ventilation and hot water supply. Passive measures, particularly natural or hybrid ventilation rather than air-conditioning, can dramatically reduce primary energy consumption. However, exploitation of renewable energy in buildings and agricultural greenhouses can, also, significantly contribute towards reducing dependency on fossil fuels. Therefore, promoting innovative renewable applications and reinforcing the renewable energy technologies market will contribute to preservation of the ecosystem by reducing emissions at local and global levels. This will also contribute to the amelioration of environmental conditions by replacing conventional fuels with renewable energies that produce no air pollution or greenhouse gases.

There is strong scientific evidence that the average temperature of the earth's surface is rising. This is a result of the increased concentration of carbon dioxide and other GHGs in the atmosphere as released by burning fossil fuels. This global warming will eventually lead to substantial changes in the world's climate, which will, in turn, have a major impact on human life and the built environment. Therefore, effort has to be made to reduce fossil energy use and to promote green energies, particularly in the building sector. Energy use reductions can be achieved by minimising the energy demand, by rational energy use, by recovering heat and the use of more green energies. This study was a step towards achieving that goal. The adoption of green or sustainable approaches to the way in which society is run is seen as an important strategy in finding a solution to the energy problem. The key factors to reducing and controlling CO_2, which is the major contributor to global warming, are the use of alternative approaches to energy generation and the exploration of how these alternatives are used today and may be used in the future as green energy sources (Omer, 2009b). Even with modest assumptions about the availability of land, comprehensive fuel-wood farming programmes offer significant energy, economic and environmental benefits. These benefits would be dispersed in rural areas where they are greatly needed and can serve as linkages for further rural economic development. The nations as a whole would benefit from savings in foreign exchange, improved energy security, and socio-economic improvements. With a nine-fold increase in forest – plantation cover, a nation's resource base would be greatly improved. The international community would benefit from pollution reduction, climate mitigation, and the increased trading opportunities that arise from new income sources.

The non-technical issues, which have recently gained attention, include: (1) Environmental and ecological factors, e.g., carbon sequestration, reforestation and revegetation. (2) Renewables as a CO_2 neutral replacement for fossil fuels. (3) Greater recognition of the importance of renewable energy, particularly modern biomass energy carriers, at the policy and planning levels. (4) Greater recognition of the difficulties of gathering good and reliable renewable energy data, and efforts to improve it. (5) Studies on the detrimental health efforts of biomass energy particularly from traditional energy users. The renewable energy resources are particularly suited for the provision of rural power supplies and a major advantage is that equipment such as flat plate solar driers, wind machines, etc., can be constructed using local resources and without the advantage results from the feasibility of local maintenance and the general encouragement such local manufacture gives to the buildup of small- scale rural based industry. This study gives some examples of small-scale energy converters, nevertheless it should be noted that small conventional, i.e., engines are currently the major source of power in rural areas and will continue to be so for a long time to come. There is a need for some further development to suit local conditions, to minimise spares holdings, to maximise interchangeability both of engine parts and of the engine application. Emphasis should be placed on full local manufacture.

The renewable energy resources are particularly suited for the provision of rural power supplies and a major advantage is that equipment such as flat plate solar driers, wind machines, etc., can be constructed using local resources and without the high capital cost of more conventional equipment. Further advantage results from the feasibility of local maintenance and the general encouragement such local manufacture gives to the buildup of small-scale rural based industry. Table 1 lists the energy sources available.

Table 1. Sources of energy

Energy source	Energy carrier	Energy end-use
Vegetation	Fuel-wood	Cooking Water heating Building materials Animal fodder preparation
Oil	Kerosene	Lighting Ignition fires
Dry cells	Dry cell batteries	Lighting Small appliances
Muscle power	Animal power	Transport Land preparation for farming Food preparation (threshing)
Muscle power	Human power	Transport Land preparation for farming Food preparation (threshing)

Table 2. Renewable applications

Systems	Applications
Water supply	Rain collection, purification, storage and recycling
Wastes disposal	Anaerobic digestion (CH_4)
Cooking	Methane
Food	Cultivate the 1 hectare plot and greenhouse for four people
	Wind generator
Electrical demands	Solar collectors
Space heating	Solar collectors and excess wind energy
Water heating	Ultimately hardware
Control system	Integration of subsystems to cut costs
Building fabric	

Currently the 'non-commercial' fuels wood, crop residues and animal dung are used in large amounts in the rural areas of developing countries, principally for heating and cooking; the method of use is highly inefficient. Table 2 presented some renewable applications.

Table 3. Energy needs in rural areas

Transport, e.g., small vehicles and boats
Agricultural machinery, e.g., two-wheeled tractors
Crop processing, e.g., milling
Water pumping
Small industries, e.g., workshop equipment
Electricity generation, e.g., hospitals and schools
Domestic, e.g., cooking, heating, and lighting
Water supply, e.g., rain collection, purification, and storage and recycling
Building fabric, e.g., integration of subsystems to cut costs
Wastes disposal, e.g., anaerobic digestion (CH_4)

Table 4. Methods of energy conversion

Muscle power	Man, animals
Internal combustion engines	
Reciprocating	Petrol- spark ignition
	Diesel- compression ignition
	Humphrey water piston
Rotating	Gas turbines
Heat engines	
Vapour (Rankine)	
Reciprocating	Steam engine
Rotating	Steam turbine
Gas Stirling (Reciprocating)	Steam engine
Gas Brayton (Rotating)	Steam turbine
Electron gas	Thermionic, thermoelectric
Electromagnetic radiation	Photo devices
Hydraulic engines	Wheels, screws, buckets, turbines
Wind engines (wind machines)	Vertical axis, horizontal axis
Electrical/mechanical	Dynamo/alternator, motor

Table 3 lists the most important of energy needs. Table 4 listed methods of energy conversion.

Considerations when selecting power plant include the following:

- Power level - whether continuous or discontinuous.
- Cost - initial cost, total running cost including fuel, maintenance and capital amortised over life.
- Complexity of operation.
- Maintenance and availability of spares.
- Life.
- Suitability for local manufacture.

The household wastes, i.e., for family of four persons, could provide 280 kWh/yr of methane, but with the addition of vegetable wastes from 0.2 ha or wastes from 1 ha growing a complete diet, about 1500 kWh/yr may be obtained by anaerobic digestion (Omer, 2009c). The sludge from the digester may be returned to the land. In hotter climates, this could be used to set up a more productive cycle (Figure 1).

There is a need for greater attention to be devoted to this field in the development of new designs, the dissemination of information and the encouragement of its use. International and government bodies and independent organisations all have a role to play in renewable energy technologies.

Society and industry in Europe and elsewhere are increasingly dependent on the availability of electricity supply and on the efficient operation of electricity systems. In the European Union (EU), the average rate of growth of electricity demand has been about 1.8% per year since 1990 and is projected to be at least 1.5% yearly up to 2030 (Omer, 2009c). Currently, distribution networks generally differ greatly from transmission networks, mainly in terms of role, structure (radial against meshed) and consequent planning and operation philosophies.

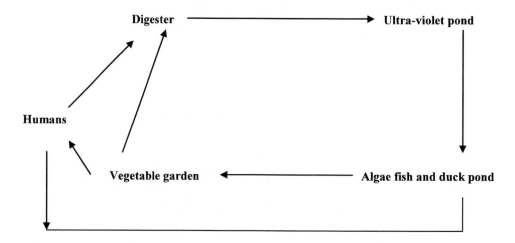

Figure 1. Biomass energy utilisation cycle.

3. ENERGY CONSUMPTION

Over the last decades, natural energy resources such as petroleum and coal have been consumed at high rates. The heavy reliances of the modern economy on these fuels are bound to end, due to their environmental impact, and the fact that conventional sources might eventually run out. The increasing price of oil and instabilities in the oil market led to search for energy substitutes.

In addition to the drain on resources, such an increase in consumption consequences, together with the increased hazards of pollution and the safety problems associated with a large nuclear fission programmes. This is a disturbing prospect. It would be equally unacceptable to suggest that the difference in energy between the developed and developing countries and prudent for the developed countries to move towards a way of life which, whilst maintaining or even increasing quality of life, reduce significantly the energy consumption per capita. Such savings can be achieved in a number of ways:

- Improved efficiency of energy use, for example better thermal insulation, energy recovery, and total energy.
- Conservation of energy resources by design for long life and recycling rather than the short life throwaway product.
- Systematic replanning of our way of life, for example in the field of transport.

Energy ratio is defined as the ratio of energy content of the food product/ energy input to produce the food.

$$Er = Ec/Ei \tag{1}$$

where Er is the energy ratio, Ec is the energy content of the food product, and Ei is the energy input to produce the food.

A review of the potential range of recyclables is presented in Table 5.

Currently the non-commercial fuelwood, crop residues and animal dung are used in large amounts in the rural areas of developing countries, principally for heating and cooking, the method of use is highly inefficient. As in the developed countries, the fossil fuels are currently of great importance in the developing countries. Geothermal and tidal energy are less important though, of course, will have local significance where conditions are suitable. Nuclear energy sources are included for completeness, but are not likely to make any effective contribution in the rural areas. Economic importance of environmental issue is increasing, and new technologies are expected to reduce pollution derived both from productive processes and products, with costs that are still unknown.

4. BIOGAS PRODUCTION

Biogas is a generic term for gases generated from the decomposition of organic material. As the material breaks down, methane (CH_4) is produced as shown in Figure 2.

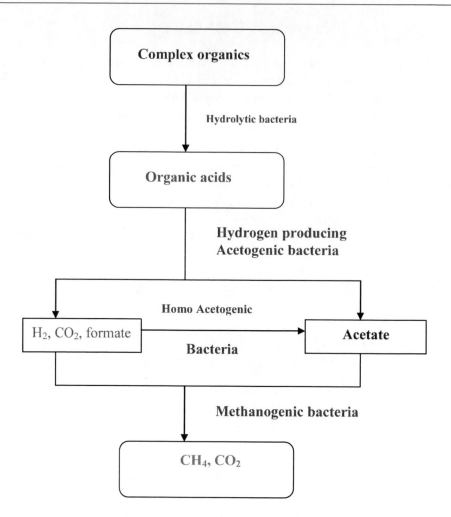

Figure 2. Biogas production process.

Sources that generate biogas are numerous and varied. These include landfill sites, wastewater treatment plants and anaerobic digesters (Omer, 2009d). Landfills and wastewater treatment plants emit biogas from decaying waste. To date, the waste industry has focused on controlling these emissions to our environment and in some cases, tapping this potential source of fuel to power gas turbines, thus generating electricity (Omer, 2009d). The primary components of landfill gas are methane (CH_4), carbon dioxide (CO_2), and nitrogen (N_2). The average concentration of methane is ~45%, CO_2 is ~36% and nitrogen is ~18% (Omer, and Yemen, 2001). Other components in the gas are oxygen (O_2), water vapour and trace amounts of a wide range of non-methane organic compounds (NMOCs). Landfill gas-to-cogeneration projects present a win-win-win situation. Emissions of particularly damaging pollutant are avoided, electricity is generated from a free fuel and heat is available for use locally.

Heat tariffs may include a number of components such as: a connection charge, a fixed charge and a variable energy charge. Also, consumers may be incentivised to lower the return temperature.

Hence, it is difficult to generalise but the heat practice for any DH company no matter what the ownership structure can be highlighted as follows:

- To develop and maintain a development plan for the connection of new consumers.
- To evaluate the options for least cost production of heat.
- To implement the most competitive solutions by signing agreements with other companies or by implementing own investment projects.
- To monitor all internal costs and with the help of benchmarking, and improve the efficiency of the company.
- To maintain a good relationship with the consumer and deliver heat supply services at a sufficient quality.

Installing DH should be pursued to meet the objectives for improving the environment through the improvement of energy efficiency in the heating sector. At the same time DH can serve the consumer with a reasonable quality of heat at the lowest possible cost. The variety of possible solutions combined with the collaboration between individual companies, the district heating association, the suppliers and consultants can, as it has been in Denmark, be the way forward for developing DH in the United Kingdom.

Table 5. Summary of material recycling practices in the construction sector (Robinson, 2007)

Construction and demolition material	Recycling technology options	Recycling product
Asphalt	Cold recycling: heat generation; Minnesota process; parallel drum process; elongated drum; microwave asphalt recycling system; finfalt; surface regeneration	Recycling asphalt; asphalt aggregate
Brick	Burn to ash, crush into aggregate	Slime burn ash; filling material; hardcore
Concrete	Crush into aggregate	Recycling aggregate; cement replacement; protection of levee; backfilling; filter
Ferrous metal	Melt; reuse directly	Recycled steel scrap
Glass	Reuse directly; grind to powder; polishing; crush into aggregate; burn to ash	Recycled window unit; glass fibre; filling material; tile; paving block; asphalt; recycled aggregate; cement replacement; manmade soil
Masonry	Crush into aggregate; heat to 900°C to ash	Thermal insulating concrete; traditional clay
Non-ferrous metal	Melt	Recycled metal
Paper and cardboard	Purification	Recycled paper
Plastic	Convert to powder by cryogenic milling; clopping; crush into aggregate; burn to ash	Panel; recycled plastic; plastic lumber; recycled aggregate; landfill drainage; asphalt; manmade soil
Timber	Reuse directly; cut into aggregate; blast furnace deoxidisation; gasification or pyrolysis; chipping; moulding by pressurising timber chip under steam and water	Whole timber; furniture and kitchen utensils; lightweight recycled aggregate; source of energy; chemical production; wood-based panel; plastic lumber; geofibre; insulation board

Three scales of CHP which were largely implemented in the following chronological order: (1) Large-scale CHP in cities (>50 MWe), industrial and small-scale CHP. (2) Small (5 kWe – 5 MWe) and medium-scale (5-50 MWe).

5. Wave Power Conversion Devices

The patent literature is full of devices for extracting energy from waves, i.e., floats, ramps, and flaps, covering channels (Swift-Hook, et al., 1975). Small generators driven from air trapped by the rising and falling water in the chamber of a buoy are in use around the world (Swift-Hook, et al., 1975). Wave power is one possibility that has been selected. Figure 3 shows the many other aspects that will need to be covered. A wave power programme would make a significant contribution to energy resources within a relatively short time and with existing technology.

Wave energy has also been in the news recently. There is about 140 megawatts per mile available round British coasts. It could make a useful contribution people needs in the UK. Although very large amounts of power are available in the waves, it is important to consider how much power can be extracted. A few years ago only a few percent efficiency had been achieved. Recently, however, several devices have been studied which have very high efficiencies. Some form of storage will be essential on a second-to-second and minute-to-minute basis to smooth the fluctuations of individual waves and wave's packets but storage from one day to the next will certainly not be economical. This is why provision must be made for adequate standby capacity.

The increased availability of reliable and efficient energy services stimulates new development alternatives. This study discusses the potential for such integrated systems in the stationary and portable power market in response to the critical need for a cleaner energy technology.

Anticipated patterns of future energy use and consequent environmental impacts (acid precipitation, ozone depletion and the greenhouse effect or global warming) are comprehensively discussed in this theme. Throughout the theme several issues relating to renewable energies, environment and sustainable development are examined from both current and future perspectives. It is concluded that renewable environmentally friendly energy must be encouraged, promoted, implemented and demonstrated by full-scale plant (device) especially for use in remote rural areas. Globally, buildings are responsible for approximately 40% of the total world annual energy consumption. Most of this energy is for the provision of lighting, heating, cooling, and air conditioning. Increasing awareness of the environmental impact of CO_2 and NO_x and CFCs emissions triggered a renewed interest in environmentally friendly cooling, and heating technologies. Under the 1997 Montreal Protocol, governments agreed to phase out chemicals used as refrigerants that have the potential to destroy stratospheric ozone. It was therefore considered desirable to reduce energy consumption and decrease the rate of depletion of world energy reserves and pollution of the environment. One way of reducing building energy consumption is to design buildings, which are more economical in their use of energy for heating, lighting, cooling, ventilation and hot water supply. Passive measures, particularly natural or hybrid ventilation rather than air-conditioning, can dramatically reduce primary energy consumption.

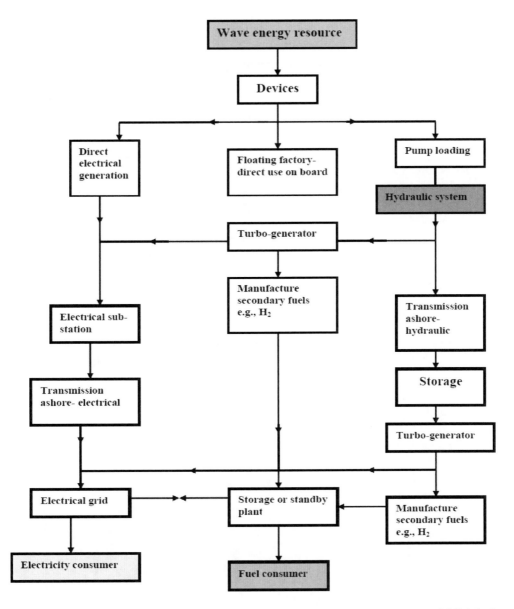

Figure 3. Possible systems for exploiting wave power, each element represents an essential link in the chain from sea waves to consumer.

However, exploitation of renewable energy in buildings and agricultural greenhouses can, also, significantly contribute towards reducing dependency on fossil fuels. Therefore, promoting innovative renewable applications and reinforcing the renewable energy market will contribute to preservation of the ecosystem by reducing emissions at local and global levels. This will also contribute to the amelioration of environmental conditions by replacing conventional fuels with renewable energies that produce no air pollution or greenhouse gases. The provision of good indoor environmental quality while achieving energy and cost efficient operation of the heating, ventilating and air-conditioning (HVAC) plants in buildings represents a multi variant problem. The comfort of building occupants is dependent on many environmental parameters including air speed, temperature, relative humidity and quality in addition to lighting and noise. The overall objective is to provide a high level of building performance (BP), which can be defined as indoor environmental quality (IEQ), energy efficiency (EE) and cost efficiency (CE).

6. ETHANOL PRODUCTION

Alternative fuels were defined as methanol, ethanol, natural gas, propane, hydrogen, coal-derived liquids, biological material and electricity production (Sims, 2007). The fuel pathways currently under development for alcohol fuels are shown in Figure 4. The production of agricultural biomass and its exploitation for energy purposes can contribute to alleviate several problems, such as the dependence on import of energy products, the production of food surpluses, the pollution provoked by the use of fossil fuels, the abandonment of land by farmers and the connected urbanisation. Biomass is not at the moment competitive with mineral oil, but, taking into account also indirect costs and giving a value to the aforementioned advantages, public authorities at national and international level can spur its production and use by incentives of different nature. In order to address the problem of inefficiency, research centres around the world have investigated the viability of converting the resource to a more useful form, namely solid briquettes and fuel gas (Sims, 2007) (Figure 5).

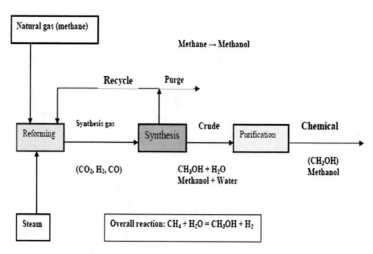

Figure 4. Schematic process flowsheet.

The main advantages are related to energy, agriculture and environment problems, are foreseeable both at regional level and at worldwide level and can be summarised as follows:

- Reduction of dependence on import of energy and related products.
- Reduction of environmental impact of energy production (greenhouse effect, air pollution, and waste degradation).

- Substitution of food crops and reduction of food surpluses and of related economic burdens, and utilisation of marginal lands and of set aside lands.
- Reduction of related socio-economic and environmental problems (soil erosion, urbanisation, landscape deterioration, etc.).
- Development of new know-how and production of technological innovation.

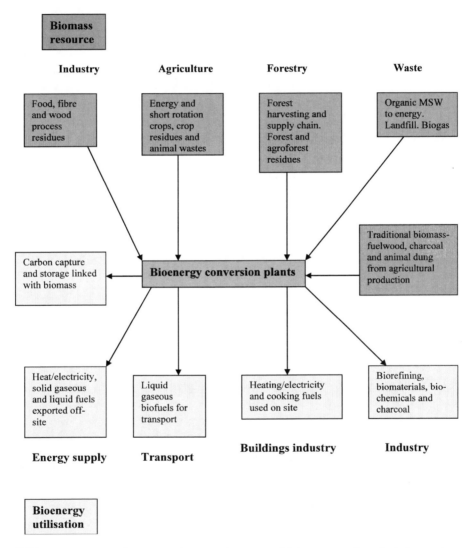

Figure 5. Biomass resources from several sources is converted into a range of products for use by transport, industry and building sectors (Sims, 2007).

Biomass resources play a significant role in energy supply in all developing countries. Biomass resources should be divided into residues or dedicated resources, the latter including firewood and charcoal can also be produced from forest residues. Ozone (O_3) is a naturally occurring molecule that consists of three oxygen atoms held together by the bonding of the oxygen atoms to each other. The effects of the chlorofluorocarbons (CFCs) molecule can last for over a century. This reaction is shown in Figure 6.

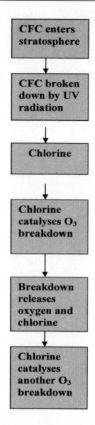

Figure 6. The process of ozone depletion (Trevor, 2007).

It is a common misconception that the reason for recycling old fridge is to recover the liquid from the cooling circuit at the back of the unit. The insulating foams used inside some fridges act as sinks of CFCs - the gases having been used as blowing agents to expand the foam during fridge manufacture. Although the use of ozone depleting chemicals in the foam in fridges has declined in the West, recyclers must consider which strategy to adopt to deal with the disposal problem they still present each year. It is common practice to dispose of this waste wood in landfill where it slowly degraded and takes up valuable void space. This wood is a good source of energy and is an alternative to energy crops. Agricultural wastes are abundantly available globally and can be converted to energy and useful chemicals by a number of microorganisms. The success of promoting any technology depends on careful planning, management, implementation, training and monitoring. Main features of gasification project are:

- Networking and institutional development/strengthening.
- Promotion and extension.
- Construction of demonstration projects.
- Research and development; and training and monitoring.

7. BIOMASS CHP

Combined heat and power (CHP) installations are quite common in greenhouses, which grow high-energy, input crops (e.g., salad vegetables, pot plants, etc.). Scientific assumptions for a short-term energy strategy suggest that the most economically efficient way to replace the thermal plants is to modernise existing power plants to increase their energy efficiency and to improve their environmental performance. However, utilisation of wind power and the conversion of gas-fired CHP plants to biomass would significantly reduce the dependence on imported fossil fuels. Although a lack of generating capacity is forecasted in the long-term, utilisation of the existing renewable energy potential and the huge possibilities for increasing energy efficiency are sufficient to meet future energy demands in the short-term.

A total shift towards a sustainable energy system is a complex and long process, but is one that can be achieved within a period of about 20 years. Implementation will require initial investment, long-term national strategies and action plans. However, the changes will have a number of benefits including a more stable energy supply than at present, and major improvement in the environmental performance of the energy sector, and certain social benefits. A national vision (Omer, 2009d) used a methodology and calculations based on computer modelling that utilised:

- Data from existing governmental programmes.
- Potential renewable energy sources and energy efficiency improvements.
- Assumptions for future economy growth.
- Information from studies and surveys on the recent situation in the energy sector.

In addition to realising the economic potential identified by the National Energy Savings Programme, a long-term effort leading to a 3% reduction in specific electricity demand per year after 2020 is proposed. This will require further improvements in building codes, and continued information on energy efficiency.

The environmental Non Governmental Organisations (NGOs) are urging the government to adopt sustainable development of the energy sector by:

- Diversifying of primary energy sources to increase the contribution of renewable and local energy resources in the total energy balance.
- Implementing measures for energy efficiency increase at the demand side and in the energy transformation sector.

The price of natural gas is set by a number of market and regulatory factors that include supply and demand balance and market fundamentals, weather, pipeline availability and deliverability, storage inventory, new supply sources, prices of other energy alternatives and regulatory issues and uncertainty. Classic management approaches to risk are well documented and used in many industries. This includes the following four broad approaches to risk:

- Avoidance includes not performing an activity that could carry risk. Avoidance may seem the answer to all risks, but avoiding risks also means losing out on potential gain.
- Mitigation/reduction involves methods that reduce the severity of potential loss.
- Retention/acceptance involves accepting the loss when it occurs. Risk retention is a viable strategy for small risks. All risks that are not avoided or transferred are retained by default.
- Transfer means causing another party to accept the risk, typically by contract.

Methane is a primary constituent of landfill gas (LFG) and a potent greenhouse gas (GHG) when released into the atmosphere. Globally, landfills are the third largest anthropogenic emission source, accounting for about 13% of methane emissions or over 818 million tones of carbon dioxide equivalent (MMTCO$_2$e) (Brain, and Mark 2007) as shown in Figures 7-9.

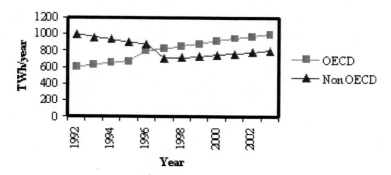

Figure 7. Global CHP trends from 1992-2003 (IEA, 2007).

1 Food, 2 Textile, 3 Pulp & paper, 4 Chemicals, 5 Refining, 6 Minerals, 7 Primary metals, and 8 others

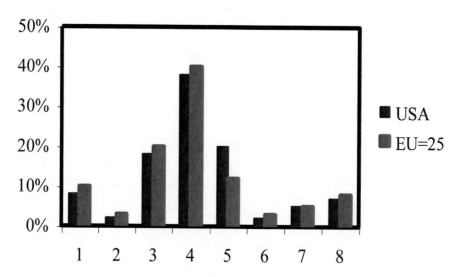

Figure 8. Distribution of industrial CHP capacity in the EU and USA (IEA, 2007).

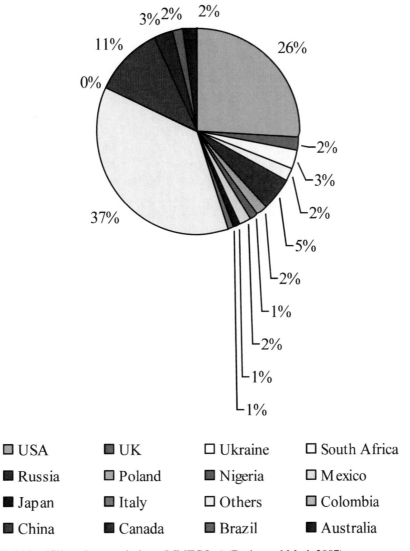

Figure 9. World landfill methane emissions (MMTCO$_2$e) (Brain, and Mark 2007).

8. GEOTHERMAL ENERGY

Geothermal steam has been used in volcanic regions in many countries to generate electricity. The use of geothermal energy involves the extraction of heat from rocks in the outer part of the earth. It is relatively unusual for the rocks to be sufficiently hot at shallow depth for this to be economically attractive. Virtually all the areas of present geothermal interest are concentrated along the margins of the major tectonic plates, which form the surface of the earth. The forced or natural circulation of water through permeable hot rock conventionally extracts heat.

There are various practical difficulties and disadvantages associated with the use of geothermal power:

Transmission: geothermal power has to be used where it is found. In Iceland it has proved feasible to pipe hot water 20 km in insulated pipes but much shorter distances are preferred.

Environmental problems: these are somewhat variable and are usually not great. Perhaps the most serious is the disposal of warm high salinity water where it cannot be reinjected or purified. Dry steam plants tend to be very noisy and there is releases of small amounts of methane, hydrogen, nitrogen, amonia and hydrogen sulphide and of these the latter presents the main problem.

The geothermal fluid is often highly chemically corrosive or physically abrasive as the result of the entrained solid matter it carries. This may entail special plant design problems and unusually short operational lives for both the holes and the installations they serve.

Because the useful rate of heat extraction from a geothermal field is in nearly all cases much higher than the rate of conduction into the field from the underlying rocks, the mean temperatures of the field is likely to fall during exploitation. In some low rainfall areas there may also be a problem of fluid depletion. Ideally, as much as possible of the geothermal fluid should be reinjected into the field. However, this may involve the heavy capital costs of large condensation installations. Occasionally, the salinity of the fluid available for reinjection may be so high (as a result of concentration by boiling) that is unsuitable for reinjection into ground. Ocasionally, the impurities can be precipitated and used but this has not generally proved commercially attractive.

World capacity of geothermal energy is growing at a rate of 2.5% per year from a 2005 level of 28.3 GW (Rawlings, 1999). The GSHPs account for approximately 54% of this capacity almost all of it in the North America and Europe (Rawlings, 1999). The involvement of the UK is minimal with less than 0.04% of world capacity and yet is committed to substantial reduction in carbon emission beyond the 12.5% Kyoto obligation to be achieved by 2012. The GSHPs offer a significant potential for carbon reduction and it is therefore expected that the market for these systems will rise sharply in the UK in the immediate years ahead given to low capacity base at present.

There are numerous ways of harnessing low-grade heat from the ground for use as a heat pump source or air conditioning sink. For small applications (residences and small commercial buildings) horizontal ground loop heat exchangers buried typically at between 1 m and 1.8 m below the surface can be used provided that a significant availability of land surrounding the building can be exploited which tends to limit these applications to rural settings.

Heat generation within the earth is approximately 2700 GW, roughly an order of magnitude greater than the energy associated with the tides but about four orders less than that received by the earth from the sun (Oxburgh, 1975).

Temperature distributions within the earth depend on:

- The abundance and distribution of heat producing elements within the earth.
- The mean surface temperature (which is controlled by the ocean/atmosphere system).
- The thermal properties of the earth's interior and their lateral and radial variation.
- Any movements of fluid or solid rock materials occurring at rates of more than a few millimetres per year.

Of these four factors the first two are of less importance from the point of view of geothermal energy. Mean surface temperatures range between 0-30°C and this variation has a small effect on the useable enthalpy of any flows of hot water. Although radiogenic heat production in rocks may vary by three orders of magnitude, there is much less variation from place to place in the integrated heat production with depth. The latter factors, however, are of great importance and show a wide range of variation. Their importance is clear from the relationship:

$$\beta = q/k \qquad (1)$$

where:

β is the thermal gradient for a steady state ($^{\circ}$C/km), q is the heat flux (10^{-6} cal cm^{-2} sec^{-1}) and k is the thermal conductivity (cal cm^{-1} sec^{-1} $^{\circ}$C^{-1}).

The first requirement of any potential geothermal source region is that β being large, i.e., that high rock temperatures occur at shallow depth. Beta will be large if either q is large or k is small or both. By comparison with most everyday materials, rocks are poor conductors of heat and values of conductivity may vary from 2 x 10^{-3} to 10^{-2} cal cm^{-1} sec^{-1} $^{\circ}$C^{-1}. The mean surface heat flux from the earth is about 1.5 heat flow units (1 HFU = 10^{-6} cal cm^{-2} sec^{-1}) (Oxburgh, 1975). Rocks are also very slow respond to any temperature change to which they are exposed, i.e., they have a low thermal diffusivity:

$$K = k/\rho C_p \qquad (2)$$

where:

K is thermal diffusivity; ρ and C_p are density and specific heat respectively.

These values are simple intended to give a general idea of the normal range of geothermal parameters (Table 6). In volcanic regions, in particular, both q and β can vary considerably and the upper values given are somewhat nominal.

Table 6. Values of geothermal parameters

Parameter	Lower	Average	Upper
q (HFU)	0.8	1.5	3.0 (non volcanic) $\approx$100 (volcanic)
k =cal cm^{-2} sec^{-1} $^{\circ}$C^{-1}	2x10^{-3}	6x10^{-3}	12x10^{-3}
β =$^{\circ}$C/km	8	20	60 (non volcanic) $\approx$300 (volcanic)

9. LANDFILL GAS

Landfill gas (LFG) is currently extracted at over 1200 landfills worldwide for a variety of energy purposes (Table 7), such as:

- Creating pipeline quality gas or an alternative fuel for vehicles.
- Processing the LFG to make it available as an alternative fuel to local industrial or commercial customers.

- Generation of electricity with engines, turbines, micro-turbines and other emerging technologies.

In terms of solid waste management policy, many NGOs have changed drastically in the past ten years from a mass production and mass consumption society to 'material-cycle society' (Abdeen, 2008). In addition to national legislation, municipalities are legally obliged to develop a plan for handling the municipal solid waste (MSW) generated in administrative areas. Such plans contain:

- Estimates of future waste volume.
- Measures to reduce waste.
- Measures to encourage source separation.
- A framework for solid waste disposal and the construction and management of solid waste management facilities.

Landfilling is in the least referred tier of the hierarchy of waste management options: waste minimisation, reuse and recycling, incineration with energy recovery, and optimised final disposal. The key elements are as follows: construction impacts, atmospheric emissions, noise, water quality, landscape, visual impacts, socio economics, ecological impacts, traffic, solid waste disposal and cultural heritage.

Table 7. Types of LFG implemented recently worldwide

Landfill caps	Electricity generation	Fuel production
□ Soil caps □ Clay caps □ Geo-membrane caps	□ Reciprocating engines □ Combustion turbines □ Micro-turbines □ Steam turbines □ Fuel cells	□ Medium BTU gas □ High BTU gas □ Liquefied methane **Thermal generation**
LFG destruction		□ Boilers □ Kilns
□ Flares		□ Greenhouse heaters

10. ENERGY EFFICIENCY

Energy efficiency is the most cost-effective way of cutting carbon dioxide emissions and improvements to households and businesses. It can also have many other additional social, economic and health benefits, such as warmer and healthier homes, lower fuel bills and company running costs and, indirectly, jobs. Britain wastes 20 per cent of its fossil fuel and electricity use. This implies that it would be cost-effective to cut £10 billion a year off the collective fuel bill and reduce CO_2 emissions by some 120 million tones. Yet, due to lack of good information and advice on energy saving, along with the capital to finance energy efficiency improvements, this huge potential for reducing energy demand is not being realised. Traditionally, energy utilities have been essentially fuel providers and the industry has pursued profits from increased volume of sales. Institutional and market arrangements have favoured energy consumption rather than conservation. However, energy is at the centre of the sustainable development paradigm as few activities affect the environment as much as the continually increasing use of energy. Most of the used energy depends on finite resources, such as coal, oil, gas and uranium. In addition, more than three quarters of the world's consumption of these fuels is used, often inefficiently, by only one quarter of the world's population. Without even addressing these inequities or the precious, finite nature of these resources, the scale of environmental damage will force the reduction of the usage of these fuels long before they run out.

Throughout the energy generation process there are impacts on the environment on local, national and international levels, from opencast mining and oil exploration to emissions of the potent greenhouse gas carbon dioxide in ever increasing concentration. Recently, the world's leading climate scientists reached an agreement that human activities, such as burning fossil fuels for energy and transport, are causing the world's temperature to rise. The Intergovernmental Panel on Climate Change has concluded that "the balance of evidence suggests a discernible human influence on global climate". It predicts a rate of warming greater than anyone had seen in the last 10,000 years, in other words, throughout human history. The exact impact of climate change is difficult to predict and will vary regionally. It could, however, include sea level rise, disrupted agriculture and food supplies and the possibility of more freak weather events such as hurricanes and droughts. Indeed, people already are waking up to the financial and social, as well as the environmental, risks of unsustainable energy generation methods that represent the costs of the impacts of climate change, acid rain and oil spills. The insurance industry, for example, concerned about the billion dollar costs of hurricanes and floods, has joined sides with environmentalists to lobby for greenhouse gas emissions reduction. Friends of the earth are campaigning for a more sustainable energy policy, guided by the principal of environmental protection and with the objectives of sound natural resource management and long-term energy security. The key priorities of such an energy policy must be to reduce fossil fuel use, move away from nuclear power, improve the efficiency with which energy is used and increase the amount of energy obtainable from sustainable and renewable energy sources. Efficient energy use has never been more crucial than it is today, particularly with the prospect of the imminent introduction of the climate change levy (CCL). Establishing an energy use action plan is the essential foundation to the elimination of energy waste. A logical starting point is to carry out an energy audit that enables the assessment of the energy use and determine what actions to take. The actions are best categorised by splitting measures into the following three general groups:

(1) High priority/low cost:

These are normally measures, which require minimal investment and can be implemented quickly. The followings are some examples of such measures:

- Good housekeeping, monitoring energy use and targeting waste-fuel practices.
- Adjusting controls to match requirements.
- Improved greenhouse space utilisation.
- Small capital item time switches, thermostats, etc.
- Carrying out minor maintenance and repairs.
- Staff education and training.

- Ensuring that energy is being purchased through the most suitable tariff or contract arrangements.

(2) Medium priority/medium cost:

Measures, which, although involve little or no design, involve greater expenditure and can take longer to implement. Examples of such measures are listed below:

- New or replacement controls.
- Greenhouse component alteration, e.g., insulation, sealing glass joints, etc.
- Alternative equipment components, e.g., energy efficient lamps in light fittings, etc.

(3) Long term/high cost:

These measures require detailed study and design. They can be best represented by the followings:

- Replacing or upgrading of plant and equipment.
- Fundamental redesign of systems, e.g., CHP installations.

This process can often be a complex experience and therefore the most cost-effective approach is to employ an energy specialist to help.

11. POLICY RECOMMENDATIONS FOR A SUSTAINABLE ENERGY FUTURE

Sustainability is regarded as a major consideration for both urban and rural development. People have been exploiting the natural resources with no consideration to the effects, both short-term (environmental) and long-term (resources crunch). It is also felt that knowledge and technology have not been used effectively in utilising energy resources. Energy is the vital input for economic and social development of any country. Its sustainability is an important factor to be considered. The urban areas depend, to a large extent, on commercial energy sources. The rural areas use non-commercial sources like firewood and agricultural wastes. With the present day trends for improving the quality of life and sustenance of mankind, and environmental issues are considered highly important. In this context, the term energy loss has no significant technical meaning. Instead, the exergy loss has to be considered, as destruction of exergy is possible. Hence, exergy loss minimisation will help in sustainability.

The development of a renewable energy in a country depends on many factors. Those important to success are listed below:

(1) Motivation of the Population

The population should be motivated towards awareness of high environmental issues and rational use of energy in order to reduce cost. Subsidy programme should be implemented as incentives to install biomass energy plants. In addition, image campaigns to raise awareness of renewable technology.

(2) Technical Product Development

To achieve technical development of biomass energy technologies the following should be addressed:

- Increasing the longevity and reliability of renewable technology.
- Adapting renewable technology to household technology (hot water supply).
- Integration of renewable technology in heating technology.
- Integration of renewable technology in architecture, e.g., in the roof or façade.
- Development of new applications, e.g., solar cooling.
- Cost reduction.

(3) Distribution and Sales

Commercialisation of biomass energy technology requires:

- Inclusion of renewable technology in the product range of heating trades at all levels of the distribution process (wholesale, retail, etc.).
- Building distribution nets for renewable technology.
- Training of personnel in distribution and sales.
- Training of field sales force.

(4) Consumer Consultation and Installation

To encourage all sectors of the population to participate in adoption of biomass energy technologies, the following has to be realised:
- Acceptance by craftspeople, and marketing by them.
- Technical training of craftspeople, initial and follow-up training programmes.
- Sales training for craftspeople.
- Information material to be made available to craftspeople for consumer consultation.

(5) Projecting and Planning

Successful application of biomass technologies also require:

- Acceptance by decision makers in the building sector (architects, house technology planners, etc.).
- Integration of renewable technology in training.
- Demonstration projects/architecture competitions.
- Biomass energy project developers should prepare to participate in the carbon market by:

- Ensuring that renewable energy projects comply with Kyoto Protocol requirements.
- Quantifying the expected avoided emissions.
- Registering the project with the required offices.
- Contractually allocating the right to this revenue stream.

Other ecological measures employed on the development include:

- Simplified building details.
- Reduced number of materials.
- Materials that can be recycled or reused.
- Materials easily maintained and repaired.
- Materials that do not have a bad influence on the indoor climate (i.e., non-toxic).
- Local cleaning of grey water.
- Collecting and use of rainwater for outdoor purposes and park elements.
- Building volumes designed to give maximum access to neighbouring park areas.
- All apartments have visual access to both backyard and park.

(6) Energy Saving Measures

The following energy saving measures should also be considered:

- Building integrated solar PV system.
- Day-lighting.
- Ecological insulation materials.
- Natural/hybrid ventilation.
- Passive cooling, and passive solar heating.
- Solar heating of domestic hot water.
- Utilisation of rainwater for flushing.

Improving access for rural and urban low-income areas in developing countries must be through energy efficiency and renewable energies. Sustainable energy is a prerequisite for development. Energy-based living standards in developing countries, however, are clearly below standards in developed countries. Low levels of access to affordable and environmentally sound energy in both rural and urban low-income areas are therefore a predominant issue in developing countries. In recent years many programmes for development aid or technical assistance have been focusing on improving access to sustainable energy, many of them with impressive results.

Apart from success stories, however, experience also shows that positive appraisals of many projects evaporate after completion and vanishing of the implementation expert team. Altogether, the diffusion of sustainable technologies such as energy efficiency and renewable energies for cooking, heating, lighting, electrical appliances and building insulation in developing countries has been slow.

Energy efficiency and renewable energy programmes could be more sustainable and pilot studies more effective and pulse releasing if the entire policy and implementation process was considered and redesigned from the outset. New financing and implementation processes are needed, which allow reallocating financial resources and thus enabling countries themselves to achieve a sustainable energy infrastructure. The links between the energy policy framework, financing and implementation of renewable energy and energy efficiency projects have to be strengthened and capacity building efforts are required. Energy resources are needed for societal development. Their sustainable development requires a supply of energy resources that are sustainably available at a reasonable cost and can cause no negative societal impacts.

12. ENVIRONMENTAL ASPECTS OF ENERGY CONVERSION AND USE

Environment has no precise limits because it is in fact a part of everything. Indeed, environment is, as anyone probably already knows, not only flowers blossoming or birds singing in the spring, or a lake surrounded by beautiful mountains. It is also human settlements, the places where people live, work, rest, the quality of the food they eat, the noise or silence of the street they live in. Environment is not only the fact that our cars consume a good deal of energy and pollute the air, but also, that we often need them to go to work and for holidays.

Table 8. Annual greenhouse emissions from different sources of power plants

Primary source of energy	Emissions (x 10^3 metric tones)		Waste (x 10^3 metric tones)	Area (km^2)
	Atmosphere	Water		
Coal	380	7-41	60-3000	120
Oil	70-160	3-6	negligible	70-84
Gas	24	1	-	84
Nuclear	6	21	2600	77

Table 9. Energy consumption in different continents

Region	Population (millions)	Energy (Watt/m^2)
Africa	820	0.54
Asia	3780	2.74
Central America	180	1.44
North America	335	0.34
South America	475	0.52
Western Europe	445	2.24
Eastern Europe	130	2.57
Oceania	35	0.08
Russia	330	0.29

Obviously man uses energy just as plants, bacteria, mushrooms, bees, fish and rats do. Man largely uses solar energy- food, hydropower, wood- and thus participates harmoniously in the natural flow of energy through the environment. But man also uses oil, gas, coal and nuclear power. By using such sources of energy, man is thus modifying his environment.

The atmospheric emissions of fossil fuelled installations are mosty aldehydes, carbon monoxide, nitrogen oxides, sulpher oxides and particles (i.e., ash) as well as carbon dioxide. Table 8 shows estimates include not only the releases occuring at the power plant itself but also cover fuel extraction and treatment, as well as the storage of wastes and the area of land required for operations. Table 9 shows energy consumption in different regions of the world.

13. GREENHOUSES ENVIRONMENT

Greenhouse cultivation is one of the most absorbing and rewarding forms of gardening for anyone who enjoys growing plants. The enthusiastic gardener can adapt the greenhouse climate to suit a particular group of plants, or raise flowers, fruit and vegetables out of their natural season. The greenhouse can also be used as an essential garden tool, enabling the keen amateur to expand the scope of plants grown in the garden, as well as save money by raising their own plants and vegetables. There was a decline in large private greenhouses during the two world wars due to a shortage of materials for their construction and fuel to heat them. However, in the 1950s mass-produced, small greenhouses became widely available at affordable prices and were used mainly for raising plants (John, 1993). Also, in recent years, the popularity of conservatories attached to the house has soared. Modern double-glazing panels can provide as much insulation as a brick wall to create a comfortable living space, as well as provide an ideal environment in which to grow and display tender plants.

The comfort in a greenhouse depends on many environmental parameters. These include temperature, relative humidity, air quality and lighting. Although greenhouse and conservatory originally both meant a place to house or conserve greens (variegated hollies, cirrus, myrtles and oleanders). A greenhouse today implies a place in which plants are raised while conservatory usually describes a glazed room where plants may or may not play a significant role. Indeed, a greenhouse can be used for so many different purposes. It is, therefore, difficult to decide how to group the information about the plants that can be grown inside it.

Throughout the world urban areas have increased in size during recent decades. About 50% of the world's population and approximately 76% in the more developed countries are urban dwellers (UN, 2001). Even though there is an evidence to suggest that in many 'advanced' industrialised countries there has been a reversal in the rural-to-urban shift of populations, virtually all population growth expected between 2000 and 2030 will be concentrated in urban areas of the world. With an expected annual growth of 1.8%, the world's urban population will double in 38 years (UN, 2001). This represents a serious contributing to the potential problem of maintaining the required food supply. Inappropriate land use and management, often driven by intensification resulting from high population pressure and market forces, is also a threat to food availability for domestic, livestock and wildlife use. Conversion to cropland and urban-industrial establishments is threatening their integrity. Improved productivity of peri-urban agriculture can, therefore, make a very large contribution to meeting food security needs of cities as well as providing income to the peri-urban farmers. Hence, greenhouses agriculture can become an engine of pro-poor 'trickle-up' growth because of the synergistic effects of agricultural growth such as (UN, 2001):

- Increased productivity increases wealth.
- Intensification by small farmers raises the demand for wage labour more than by larger farmers.
- Intensification drives rural non-farm enterprise and employment.
- Alleviation of rural and peri-urban poverty is likely to have a knock-on decrease of urban poverty.

Despite arguments for continued large-scale collective schemes there is now an increasingly compelling argument in favour of individual technologies for the development of controlled greenhouses. The main points constituting this argument are summarised by (UN, 2001) as follows:

- Individual technologies enable the poorest of the poor to engage in intensified agricultural production and to reduce their vulnerability.
- Development is encouraged where it is needed most and reaches many more poor households more quickly and at a lower cost.
- Farmer-controlled greenhouses enable farmers to avoid the difficulties of joint management.

Such development brings the following challenges (UN, 2001):

- The need to provide farmers with ready access to these individual technologies, repair services and technical assistance.
- Access to markets with worthwhile commodity prices, so that sufficient profitability is realised.
- This type of technology could be a solution to food security problems. For example, in greenhouses, advances in biotechnology like the genetic engineering, tissue culture and market-aided selection have the potential to be applied for raising yields, reducing pesticide excesses and increasing the nutrient value of basic foods.

However, the overall goal is to improve the cities in accordance with the Brundtland Report (WCED, 1987) and the investigation into how urban green could be protected. Indeed, greenhouses can improve the urban environment in multitude of ways. They shape the character of the town and its neighbourhoods, provide places for outdoor recreation, and have important environmental functions such as mitigating the heat island effect, reduce surface water runoff, and creating habitats for wildlife. Following analysis of social, cultural and ecological values of urban green, six criteria in order to evaluate the role of green urban in towns and cities were prescribed (WCED, 1987). These are as follows:

- Recreation, everyday life and public health.
- Maintenance of biodiversity - preserving diversity within species, between species, ecosystems, and of landscape types in the surrounding countryside.

- City structure - as an important element of urban structure and urban life.
- Cultural identity - enhancing awareness of the history of the city and its cultural traditions.
- Environmental quality of the urban sites - improvement of the local climate, air quality and noise reduction.
- Biological solutions to technical problems in urban areas - establishing close links between technical infrastructure and green-spaces of a city.

The main reasons why it is vital for greenhouses planners and designers to develop a better understanding of greenhouses in high-density housing can be summarised as follows (WCED, 1987):

- Pressures to return to a higher density form of housing.
- The requirement to provide more sustainable food.
- The urgent need to regenerate the existing, and often decaying, houses built in the higher density, high-rise form, much of which is now suffering from technical problems.

The connection between technical change, economic policies and the environment is of primary importance as observed by most governments in developing countries, whose attempts to attain food self-sufficiency have led them to take the measures that provide incentives for adoption of the Green Revolution Technology (Herath, 1985). Since, the Green Revolution Technologies were introduced in many countries actively supported by irrigation development, subsidised credit, fertiliser programmes, and self-sufficiency was found to be not economically efficient and often adopted for political reasons creating excessive damage to natural resources. Also, many developing countries governments provided direct assistance to farmers to adopt soil conservation measures. They found that high costs of establishment and maintenance and the loss of land to hedgerows are the major constraints to adoption (Herath, 1985). The soil erosion problem in developing countries reveals that a dynamic view of the problem is necessary to ensure that the important elements of the problem are understood for any remedial measures to be undertaken. The policy environment has, in the past, encouraged unsustainable use of land (Herath, 1985). In many regions, government policies such as provision of credit facilities, subsidies, price support for certain crops, subsidies for erosion control and tariff protection, have exacerbated the erosion problem. This is because technological approaches to control soil erosion have often been promoted to the exclusion of other effective approaches. However, adoption of conservation measures and the return to conservation depend on the specific agro-ecological conditions, the technologies used and the prices of inputs and outputs of production.

13.1. Types of Greenhouses

Choosing a greenhouse and setting it up are important, and often expensive, steps to take. Greenhouses are either freestanding or lean-to, that is, built against an existing wall. A freestanding greenhouse can be placed in the open, and, hence, take advantage of receiving the full sun throughout the day. It is, therefore, suitable for a wide range of plants. However, its main disadvantage when compared to a lean-to type is that more heat is lost through its larger surface area. This is mainly why lean-to greenhouses have long been used in the walled gardens of large country houses to grow Lapageria rosea and other plants requiring cool, constant temperature, such as half-hardly ferns. However, generally, good ventilation and shading in the spring and summer to prevent overheating are essential for any greenhouse. The high daytime temperatures will warm the back wall, which acts as a heat battery, releasing its accumulated heat at night. Therefore, plants in a greenhouse with this orientation will need the most attention, as they will dry out rapidly.

Also, greenhouses vary considerably in their shapes and internal dimensions. Traditional greenhouses have straight sides, which allow the maximum use of internal space, and are ideal for climbers (Herath, 1985). On the other hand, greenhouses with sloping sides have the advantage of allowing the greatest penetration of sunlight, even during winter (Herath, 1985). The low winter sun striking the glass at $90^{\circ}C$ lets in the maximum amount of light. Where the sun strikes the glass at a greater or lesser angle, a proportion of the light is reflected away from greenhouse. Sloping sides, also, offer less wind resistance than straight sides and therefore, less likely to be damaged during windy weather. This type of greenhouse is most suitable for short winter crops, such as early spring lettuce, and flowering annuals from seed, which do not require much headroom.

A typical greenhouse is shown schematically in Figure 10. However, there are several designs of greenhouses, based on dimensions, orientation and function. The following three options are the most widely used:

- A ready-made design.
- A designed, which is constructed from a number of prefabricated modules.
- A bespoke design.

Of these, the prefabricated ready-made design, which is utilised to fit the site, is the cheapest greenhouses and gives flexibility. It is, also, the most popular option (Herath, 1985).

Specific examples of commercially available designs are numerous. Dutch light greenhouses, for example, have large panes of glass, which cast little shade on the plants inside. They are simple to erect, consisting of frames bolted together, which are supported on a steel framework for all but the smallest models. They are easy to move and extra sections can be added on to them, a useful attraction (Herath, 1985). Curvilinear greenhouses, on the other hand, are designed primarily to let in the maximum amount of light throughout the year by presenting at least one side perpendicular to the sun. This attractive style of greenhouse tends to be expensive because of the number of different angles, which require more engineering (Herath, 1985). Likewise, the uneven span greenhouses are designed for maximum light transmission on one side. These are generally taller than traditional greenhouses, making them suitable for tall, early season crops, such as cucumbers (Herath, 1985). Also, the polygonal greenhouses are designed more as garden features than as practical growing houses, and consequently, are expensive. Their internal space is somewhat limited and on smaller models over-heading can be a problem because of their small roof ventilations. They are suitable for growing smaller pot plants, such as pelargoniums and cacti (Herath, 1985).

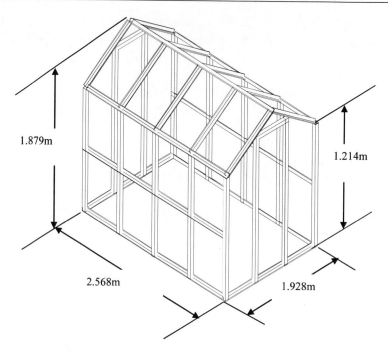

Figure 10. Greenhouse and base with horticultural glass.

Another example is the solar greenhouses. These are designed primarily for areas with very cold winters and poor winter light. They take the form of lean-to structures facing the sun, are well insulated to conserve heat and are sometimes partially sunk into the ground. They are suitable for winter vegetable crops and early-sown bedding plants, such as begonias and pelargoniums (Herath, 1985). Mini lean-to greenhouses are suitable for small gardens where space is limited. They can, also, be used to create a separate environment within larger greenhouses. The space inside is large enough to grow two tomato or melon plants in growing bags, or can install shelves to provide a multi-layered growing environment, ideal for many small potted plants and raising summer bedding plants (Herath, 1985).

13.2. Construction Materials

Different materials are used for the different parts. However, wood and aluminium are the two most popular materials used for small greenhouses. Steel is used for larger structures and UPVC for conservatories (Jonathon, 1991).

13.3. Ground Radiation

Reflection of sunrays is mostly used for concentrating them onto reactors of solar power plants. Enhancing the insolation for other purposes has, so far, scarcely been used. Several years ago, application of this principle for increasing the ground irradiance in greenhouses, glass covered extensions in buildings, and for illuminating northward facing walls of buildings was proposed (Achard and Gicqquel, 1986). Application of reflection of sun's rays was motivated by the fact that ground illuminance/irradiance from direct sunlight is of very low intensity in winter months, even when skies are clear, due to the low incident angle of incoming radiation during most of the day. This is even more pronounced at greater latitudes. As can be seen in Figure 11, which depicts a sunbeam split into its vertical and horizontal components, nearly all of the radiation passes through a greenhouse during most of the day.

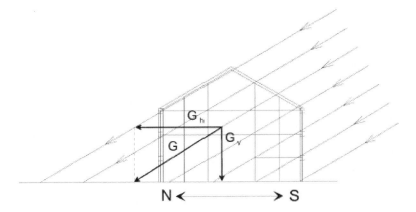

Figure 11. Relative horizontal and vertical components of solar radiation.

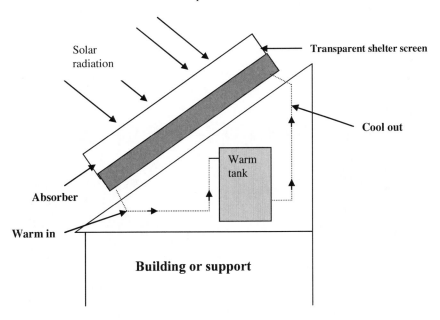

Figure 12. Solar heaters for hot water.

Large-scale, conventional, power plant such as hydropower has an important part to play in development. It does not, however, provide a complete solution. There is an important complementary role for the greater use of small-scale, rural based and power plant (device). Such plant can be used to assist development since it can be made locally using local resources, enabling a rapid built-up in total equipment to be made without a corresponding and unacceptably large demand on central funds. Renewable resources are particularly suitable for providing the energy for such equipment and its use is also compatible with the long-term aims. It is possible with relatively simple flat plate solar collectors (Figure 12) to provide warmed water and enable some space heating for homes and offices, which is particularly useful when the buildings are well insulated and thermal capacity sufficient for the carryover of energy from day to night is arranged.

13.4. Greenhouse Environment

It has been known for long time now that urban centres have mean temperatures higher than their less developed surroundings. The urban heat increases the average and peak air temperatures, which, in turn, affect the demand for heating and cooling. Higher temperatures can be beneficial in the heating season, lowering fuel use, but they exacerbate the energy demand for cooling in summer time. In temperate climates neither heating nor cooling may dominate the fuel use in a building, and the balance of the effect of the heat is less. The solar gains, however, would affect the energy consumption. Therefore, lower or higher percentage of glazing, or incorporating of shading devices might affect the balance between annual heating and cooling load. As the provision of cooling is expensive with higher environmental cost, ways of using innovative alternative systems like mop fans will be appreciated (Figures 13-15). Indeed, considerable research activities have been devoted to the development of alternative methods of refrigeration and air-conditioning. The mop fan is a novel air-cleaning device that fulfils the functions of de-dusting of gas streams, and removal of gaseous contaminations from gas streams and gas circulation (Erlich, 1991). Hence, the mop fan seems particularly suitable for applications in industrial, agricultural and commercial buildings and greenhouses.

Figure 13. Mop fan systems.

Indoor conditions are usually fixed by comfort conditions, with air temperatures ranging from 15°C to 27°C, and relative humidities ranging from 50% to 70% (Abdeen, 2008). The system coefficient of performance (COP) is defined as the ration between the cooling effect in the greenhouse and the total amount of air input to the mop fan. Hence,

COP = cooling delivered/air input to the mop fan   (3)

Figure 14. Mop fan in greenhouse.

Figure 15. Plants in greenhouse.

Therefore, system performance (COP) varies with indoor and outdoor conditions. A lower ambient temperature and a lower ambient relative humidity lead to a higher COP. This means that the system will be, in principle, more efficient in colder and drier climates. The effect of indoor (greenhouse) conditions and outdoor (ambient) conditions (temperature and relative humidity) on system performance is illustrated in Figure 16.

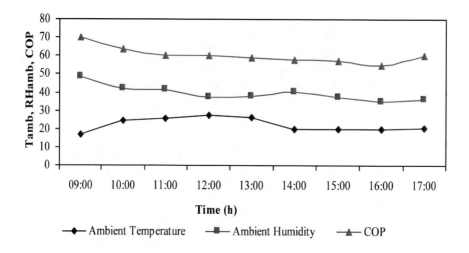

Figure 16. Ambient temperature, relative humidity and COP.

CONCLUSION

There is strong scientific evidence that the average temperature of the earth's surface is rising. This is a result of the increased concentration of carbon dioxide and other GHGs in the atmosphere as released by burning fossil fuels. This global warming will eventually lead to substantial changes in the world's climate, which will, in turn, have a major impact on human life and the built environment. Therefore, effort has to be made to reduce fossil energy use and to promote green energies, particularly in the building sector. Energy use reductions can be achieved by minimising the energy demand, by rational energy use, by recovering heat and the use of more green energies. This study was a step towards achieving that goal. The adoption of green or sustainable approaches to the way in which society is run is seen as an important strategy in finding a solution to the energy problem. The key factors to reducing and controlling CO_2, which is the major contributor to global warming, are the use of alternative approaches to energy generation and the exploration of how these alternatives are used today and may be used in the future as green energy sources. Even with modest assumptions about the availability of land, comprehensive fuel-wood farming programmes offer significant energy, economic and environmental benefits. These benefits would be dispersed in rural areas where they are greatly needed and can serve as linkages for further rural economic development. The nations as a whole would benefit from savings in foreign exchange, improved energy security, and socio-economic improvements. With a nine-fold increase in forest – plantation cover, a nation's resource base would be greatly improved. The international community would benefit from pollution reduction, climate mitigation, and the increased trading opportunities that arise from new income sources. The non-technical issues, which have recently gained attention, include: (1) Environmental and ecological factors, e.g., carbon sequestration, reforestation and revegetation. (2) Renewables as a CO_2 neutral replacement for fossil fuels. (3) Greater recognition of the importance of renewable energy, particularly modern biomass energy carriers, at the policy and planning levels. (4) Greater recognition of the difficulties of gathering good and reliable renewable energy data, and efforts to improve it. (5) Studies on the detrimental health efforts of biomass energy particularly from traditional energy users. Two of the most essential natural resources for all life on the earth and for man's survival are sunlight and water. Sunlight is the driving force behind many of the renewable energy technologies. The worldwide potential for utilising this resource, both directly by means of the solar technologies and indirectly by means of biofuels, wind and hydro technologies is vast. During the last decade interest has been refocused on renewable energy sources due to the increasing prices and fore-seeable exhaustion of presently used commercial energy sources. Plants, like human beings, need tender loving care in the form of optimum settings of light, sunshine, nourishment, and water. Hence, the control of sunlight, air humidity and temperatures in greenhouses are the key to successful greenhouse gardening. The mop fan is a simple and novel air humidifier; which is capable of removing particulate and gaseous pollutants while providing ventilation. It is a device ideally suited to greenhouse applications, which require robustness, low cost, minimum maintenance and high efficiency. A device meeting these requirements is not yet available to the farming community. Hence, implementing mop fans aides sustainable development through using a clean, environmentally friendly device that decreases load in the greenhouse and reduces energy consumption.

REFERENCES

Abdeen, M. O. (2008). Chapter 10: Development of integrated bioenergy for improvement of quality of life of poor people in developing countries, In: *Energy in Europe: Economics, Policy and Strategy- IB*, Editors: Flip L. Magnusson and Oscar W. Bengtsson, 2008 NOVA Science Publishers, Inc., p.341-373, New York, USA.

Achard, P., and Gicqquel, R. (1986). European passive solar handbook. Brussels: Commission of the European Communities.

Bos, E., My, T., Vu, E., and Bulatao R. (1994). World population projection: 1994-95. Edition, published for the World Bank by the John Hopkins University Press. Baltimore and London.

Brain, G., and Mark, S. (2007). Garbage in, energy out: landfill gas opportunities for CHP projects. *Cogeneration and On-Site Power* 8 (5): 37-45.

Erlich, P. (1991). *Forward facing up to climate change, in Global Climate Change and Life on Earth*. R.C. Wyman (Ed), Chapman and Hall, London.

Herath, G. (1985). The green revolution in Asia: productivity, employment and the role of policies. Oxford Agrarian Studies. 14: 52-71.

International Energy Agency (IEA). (2007). Indicators for Industrial Energy Efficiency and CO_2 Emissions: A Technology Perspective.

Jonathon, E. (1991). Greenhouse gardening. The Crowood Press Ltd. UK.

John, W. (1993). *The glasshouse garden*. The Royal Horticultural Society Collection. UK.

Omer, A.M., and Yemen, D. (2001). Biogas an appropriate technology. *Proceedings of the 7th Arab International Solar Energy Conference,* P.417, Sharjah, UAE, 19-22 February 2001.

Omer, A. M. (2009a). Environmental and socio-economic aspect of possible development in renewable energy use, In: *Proceedings of the 4th International Symposium on Environment,* Athens, Greece, 21-24 May 2009.

Omer, A. M. (2009b). Energy use, environment and sustainable development, In: *Proceedings of the 3rd International Conference on Sustainable Energy and Environmental Protection (SEEP 2009),* Paper No.1011, Dublin, Republic of Ireland, 12-15 August 2009.

Omer, A. M. (2009c). Energy use and environmental: impacts: a general review, *Journal of Renewable and Sustainable Energy,* Vol.1, No.053101, p.1-29, United State of America, September 2009.

Omer, A. M. (2009d). Chapter 3: Energy use, environment and sustainable development, *In: Environmental Cost Management,* Editors: Randi Taylor Mancuso, 2009 NOVA Science Publishers, Inc., p.129-166, New York, USA.

Oxburgh, E.R. (1975). *Geothermal energy.* Aspects of Energy Conversion. p. 385-403.

Rawlings, R.H.D. (1999). *Technical Note TN 18/99 – Ground Source Heat Pumps: A Technology Review.* Bracknell. The Building Services Research and Information Association.

Robinson, G. (2007) Changes in construction waste management. *Waste Management World,* p. 43-49. May-June 2007.

Sims, R.H. (2007). Not too late: IPCC identifies renewable energy as a key measure to limit climate change. *Renewable Energy World* 10 (4): 31-39.

Swift-Hook, D.T., et al. (1975). Characteristics of a rocking wave power devices. Nature 254: 504.

Trevor, T. (2007). Fridge recycling: bringing agents in from the cold. *Waste Management World* 5: 43-47.

United Nations (UN). (2001). *World Urbanisation Prospect: The 1999 Revision.* New York. The United Nations Population Division.

World Commission on Environment and Development (WCED). (1987*). Our common future.* New York. Oxford University Press.

Chapter 6

Medicines and Biology: Sustainable Management of Pharmacy, Pharmacists and Pharmaceuticals and How to Bridge the Gap in Human Resources for Health

Abstract

Worldwide there are different systems for providing pharmacy services. Most countries have some element of state assistance, either for all patients or selected groups such as children, and some private provisions. Medicines are financed either through cost sharing or full private. The role of the private services is therefore much more significant. Nationally, there is a mismatch between the numbers of pharmacists and where are they worked, and the demand for pharmacy services. The position is exacerbated locally where in some areas of poor; there is a real need for pharmacy services, which is not being met and where pharmacists have little spare capacity. Various changes within the health-care system require serious attention be given to the pharmacy human resources need. In order to stem the brain drain of pharmacists, it is, however, necessary to have accurate information regarding the reasons that make the pharmacists emigrate to the private sector. Such knowledge is an essential in making of informed decisions regarding the retention of qualified, skilled pharmacists in the public sector for long time. There are currently 3000 pharmacists registered with the Sudan Medical Council of whom only 10% are working with the government. The pharmacist: population ratio indicates there is one pharmacist for every 11,433 inhabitants in Sudan, compared to the World Health Organisation (WHO) average for industrialised countries of one pharmacist for 2,300 inhabitants. The situation is particularly problematic in the Southern states where there is no pharmacist at all. The distribution of pharmacists indicates the majority are concentrated in Khartoum state. When population figures are taken into consideration all states except Khartoum and Gezira states are under served compared to the WHO average. This mal-distribution requires serious action as majority of the population is served in the public sector. This study reveals the low incentives, poor working conditions, job dissatisfaction and lack of professional development programmes as main reasons for the immigration to the private-sector. The objective of this communication is to highlight and provide an overview of the reasons that lead to the immigration of the public sector pharmacists to the private-sector in Sudan. The survey has been carried out in September 2004. Data gathered by the questionnaires were analysed using Statistical Package for Social Sciences (SPSS) version 12.0 for windows. The result have been evaluated and tabulated in this study. The data presented in this theme can be considered as nucleus information for executing research and development for pharmacists and pharmacy. More measures must be introduced to attract pharmacists into the public sector. The emerging crisis in pharmacy human resources requires significant additional effort to gather knowledge and dependable data that can inform reasonable, effective, and coordinated responses from government, industry, and professional associations.

Keywords: Sudan, healthcare, pharmacy, pharmacist retention, private-sector, public-sector

1. INTRODUCTION

Sudan is geo-politically well located, bridging the Arab world to Africa. Its large size and extension from south to north provides for several agro-ecological zones with a variety of climatic conditions, rainfall, soils and vegetation (Appendix 1). Sudan is one of the most diverse countries in Africa, home to deserts, mountain ranges, swamps and rain forests. The present policy of the national health–care system in Sudan is based on ensuring the welfare of the Sudanese inhabitants through increasing national production and upgrading the productivity of individuals. A health development strategy has been formulated in a way that realises the relevancy of health objectives to the main goals of the national development plans. The strategy of Sudan at the national level aims at developing the Primary Health Care (PHC) services in the rural areas as well as urban areas (GOS, 2002). Methods of preventing and controlling health problems are the following:

- Promotion of food supply and proper nutrition
- An adequate supply of safe water and basic sanitation
- Maternal and child health care
- Immunisation against major infectious diseases
- Preventing and control of locally endemic diseases, and
- Provision of essential drugs

This will be achieved through a health system consisting of three levels (state, provincial and localities), including the referral system, secondary and tertiary levels.

Poverty and the accompanying ignorance (lacking knowledge, generally do not have many options often than exploiting their local environment) of natural resource degradation present major obstacles to sustainable development. In Sudan about 75% of the population live in poor conditions (scarcity of food, water, clothes, health services, education, etc.), while 20% live in abject poverty. Small holders and pastoral groups have intensified exploitation of the land, contributing to widespread soil erosion (Omer, 1994). The economic dividend of a full peace settlement could be great. Sudan has large areas of cultivatable land, as well as gold and cotton. Its oil reserves are ripe for further exploitation.

In Sudan, with more than ten million people do not have adequate access to health care; twenty million inhabitants are without access to pharmacy, and a very low proportion of people being treated in hospitals. The investment, which is needed to fund the extension and improvement of these services, is substantial. Most governments in developing countries are ready to admit that they lack the financial resources for proper health and pharmacy schemes. Moreover, historically, bilateral and multilateral funding accounts for less than 10% of total investment needed. Thus the need for private financing is imperative.

Many healthy utilities in developing countries need to work in earnest to improve the efficiency of operations. These improvements will not only lead to better services but also to enhanced net cash flows that can be re-invested to improve the quality of service. Staff productivity is another area where significant gains can be achieved. Failure of subsidies to reach intended objectives is due, in part, to lack of transparency in their allocation. Subsidies are often indiscriminately assigned to support investment programmes that benefit more middle and high-income families, which are already receiving acceptable service. Consumption subsidies often benefit upper-income domestic consumers' substantially more than low-income ones. Many developing countries (Sudan is not an exception) are encouraging the participation of the private sector as a means to improve productivity in the provision of health and pharmacies services. Private-sector involvement is also needed to increase financial flows to expand the coverage and quality of services. Many successful private-sector interventions have been under taken. Private operators are not responsible for the financing of works, nonetheless they can bring significant productivity gains, which would allow the utility to allocate more resources to improve and extend services. Redressing productivity, subsidy and cross-subsidy issues before the private-sector is invited to participate, has proven to be less contentious (Show and Griffin, 1995).

Despite the constraints, over the last decade the rate of implementation of rural and peri-urban pharmacy supplies and healthy programmes has increased considerably, and many people are now being served more adequately. The following are Sudan experience in pharmacy supply and healthy projects:

At community level:

- Participatory approaches in planning, implementation and monitoring.
- Establishment and training of reliable financial and maintenance management.
- Sensitive timing of health and hygiene education.

At state and national level:

- Integrated multi-sectoral approach development.
- Training approach and material development for state and extension staff.
- Continuing support from integrated multi-sectoral extension team.
- Establishment of technical support system.
- Multi-sectoral advisory group including training and research institutions.
- Development and dissemination of relevant information for state and extension staff.

Coordination and integration of various aspects of health, and pharmacy management with other related resources is societal concern. The following are recommended:

- Community must be the focus of benefits accruing from restructures, legislature to protect community interest on the basis of equity and distribution, handover the assets to the community should be examined; and communities shall encourage the transfer the management of health schemes to a professional entity.

- The private-sector should be used to mobilise, and strengthen the technical and financial resources, from within and without the country to implement the services, with particular emphasis on utilisation of local resources.
- The government should provide the necessary financial resources to guide the process of community management of pharmacy supplies. The government to divert from provision of services and be a facilitator through setting up standards, specifications and rules to help harmonise the private sector and establish a legal independent body by an act of parliament to monitor and control the providers. Government to assist the poor communities who cannot afford service cost, and alleviate social-economic negative aspects of privatisation.
- The sector-actors should create awareness to the community of the roles of the private-sector and government in the provision of health and pharmacy services.
- Support agencies assist with the financial and technical support, the training facilities, coordination, development and dissemination of health projects, and then evaluation of projects.

Health system in Sudan is characterised by heavily reliance on charging users at the point of access (private expenditure on health is 79.1 percent (WHO, 2004)), with less use of prepayment system such as health insurance. The way the health system is funded, organised, managed and regulated affects health workers' supply, retention, and the performance. The contested policies of public health sector reform can be construed as attempts to craft the incentive environment to produce improved performance (Hongoro and McPake, 2004). The migration of doctors to Gulf States and more recently to the UK leaves easily noticeable gap in health care system in Sudan. The loss of pharmacists from public-sector mainly to private sector could be equally detrimental.

Primary Health Care was adopted as a main strategy for health care provision in Sudan and new strategies were introduced during the last decade, include:

- Health area system
- Polio eradication in 1988
- IMCI initiative
- Rollback malaria strategy
- Basic developmental need approach in 1997
- Safe motherhood, making pregnancy safer initiative, eradication of harmful traditional practices and emergency obstetrics' care programmes

The 25 years pharmacy strategy aims to help people maintain their health, manage common ailments, make the best use of prescribed medicines and manage long-term medication needs by providing a service which is easily accessible to all, tailored to individual needs, efficient, co-coordinated with other professionals, and of a quality that satisfy customers (MOH, 2003). The ability of the pharmacy profession to provide patients with more support in using medicines and to make them more confident in advice they are given depends entirely on the quality and quantity of Pharmacy Human Resources (PHRs) available to do the job. The PHRs in public-sector are a critical component in the National Drug Policy (NDP) and the 25 years pharmacy strategy. Implementing of the pharmacy strategy and achieving its objectives depend upon people. It requires high qualified and experienced professionals, including policy-makers, pharmacists, doctors, pharmacy technicians, and paramedical staff, economists and researchers. The goals of the 25 years pharmacy strategy will not be achieved without increasing the number and quality of pharmacists working in the public-sector. The brain drain will affect the pharmacists' key role in the implementation of NDP and 25 years pharmacy strategy (MOH, 1997). Pharmacists will implement the strategy only if they understand its rationale and objectives, when they are trained to do their jobs well, paid adequate wages, and motivated to maintain high standard. Lack of appropriate expertise has been a decisive factor in the failure of some countries to achieve the objectives of national drug policy.

Although substantial new resources such as oil production, peace agreement and increased Revolving Drug Funds (RDFs) coverage are promised to health system, many of the constraints cannot be easily resolved by money alone (Appendix 2). Worldwide there are different systems for providing pharmacy services. However, viewed across a variety of characteristics, the pharmacists' profession is clearly in transition. Where this evolution is leading is not clear. Increased numbers of drug therapies, an aging but more knowledgeable and demanding population, and deficiencies in other areas of the health care system seem to be driving increased demand for the clinical counselling skills of the pharmacists. Given the growing evidence of drug related complications, however, the well documented ability of pharmacists to anticipate and forestall many of these problems. A more likely scenario is that pharmacists will be increasingly valued and demanded for their knowledge, skills and cost effectiveness contribution to the health care system (CPA, 2001). The shortage of pharmacists at points of drug dispensing deprives the population of vital expertise in the management of medicine related problems in both community and hospital setting (Matowe, et al., 2004).

The drug distribution network in Sudan consists of open market, drug vendors (known as home drug store), community (private) pharmacies, people's pharmacies, private and public hospitals, doctors' private clinics, non-governmental organisations (NGOs) clinics, private medicines importers (wholesalers), public wholesalers (i.e., Central Medical Supplies and Khartoum State Revolving Drug Fund) and local pharmaceutical manufacturers. The states' departments of pharmacy statutorily license community and Peoples' pharmacies. A superintending pharmacist, who is permanently registered with Sudan Medical Council and licensed, oversees the pharmacy any time it is opened for business (The Act, 2001). With such pharmacies there should not be any serious of the sale of fake drugs. Unfortunately however, there are many pharmacies working without qualified pharmacists (MOH, 2003).

During the last decades, the pharmacy workforces have witnessed a significant increase in the number of pharmacies, drug importing companies and pharmaceutical manufacturers as shown in Table 1.

In the public-sector, adoption of cost sharing policy as a mechanism of financing for essential medicines at full price cost requires far more expertise than simply distributing free medicines. This policy increases the demand for pharmacists in hospitals. The new concept of pharmaceutical care and recognition pharmacists as health care team members will boost the demand for the skilled PHRs. The Federal Ministry of Health (MOH) faces two major issues with the PHRs. First, there is shortage of current pharmacists in the public sector. Secondly, the future role of pharmacists within the health cares system.

Table 1. Pharmacists' labour market (MOH, 2003)

Institutions	1989	2003	Increase in (%)
Faculties of Pharmacy	1	7	600%
Registered Pharmacists	1505	2992	99%
Public Sector Pharmacists	162	300	85%
Hospital Pharmacies	205	304	48%
Community Pharmacies	551	779	41%
Drug Importing Companies	77	175	127%
Drug Manufacturers	5	14	180%

As well as involving several of Sudan's neighbours, the civil war has proved costly; with the result many Sudanese have seen a fall in living standards. The political upheaval and economic meltdown in public-sector play an important role in driven pharmacists out. This will render the public sector remains unattractive compared with elsewhere, and the private sector will continue to suck in qualified pharmacists in increasing numbers, and the public sector will continue to finance it.

There are considerable published works about brain drain of health professionals (mainly doctors and nurses) from developing countries to the developed ones (e.g., Lerberghe et al., 2002; and Hongoro and McPake, 2004). But, there are no many empirical studies that examining the same questions about the brain drain of pharmacists from government institutions to the private-sector. The findings of this study will demonstrate factors and explain the reasons behind the brain drain of pharmacists. The data are meant to provide health officials with evidence-based information about the causes of the pharmacists' attrition. Such information is necessary for formulating appropriate policies for the retention of pharmacists in Sudanese public-sector. The study can benefit other developing countries with similar situation especially in Sub-Saharan Africa. It will also encourage human resources (planners and policy-makers) to be open to the application of business instrument when dealing with pharmacy manpower within the public sector.

2. AIMS AND OBJECTIVES

The main purpose of this research is to determine and analyse the reasons of pharmacists' brain drain from public to private-sectors in Sudan and to set a recommendation to remedy this situation. The specific objectives are to answer the following questions:

- Why do pharmacists leave the public-sectors and what are the most important reasons, which encourage them to join the private-sector?
- What are the main reasons that make public sector pharmacists have intention to quit from civil service?
- What are the encouraging factors, which retain pharmacists in the public-sector?

3. METHOD

The logical target was a small sample that can describe a population group; however, the survey did not attempt to characterise the entire pharmacists working in Sudan. Thus, the objective of this study is not to generate statistically significant findings, but to explore the reasons of brain drain of pharmacists from public to private-sectors and had been sized to be feasible in the time and resources available. The information necessary to explore the reasons of brain drain of pharmacists from public to private-sectors were collected from 54 pharmacists working for private sectors (32 community pharmacies and 22 from drug importing companies) and 26 working with public sectors. All the above pharmacies were registered with the Sudan Medical Council. These samples were obtained from registered pharmacists. The samples are nevertheless, thought to be sufficient to valid the conclusion drawn from this research.

Data were collected through the use of two self-completing questionnaires: one addressed to pharmacists working with government institutions (Appendix 3), and the other from those who working with the private sector (Appendix 4). The questionnaires using close-ended questions were phrased in such a way that a limited range of response was obtained and to get reliable and consistent information. The questionnaires then pre-coded. The questionnaires were translated back into clear Arabic language. Since, the ambiguous questions would lead to responses that do not accurately capture respondents' views or not bothering to respond (Boynton, et al., 2004).

Each questionnaire was tested at the field to make sure that all relevant issues were covered and pre-codes were correct. Four pharmacists working with private-sector (two from community pharmacies and the other two from drug importing companies) were asked to fill the questionnaire and feed back the authors (How long did it take them to answer the questions and whether there was unclear question(s) or not?). The same scenario was repeated with two public-sector pharmacists to test the questionnaire designed to address those working with the government institutions. The pilot survey participants were not in the selected study samples. The responses were positive, though minor changes were made to both questionnaires (mainly in a formatting). A category "Others (please specify)" was added after certain questions to accommodate any response not listed. The questionnaires took the respondents from 6 to 8 minutes to be answered carefully.

The participants from drug importing companies were selected by using systematic sampling methods. The author agreed to select the first name appears in the list of the medicines importing companies' responsible pharmacists after their ascending sorting. Thereafter, every eighth pharmacists (the total number of drug companies is 175) on the list is to complete the sample size of 22. The respondents from community pharmacies were selected from the list of licensed community pharmacies responsible pharmacists using the same procedure as in the case of drug importing companies. After the selection of the first name, every twenty-fifth pharmacists (the total number is 779 pharmacies) on the list to complete the sample of 32 participants. The electronic lists were obtained from the General Directorate of Pharmacy, and the Federal Ministry of Health – Khartoum.

Member of supportive staff within the Directorate of Pharmacy distributed the questionnaire to the pharmacies and drug companies at Khartoum State. After one week latter, all questionnaires were collected with 100% response rate. Those who work with the

Federal and Khartoum State Department of Pharmacies were asked to fill in a questionnaire specially designed for those who work with the government. The questionnaire was distributed to pharmacists using internal mail system (i.e., cirque). 26 responses representing 87% of the study population were received. This study was carried out between 10th and 15th of September 2004. The questionnaire was translated back into English in order to ensure there is no loss or change in meanings. Data gathered by the questionnaires were electronically analysed using Statistical Package for Social Sciences (SPSS) version 12.0 for windows.

4. RESULTS

4.1. Public Sector Pharmacists

The total number of respondents from public-sector was 26 pharmacists (53.8%) of them were males. The majority (73%) of respondents graduated within or after 1991. Most (69%) of them had studied in Sudan. Surprisingly, (57%) of pharmacists (53.8% male) were employed in the private at some time in the past before joining the public-sector. This is due to the fact some of the current pharmacy managers in the Federal and Khartoum State Departments of Pharmacy had private-sector experience. The top three reasons that de-motivate pharmacists who had experience with the private were lack of ownership feeling (21.4%), sense of working for specific person (21.4%) and job dissatisfaction (14.3%). Most (80.8%) of respondents joined the public-sector due to job satisfaction and feeling of ownership (65.4%) as illustrated in Table 2 and Figure 1. In answering the question: 'Do you have intention to leave the public-sector at some time in the future?' (61.5%) of respondents answered 'Yes'. The vast majority (87.5%) of them owing their intention to leave for better benefits in the private-sector compared with the public sector as given in Table 3. Table 4 shows (69.2%) of respondents mentioned monetary issues as one of the reasons discourages them from continuing with public-sector.

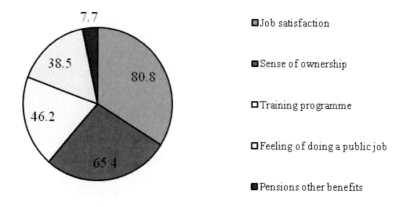

Figure 1. Reasons for joining public sector.

Table 2. Reasons for choosing public-sector (N = 26)

Reasons	Percent
Job satisfaction	80.8
Sense of ownership	65.4
Training programme	46.2
Feeling of doing a public job	38.5
Pensions and other benefits	7.7

Table 3. 'Why are you intending to leave the public-sector?' (N= 16)

Reasons	Percent
High wages and incentives in the private-sector	87.5
Private sector offers vehicles	56.3
The private give full treatment when feeling ill	50.0
Job satisfaction in the private	6.3

Table 4. Reasons discourage you to continue with public-sector (N = 26)

Reasons	Percent
Monetary issues	69.2
Lack of recognition of what I have done	57.7
Dim vision	53.8
Sense of instability	53.8
Those who work and those who do not are equal	53.8
Policy-makers do not care about pharmacy	53.8
Lack of job satisfaction	34.6
Political issues	15.4

The respondents recommended continuing pharmacy professionals' development to assure the role of the pharmacists in the health care, creation of new jobs, increase the salaries of public sector pharmacists and activation of federal pharmacy and poisons board.

4.2. Private Sector Pharmacists

The number of respondents from the private-sector was 54 (80%) of them were male. (77.8%) had studied in Sudan and the majority (74%) graduated during or after 1991. 32 (59.3%) of the respondents worked with community pharmacy whereas, 22 (40.7%) were drug companies employees. Salaries in the private-sector ranged from LS 500,000 to LS 2,500,000 Sudanese pound (LS) or more (1 US$ = LS 2500). 35 (65%) pharmacists had previous public-sector experience. In answering the question 'Why did you leave the public sector?' (51.4%) of respondents had left the public sector because policy-makers did not care of pharmacy (Table 5). The main reasons for choosing the private-sectors mentioned by respondents are the salaries (61.8%); the job satisfaction (52.9%) and the vehicle (26.5%) are shown in Table 6 and Figure 2.

Substantial percentage (78.4%) of the respondents answer "yes" to the question: thinking about your own job; could you leave the private and join the public sector at some time in the future? Table 7 shows the reasons, which encourage pharmacists who were in the private sector (at the time of the study), and are willing to join the public-sector.

Table 5. 'Why do you leave the public-sector?' (N = 35)

Reasons	Percent
Policy makers do not care of pharmacy	51.4
Those who work and those who do not are equal	42.9
Low salaries and incentives	42.9
Lack of recognition of what I have done	31.4
Instability feeling	28.6
Lack of job satisfaction	28.6
Dim vision	25.7
Political issues	17.1
Others*	28.6

*No training, hospitals are without medicines and domination of doctors.

Table 6. Reasons for choosing the private-sector (N = 34)

No.	Reasons	Percent
1	Salaries are better than public sector	61.8
2	Job satisfaction	52.9
3	Private sector offers vehicles	26.5
4	Full treatment when feeling ill	14.7
5	Others*	23.5

*No jobs available in the public sector and mismanagement. It is easy to have a private job to increase the income and flexibility of working environment.

Table 7. 'What encourages you to join the public-sector?' (N = 43)

Reasons	Percent
Job satisfaction	69.8
No feeling of working for specific person	62.8
Overseas training	62.8
Internal training	55.8
Feeling of ownership	48.8
Better salaries	27.9
Others*	18.6

*Public sector reserves rights when ill, job satisfaction, stability and fair competition.

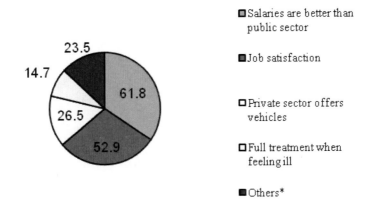

Figure 2. Reasons for preferring the private-sector.

The most important reasons discourage the pharmacists who were in the private-sector during the study period from joining the public-sector are presented in Table 8.

Table 8. 'Why did not some private-sector pharmacists like to join the public sector?' (N = 31)

Reasons	Percent
Monetary issues	64.5
Dim vision	51.6
Instability feeling	38.7
Lack of job satisfaction	19.4
Political issues	12.9
Others*	1.5

*Government neglects pharmacists and the domination of doctors.

5. DISCUSSIONS

5.1. Public Sector Pharmacy Workforces

The public health sector reform seems to have undermined pharmacy human resources in health sector as often as making a positive contribution. Without motivated, competent, and well-funded pharmacy workforces, there is a dangerous infusion of money for establishing drugs revolving funds in different states. To address the national problem of access to essential medicines will be either misused or wasted, or both.

Challenges with respect to pharmacy human resources vary greatly between and within states, and associated with the political commitment of the states` government and their ministers of health. The public sectors' pharmacists in many states are adversely affected by sever under investment from the states and national funds, as well as external sources. For example, pharmacy budget from World Health Organisation (WHO) reduced from US$ 200,000 in 2003 to only US$ 93,000 in 2004 (MOH, 2004). Driven by financial limitations, pharmacy workforces planning at federal and states ministries of health has been unable to match pharmacists requirements, the needs of community and the health system as whole.

It has been quite evident the civil service management system is detrimental to the retention of skilled pharmacists. Like other disciplines, the service affair authority determines the number of pharmacists` jobs in the public-sector. It also sets salary scale and other incentives schemes in coordination with the Federal Ministry of Finance and Economic Planning. Although health professionals in hospitals tend to work in shifts and have to face different working conditions, the incentive system was not flexible enough to cope with differences between health professionals and other civil servants. Notably, the gap in the pharmacy workforces do not generally relate to pharmacists, but to pharmacy assistants who constitute the bulk of the workforces. The difficulties caused by low pharmacy staff numbers are compounded by morale problems, skill balances and geographical mal-distribution, most of which are related to poor human resources management (Narasimhan, et al., 2004). How can the ministry of health grapple successfully with the demands of pharmaceutical care crises and the requirements of transformed pharmacy profession, if it lacks the very foundation of pharmacy care – motivated, trained and supported pharmacists.

5.2. Mal-Distribution

Around 3000 pharmacists are registered in Sudan. Only 300 (10%) works with the public sector. 25, 25, 20 pharmacists were employed in Khartoum, Khartoum North and Omdurman hospitals respectively. Some states (e.g., Southern states has only 2 pharmacists) were not included in Table .9. This anomaly seems to imply the number of pharmacists in the public sector (has not only been insufficient in absolute terms, but also has been inefficient in its distribution). This number will be depleted and the situation may be getting worse. One reason is migration to the private-sector.

5.3. Working Conditions

Sudan like many developing countries, the essential working conditions is not met. Social or personal development opportunities are limited. Therefore, it is difficult for health professionals in general and pharmacists in particular to remain satisfied. The poor working conditions, remuneration and other factors pushed pharmacists out of the public-sector.

5.4. Incentives and Remuneration

The question 'why public-sector pharmacists intending to leave?' showed the issues of salary and remuneration dominated. (87.5%) of pharmacists at Khartoum area stated in surveyed high wages and incentives in the private-sector. (56.3%) stated vehicle as the reason for intending to leave the public-sector. The study revealed the salaries of the majority (78%) of private sector pharmacists are more than 3 times the salaries of the public-sector pharmacists. On average, the private pharmacists earn L.S 2 million compared with around L.S 600,000 for the public-sector pharmacists. It is not uncommon to the public pharmacists to engage in dual practice (such as night shifts or working with the drug companies as fulltime at the same time) or solicit informal payments (such as registration of a pharmacy or a drug company without even visiting it) to supplement their income. Hence, this causes various further difficulties in accountability and equity of access.

Table 9. Pharmacists' distribution at state levels

State	Number of pharmacists
Department of Pharmacy (DOP) Khartoum State[*]	8
DOP-North Darfur	7
DOP-Sennar	2
DOP-North Kordofan	5
DOP-South Kordofan	7
DOP-White Nile	2
DOP-Kassala	7
DOP-River Nile	3
DOP-Northern State	3
DOP-Al Gezira[*]	6
Total[**]	50

[*]The Pharmacists who work with Revolving Drug Funds are not included. [**]Information about other States is not available (10 Southern states, 2 Darfur states, 2 Eastern states, 1 Blue Nile state, and 1 West Kordofan state).

There are some reported evidences for using of provider incentives and enablers can improve the performance under specific circumstances. For example, Eichler and colleagues (2001) showed the indicators of the achievement used to establish bonus payments improved when a bonus system was introduced in Haiti. The use of financial incentives was also reported positively to change health worker behaviour in terms of heightened productivity in Cambodia (Van Damme, et al., 2001). The findings consolidate strategies implemented by Abdullah Seedahmed at Khartoum State (Federal Ministry of Health. He was the Minister of Health, Khartoum State during 1993 –2001) and Elsadig Gasmalla, at the Red Sea, Northern, Algadarif and Al Gezira States (Minister of Health, Gezira State. He was the Minister of Health during 1996 – 2000) in attracting pharmacists to work in public-sector especially, at the Ministry of Health during their time. These strategies comprised in additional to financial incentives, the full delegation of power to the pharmacy managers, political support and motivation.

5.5. Job Satisfaction

Without professional, personal job satisfaction, and the ability to carry out a job as well as possible, the staff can become disillusioned and leave the vacancy (Hughes, 2004). In pharmacy, where practice only remotely resembles what students are taught, this makes students frustrated and disgruntled on qualification (Matowe, et al., 2004). Thus, it is not surprising the young pharmacists (74%) of private-sector pharmacists in the study graduated during or after 1991 seek better career opportunities in the private-sector, where they are offered at least better remuneration. The study revealed (69.8%) of the respondents might be encouraged to join the public service, due to job satisfaction, if other obstacles are solved.

5.6. Training Strategies

A lack of professional development can result in low staff morale (Shepherd, 1995). Training strategies that fail to emphasise continued pharmacy professional development affect not only the numbers of pharmacists in the public-sector as shown in Table .8, but also their quality and performance. In addition, poor job satisfaction, working conditions, and remuneration. These dimensions are not captured in the data to enable international comparison. But, they are widely understood to be at least as important as more quantifiable factors in explaining the performance of the health care (Hongoro and McPake, 2004).

Although there is an imperative to retain staff, and there is a link between increased retention, personal development plan and appraisal (Gould, 2004). The strategy of bonding pharmacists to government after pre-registration training has largely failed because pharmacists easily find ways to quit from the public sector (One-year houseman-ship strategy was adopted in early 1990s). This failure is partly explained in the absence of punitive action and capacity to enforce penalties (if any) and availability of buy-out options (such as attractive drug companies).

6. PHARMACEUTICALS FINANCING REFORMS

In developing countries, pharmaceuticals generally account for a more significant share of overall health expenditures than in developed countries (15%). In several African countries, it is believed to exceed 50%. In developing countries 50-90% of the overall pharmaceuticals expenditures are privately financed, which is considerably higher than in developed countries (median is 34%) (Velasquez, et al., 1998).

Financing of pharmaceutical is crucial issue for several reasons. First, because drugs are save lives and improve health, it is important that drug financing ensures access to essential drugs for all segments of the population. Second, drugs are costly. For most ministries of health, and drugs represent the largest expenditure after staff salaries. In some countries, up to 80% of household's health-related spending is on drugs. In developing countries, drugs commonly represent from 25 to 50% of total public and private health expenditures (Quick et al., 1997). Third, inadequate funding for drugs means expenditures for staff salaries and other care costs may be used inefficiently or simply wasted. Fourth, the availability and effectiveness of drugs are key factors in generating and maintaining public interest and participation in health related activities.

To be successful, user fee mechanisms must generally be accompanied by perceived quality improvements in services. The World Bank suggests the improvement in the quality of services would compensate the negative impact of prices. This implies that improved supply mechanisms for drugs are both prerequisites and outputs of successful programmes. The properly designed cost recovery programmes can encourage higher demand for modern health care and, as a result, higher level of utilisation (Hotchkiss, 1998). If all are true, it is unsurprising the utilisation of Sudan health services in the public-sector was low during the 1980s and personnel, especially in peripheral health facilities, idle most of the time. In 1992, Sudan had introduced cost recovery measures as a part of its programme of economic reforms, following a course taken by many developing countries. During the 1990's, Sudan initiated number of initiatives to establish medicine financing mechanisms as part of health reform process and decentralised decision-making at state level. In 1992, the government abolished the constitutional right of free health care. There is interest by the states to introduce a medicine financing mechanism based on the Revolving Drug Fund (RDF) experience of Khartoum State (KS).

Given the fact that less than 50% of the population has regular access to the essential medicines (Quick et al., 1997), and the highest availability of essential medicines at affordable prices in Khartoum state. The government decided to replicate the RDF to other states. Since 2001 the Central Medical Supplies Public Corporation (CMSPC) is involved in the development of the RDF in Seventeen states. The RDF has the highest level of political support as the president of Sudan himself inaugurated it.

7. RECOMMENDATIONS

The public sector is rigid, bureaucratic personnel-management practices, low incentives, poor job satisfaction and unsupportive work environment compared to the private sector. Such situation demoralised pharmacists and encourages them to join the private sector. Many (65%) of surveyed private-sector pharmacists claimed they were public sector pharmacists migrated to the private sector. Although information on migration is sparse, anecdotal evidence persuasively underscores the problem. An internal flow of pharmacists plagues all states, since pharmacists move from poorer states to wealthier ones and from the public sector to the private. Strategies to meet current and future challenges in pharmacy human resources are urgently needed. Approaches that focus on the training of individuals, which do not take into account the job satisfaction (i.e., the nature of the work itself) and pharmacists' mobility, can enjoy only limited success. Increased production alone cannot compensate for weak motivation, high attrition and increasing mobility. To reverse decades of neglect, policy-makers in both (state and federal level) should begin now, first by recognising the problem and secondly by fixing it through the immediate implementation of potentially effective strategies. Although, we do not advocate the creation of new barriers to the movement of pharmacists between private and public-sectors, steps should be taken to redress the unbalanced situation. Ten immediate steps are recommended:

- Large-scale advocacy is needed to achieve heightened political awareness within states and at federal level. One potential outcome of large-scale movement would be the beginnings of introduction of pharmacy care concept, which reshapes the pharmacy services around the patients in hospitals and community pharmacies. This concept will benefit the health care system users and motivate pharmacists to do a good job to their clients and employers. The employers need to foster an organisational culture that recognises and values staff contribution. Central to the delivery of effective recognition are employees' immediate bosses, where a participative and considerate management style is shown as a major predictive factor of retention.

- The Federal Ministry of Health (FMOH) needs to learn from the past experience of Khartoum, Red Sea, Northern, and Algadarif States and current Gezira State then, identify success stories. Pharmacists and their organisations, and Ministries of Health have not remained passive in confronting the crisis in pharmacy workforces. The goodwill and commitment of public-sector pharmacists to provide quality care despite low wages (30% of the average private salary) and medicines supply shortages at times of appalling conditions should not be overlooked.

- *Pharmacist job satisfaction:* Job satisfaction is how people feel about their jobs. Experiencing job dissatisfaction leads to withdrawal cognition and employee turnover. Job dissatisfaction can be caused in many ways, including high centralisation, routinisation, low integration, low communication and policy knowledge. Pharmacy education has a key role to prepare pharmacy student for practice and must anticipate the changing professional role. New strategies need to be developed with the participation of pharmacy professionals associations, unions, universities and ministries of health and higher education representatives to meet both; the short-term and long-term needs of pharmacists as pharmacy care providers. Technology will, no doubt give opportunity to join postgraduate studies (e.g., P.G. diploma or M.Sc. courses) from overseas via e-learning or continuing pharmacy professional development programmes.

- *Salaries and incentives structure*: This includes the process of creating new jobs, addressing low wages, as well as developing incentives structure that supports pharmacists over the course of their working lives. In order to stem the flow of pharmacists to the private-sector and increase their performance, the Ministry of Health needs to pay incentives to its pharmacy staff on a semi-private basis. Introduction of the employment contract and the application of the incentive budget line opposite performance proved to be effective in Khartoum State experience (Mohamed, 2000). The obligations of each part (employer and employee) should be written in non-ambiguous language and transparent reward system should be in place. When transparency of reward system is poor, its credibility will be questioned and pharmacists might not respond to the explicit incentive system at all. IDS, 2000 pointed the lack of training and potential career development is a particularly important contributor to voluntary resignations. Uncompetitive pay is often debated as a reason for employee turnover (IDS, 2000). The perception of receiving a fair salary is a determinant of retention. It seems to be important both at the recruitment stage and subsequently as a determinant of retention rates is the perception that employees are receiving a fair salary. It is important to note this does not necessarily equate to a large salary, since people often compare themselves with peers in the same occupations or with friends and family rather than with better paid or higher skilled workers. Also, when promises are broken and expectations are perceived (have not been met), employees take actions to withdraw from the organisation, which may include actually quitting jobs.

- *Pharmacy staff motivation:* In addition to financial incentives, Ministry of Health should continue to invest in improving the working conditions to ensure the suitable qualified and skilled pharmacists are retained for longer periods. Recruitment of qualified pharmacists (which may include looking outside the public services). A clear definition of job assignments (staff at hospitals` level enter into written contracts to perform according to the MOH guidelines) and regular supervision will assist MOH to achieve a good staff performance. The MOH should provide transport to pharmacists (senior and specialised pharmacists could be offered private vehicles) from their residence to the place of work to increase their motivation. Company-paid private medical insurance, and a company car for senior staff, child day care facilities, pension and retirement plans are the most desired and lead to employee retention.

- *Redistribution of Pharmacy workforces:* To address the problems of pharmacy profession in Sudan, an increase in access to essential medicines is insufficient. Far more important is the need to strengthen the pharmacy workforce in localities, states and federal health institutions to address the challenges and to use the resources and interventions for provision of effective pharmaceutical services.

- *Small staff and efficient teamwork:* The pharmacy workforces are divided into two levels (1) Department of pharmacies at Ministry of Health, and (2) Hospitals. The Department of pharmacy at state level should consist of 6 pharmacists at maximum and 25 at federal department of pharmacy including drug analysis laboratory. The hospitals` department of pharmacies classified as follows:

(i) Group A includes big hospitals (e.g., Khartoum and Omdurman hospitals). The numbers of pharmacists in Group A hospitals are 15 pharmacists in addition to pharmacy assistants and other supportive staff to cover all shifts. One manager, 3 pharmacist work in Drug Information Centre, three for internal hospital pharmacy, two in outpatient pharmacy, three in people pharmacy and one in clinical pharmacist;

(ii) Group B includes medium hospitals and capital cities hospitals (e.g., Ibrahim Malik, and Medani Hospitals). The Hospital Pharmacy Department (HPD) this group managed by 4 to 6 pharmacists;

(iii) Group C includes small and rural hospitals. Two pharmacists could run the HPD in these hospitals. Paying attention to create more flexible and efficient system for the PHRs management in the government institutions might help improve the condition of shortages of pharmacists in the public sector. The advantages of small staff can be easily managed, trained and financed, and teamwork could be developed. This also improves the performance and productivity of the public sector pharmacists thereby reduces the number of the PHRs needed to provide satisfactory pharmaceutical services in the public-sector institutions. The best indicators of staff retention are the fostering of friendships at work, and managers in health cares should take time to get knowing people and foster opportunities for friendship and socialising.

- National leadership at the highest level is essential and will only come to heighten the awareness of the fundamental importance of pharmacists in health care in general and in the pharmaceutical care in particular, and the development of new methods and strategies.

- Continuing pharmacy professional development: The most important element of National Drug Policy (NDP) and 25 years pharmacy strategy has yet to be tackled. The MOH should fully recognise its 25 years pharmacy strategy goals could be achieved through people's (especially pharmacists) expertise. Appropriate training and development is the key to reach those goals and make strategy visions become reality. A wide variety of external (e.g., distance or e-learning in the developed world) and internal training and development programmes for pharmacists should be introduced. A pharmacist's career or pathway should be developed. A policy for active selection of training fields should be formulated according to the priorities of health care needs. The career development relies on individual training and development to enable employees to move into more challenging roles and can provide enhanced rewards for those who are promoted.
- Pharmacy staff discipline and accountability system: Disciplinary procedures, which provide a range of possible responses (from warnings through dismissal, depending on the severity and frequency of the offence should be clearly stated in the new work contracts). Pharmacy managers and team leaders in different settings (administration or care providing, at both state and federal levels) should be trained to invoke disciplinary procedures and to bring criminal charges when necessary.

CONCLUSION

Improving effectiveness of the public pharmacy is by switching resources towards areas of need, reducing inequalities and promoting better health. Unless there are clear incentives for pharmacists, they can move away from public-sector.

Findings innovative approaches to stop brain drain of the pharmacists from the public sector and to increase their productivity and performance might be more appropriate strategies to solve the problem in Sudan. These strategies comprise, for instance, monitory incentives, continuing professional development, working condition and job satisfaction of civil service PHRs.

The study may help the Ministry of Health to better look at the real issues of the PHRs in the public-sector and formulate more relevant and useful policies and plans to retain qualified and skilled pharmacists in the public-sector on a solid evidence base. Monitoring and evaluation of information provided to the MOH.

The data must be accurate and up to date. The study revealed low salaries, job dissatisfaction in relation to the pharmacy practice and bureaucracy, working conditions, lack of recognition for contribution at work, and lack of professional development training programmes are the main factors influencing the brain drain of the PHRs.

These factors affect the PHRs immigration and retention concurrently rather than in insolation.

Given the time constraints required to get the new contracting arrangements in place, there is a risk that good practice developments in options for change for change field sites may not be used effectively (continue to evaluate and disseminate the lessons that emerge from these sites).

ETHICAL CLEARANCE AND DATA PROTECTION CONSENT

Before starting the data collection, ethical clearance is to be obtained from the Federal Ministry of Health Research Ethics Committee. The researchers have not time, so no ethical clearance was obtained. The respondents were informed all the data will be used for the academic research purposes only, and the data processing would not be used to support decision-making and would not cause substantial damage or distress.

Research Limitations

In the short survey pharmacists working with public and private-sectors in Sudan regarding the immigration of pharmacists from public to private-sector was meant to explore factors that discourage or shorten the pharmacists stay in public-sector. The design of the research itself may be considered inadequate with regard to size and selection process. However, the researchers believe it provides enough information about why pharmacists leave the public-sector. Furthermore research should be carried out to understand the scope, magnitude directions of the migratory flows, within and outside the country, as well as the characteristics and skills of the emigrated pharmacists.

This will be achieved through a health system consisting of three levels (state, provincial and localities), including the referral system, secondary and tertiary levels.

Pharmacy management should be coordinated and integrated with other various aspects of health. The following are recommended:

- Community must be the focus of benefits accruing from restructures, legislature to protect community interest on the basis of equity and distribution, handover the assets to the community should be examined; and communities shall encourage the transfer the management of health schemes to a professional entity.
- The private sector should be used to mobilise, and strengthen the technical and financial resources, from within and without the country to implement the services, with particular emphasis on utilisation of local resources.
- The government should provide the necessary financial resources to guide the process of community management of pharmacy supplies. The government to divert from provision of services and be a facilitator through setting up standards, specifications and rules to help harmonise the private sector and establish a legal independent body by an act of parliament to monitor and control the providers. Governments to assist the poor communities who cannot afford service cost, and alleviate social-economic negative aspects of privatisation.
- The sector actors should create awareness to the community of the roles of the private sector and government in the provision of health and pharmacy services.
- Support agencies assist with the financial and technical support, the training facilities, coordination, development and dissemination of health projects, and then evaluation of projects.

APPENDIX 1. FACTS ABOUT SUDAN

Full country name	Republic of the Sudan
Total area	One million square miles (2.5 x 10^6 square kilometres). Land 2.376 x 10^6 square kilometres
Population	34.3 x 10^6 inhabitants (UN, 2004)
Capital city	Khartoum (population 5 million)
Language	Arabic (official), English, Nubian, Ta Bedawie, diverse dialects of Nilotic, Nilo – Hamitic, Sudanic languages
Religions	Sunni Muslim 70% (in north), indigenous beliefs 25%, Christian 5% (mostly in south and Khartoum)
GDP per head	US $ 460 (World Bank, 2003)
Annual growth	4% (1997 est.)
Inflation	23% (1998 est.)
Monetary unit	1 Dinar = 10 Sudanese pounds (1 US $ = 250 Dinar)
Ethnic groups	Black 52%, Arab 39%, Beja 6%, Foreigners 2%, others 1%
Life expectancy	54 years (men), 57 years (women) (UN)
Main exports	Oil, cotton, sesame, livestock and hides, gum Arabic
Agricultures	Agriculture is the backbone of economic and social development. 62% of the populations are employed in agriculture. Agriculture contributes 33% of the gross national products (GNP), and 95% of all earnings.
Animal wealthy	35 x 10^6 head of cattle. 35 x 10^6 head of sheep. 35 x 10^6 head of goats. 3 x 10^6 head of camels. 0.6 x 10^6 head of horses and donkeys. Fish wealth 0.2 x 10^6 tonnes of food annually. Wildlife, birds and reptiles.
Population access to safe water (%)	73 % (UNICEF, 1999)
Population access to adequate sanitation (%)	51% (UNICEF, 1999)
Population access to health services (%)	51% (UNICEF, 1999)
Under five mortality rate	115 (per 1000 live births) (UNICEF, 1999)
Environment	Inadequate supplies of potable water, wildlife populations threatened by excessive hunting, soil erosion, and desertification.
International agreements	Party to: Biodiversity, climate change, desertification, endangered species, law of the sea, nuclear test ban, and ozone layer protection.

APPENDIX 2. SUMMARY OF INHERITED PROBLEMS FOR HEALTH SERVICES IN SUDAN

Health services	Personnel
<ul><li>Absence of referral systems</li><li>Lack of means of patients transport and ambulances</li><li>Lack of work standards</li><li>Service is not based on the concept of client satisfaction</li><li>Weak infrastructure and distribution</li><li>Lack of clear vision, mission and plans</li><li>Many health facilities are not constructed according to the recommended standards for its location, buildings, etc.</li><li>Low quality of tertiary services leading to patients seeking treatment abroad</li></ul>	<ul><li>Imbalance in training of different health care especially technical and nursing</li><li>Shortage in certain specialisations such as surgery, pathology, general practioners and family physician</li><li>High attrition rate</li><li>Lack of continuing education programmes</li><li>Poor distribution of health manpower</li><li>The standard of auxiliary workers does not meet the required level</li><li>Low personnel morale, satisfaction, ownership feelings, motivation, respect to work values and attitude towards patients and colleagues</li><li>Poor culture of evidence based practice</li><li>Absence of clear guidelines for medical practice and service protocols</li></ul>

The overall goal of the central medical supplies (CMS) ownership privatisation is to improve access to essential medicines and other medical supplies in order to improve health status of the inhabitants particularly in far states (e.g., Western and Southern States).

Establishment of alternative ownership for the CMS can be achieved by selling the majority of shares to the private sector. This will achieve the following objectives:

- High access to essential medicines of good quality and affordable prices to the states' population and governments.
- Efficiency and effectiveness in drug distribution system to avoid the serious pitfalls and incidences that reported during the last ten years in the CMS.
- Equity by reaching all remote areas currently deprived from the formal drug distribution channels.
- Improvement of the quality and quantity of delivery of medicines to the public health facilities.

The above objectives are expected to:

- Increase geographical and economic access to essential medicines in all states (i.e., in both rural and urban areas) to reach at least 80% of the population (currently less than 50% of population have access to essential medicines).

- The tax collection from the new business becomes more efficient and will increase after privatisation. The tax revenues could be used to finance other health-care activities.
- If the government reserves some shares (not more than 50%) in the new business, then its shares' profit could be used to finance free medicines project in hospitals outpatients' clinic, and other exempted medicines, e.g., renal dialysis and haemophilic patients' treatment.

APPENDIX 3. PUBLIC SECTOR PHARMACISTS BRAIN DRAIN QUESTIONNAIRE

Public Sector Pharmacists' Questionnaire

Date ....
Department..............................
Serial No..................................

Please mark the best answer with an X
1. Are you: 1. Male ☐ 2. Female☐

2. When did you graduate?

During or before 1965		1
During 1966 - 1970		2
During 1971 - 1975		3
During 1976 - 1980		4
During 1981 - 1985		5
During 1986 - 1990		6
During 1991 - 1995		7
During 1996 - 2000		8
After 2000		9

3. Country of graduation
....

4. Did you experience any private job at some time in the past before joining the public sector?

1. Yes ☐ (Go to question 5) 2. No ☐ (Go to question 6)

5. Why did you decide to leave the private-sector?

Please rate each of the following reasons BY CIRCLING ONE NUMBER ON EACH LINE

1= most important, 5 = least important					
Job dissatisfaction	1	2	3	4	5
Feeling of working for specific person	1	2	3	4	5
Low salaries and incentives	1	2	3	4	5
Lack of ownership	1	2	3	4	5
Others (Please specify)..					

6. Why did you choose the public sector?

Please rate each of the following reasons BY CIRCLING ONE NUMBER ON EACH LINE.

1= most important, 5 = least important					
Job satisfaction	1	2	3	4	5
Feeling of doing a public job	1	2	3	4	5
Salaries are better than the private	1	2	3	4	5
Feeling of ownership	1	2	3	4	5
Locally short and long training courses	1	2	3	4	5
Short and long training abroad	1	2	3	4	5
Others (Please specify)					

7. What are the reasons that encourage you to work with public sector?

Please rate each of the following reasons BY CIRCLING ONE NUMBER ON EACH LINE.

1= most important, 5 = least important					
Job satisfaction	1	2	3	4	5
Feeling of doing a public job	1	2	3	4	5
Salaries are better than the private	1	2	3	4	5
Feeling of ownership	1	2	3	4	5
Pensions and other benefits	1	2	3	4	5
Short and long training abroad	1	2	3	4	5
Others (Please specify)					

8. Do you have intention to leave the public-sector at some time in the future?

1. Yes ☐ (Go to questions 9, 10, 11) 2. No ☐ (Go to question 11)

9. Why are you intending to leave the public-sector?

Please rate each of the following reasons BY CIRCLING ONE NUMBER ON EACH LINE.

1= most important, 5 = least important					
Job satisfaction in the private sector	1	2	3	4	5
Private salaries are better than the public	1	2	3	4	5
Private sector offers me vehicle	1	2	3	4	5
The private gives me full treatment when feeling ill	1	2	3	4	5
Others (Please specify)					

10. What are the reasons that discourage you to continue with public sector?

Please rate each of the following reasons BY CIRCLING ONE NUMBER ON EACH LINE.

1= most important, 5 = least important					
Lack of recognition of what I have done	1	2	3	4	5
Monetary issues	1	2	3	4	5
Dim vision	1	2	3	4	5
Sense of instability	1	2	3	4	5
Lack of job satisfaction	1	2	3	4	5
Those who work and those who do not are equal	1	2	3	4	5
Policy-makers do not care about pharmacy	1	2	3	4	5
Political issues	1	2	3	4	5
Others (Please specify)					

11. If you have any other comments concerning the retention of public sector pharmacy human resources. Please do not hesitate to report them.
....
.... We
would like to thank you very much for your participation in our research. If you do not mind, we might need your telephone number to contact you for further clarification.

Telephone Number:..

APPENDIX 4. PUBLIC SECTOR PHARMACISTS BRAIN DRAIN QUESTIONNAIRE

Private Sector Pharmacists' Questionnaire

Date ....
Serial No....................................

Please mark the best answer with an X
1. Are you: 1. Male ☐ 2. Female☐
2. When did you graduate?

During or before 1965		1
During 1966 - 1970		2
During 1971 - 1975		3
During 1976 - 1980		4
During 1981 - 1985		5
During 1986 - 1990		6
During 1991 - 1995		7
During 1996 - 2000		8
After 2000		9

3. Country of graduation...

4. What is your current employer within the private-sector?

Medical representative in a drug company	1
Drug information pharmacist in a drug company	2
Community pharmacy Pharmacists	3
Non Governmental Organisation	4
Others (please specify)	
....	
....	

5. What is your approximate monthly salary *(IN SUDANESE POUNDS)?*

Less than 500,000		1
500,000 - 999,999		2
1,000,000 - 1,499,999		3
1,500,000 - 1,999,999		4
2,000,000 - 2,499,999		5
2,500,000 - 2,999,999		6
3,000,000 or more		7

6. Did you work with public-sector at some time in the past before joining the private sector?

1. Yes ☐ (Go to question 7) 2. No ☐ (Go to question 8)

7. Why did you decide to leave the public-sector?

Please rate each of the following reasons BY CIRCLING ONE NUMBER ON EACH LINE.

1= most important, 5 = least important					
Lack of recognition of what I had done	1	2	3	4	5
Low salaries and incentives	1	2	3	4	5
Dim vision	1	2	3	4	5
Feeling of instability	1	2	3	4	5
Lack of job satisfaction	1	2	3	4	5
Those who work harder and those who do not are equal	1	2	3	4	5
Policy-makers do not care about pharmacy	1	2	3	4	5
Political issues	1	2	3	4	5
Others (Please specify)					
....					
....					

8. Why did you choose the private-sector?

Please rate each of the following reasons BY CIRCLING ONE NUMBER ON EACH LINE

1= most important, 5 = least important					
Job satisfaction in the private sector	1	2	3	4	5
Private salaries are better than the public	1	2	3	4	5
Private sector offers me vehicle	1	2	3	4	5
The private gives me full treatment when feeling ill	1	2	3	4	5
Others (Please specify)					
....					
....					

9. Do you have any intention to leave the private-sector at some time in the future?

1. Yes ☐ (Go to questions 10) 2. No ☐ (Go to question 11)

10. What are the reasons that encourage you to join the public sector?

Please rate each of the following reasons BY CIRCLING ONE NUMBER ON EACH LINE.

1= most important, 5 = least important					
Job satisfaction in the public sector	1	2	3	4	5
No feeling of working for specific person	1	2	3	4	5
Better salaries	1	2	3	4	5
Feeling of ownership	1	2	3	4	5
Overseas training	1	2	3	4	5
Local training	1	2	3	4	5
Others (Please specify)					
....					
....					

11. What are the most important reasons that discourage you from joining the public sector at some time in the future?

Please rate each of the following reasons BY CIRCLING ONE NUMBER ON EACH LINE.

1= most important, 5 = least important					
Monetary issues	1	2	3	4	5
Dim vision	1	2	3	4	5
Sense of instability	1	2	3	4	5
Lack of job satisfaction	1	2	3	4	5
Political issues	1	2	3	4	5
Others (Please specify)					
....…...					
....…...					

12. If you can move to the public-sector, which of the following areas you are interested in?

Please rate each of the following reasons BY CIRCLING ONE NUMBER ON EACH LINE.

1= most important, 5 = least important					
Hospitals	1	2	3	4	5
Inspection department	1	2	3	4	5
Drug supply department	1	2	3	4	5
Drug information centre	1	2	3	4	5
Others (Please specify)					
....					
....					

13. If you have any other comments concerning the retention of public-sector pharmacy human resources, please do not hesitate to report them.

....

....

....

....

We would like to thank you very much for your participation in our research. If you do not mind, we might need your telephone number to contact you for further clarification.

Telephone Number:..

ACKNOWLEDGMENTS

The author wishes to thank Gamal Ali for their valuable comments and help in setting the research instruments. Dr. Abdelgadir Head of Khartoum State Ministry of Health, Research Department and his team who assisted with data collection are gratefully acknowledged.

REFERENCES

Boynton, P.M., Wood, G.W., and Greenhalgh, T. 2004. Hands-on guide to questionnaire research: reaching beyond the white middle classes. *British Medical Journal;* 328: 1433-1436.

Canadian Pharmacists Association (CPA). 2001. A situation analysis of human Resources issues in the pharmacy profession in Canada. Detailed report prepared by Peartree Solutions Inc., for human resources development in Canada.

Eichler, R., Auxila, P., and Pollock, J. 2001. Output based health care: paying for performance in Haiti. Public Policy for the Private Sector. Note No. 236. *World Bank.* Washington DC: USA.

Gould, D. 2004. Training needs analysis: an evaluation framework. *Nursing Standards* 18: 33-36.

Government of Sudan (GOS). 2002. National 25 years Strategic Plan (2002-2027). Khartoum: Sudan.

Hongoro, C., and McPake, B. 2004. How to bridge the gab in human resources for health. *The Lancet* 364:1451-1459.

Hotchkiss, D.R. 1998. The trade-off between price and quality of services in the Philippines. *Soc. Sci. Med.* 46(2): 227-242.

Hughes, A. 2004. Devising training needs analysis toolkit. *Hospital Pharmacist* 11:385-88.

IDS. 2000. Personnel policy and practice: improving staff retention. Income Data Services Ltd. London: UK.

Lerberghe, WV, Ferrinho, P, Omar, MC, Blaise, P, and Bugalho, AM. 2002. When staff is underpaid: dealing with the individual coping strategies of health personnel. *Bulletin of the World Health Organization* 80(7): 581-584.

Matowe, L., Duwiejua, M., and Norris, P. 2004. Is there a solution to the pharmacists' brain drain from poor to rich countries? *The Pharmaceutical Journal* 272:98-99.

Ministry of Health (MOH). 1997. *National Drug policy*. Khartoum: Sudan.

Ministry of Health (MOH). 2003. 25 years Pharmacy Strategy (2002-2027). Khartoum: Sudan. Unpublished Report.

Ministry of Health (MOH). 2004. *FGDP annual report*. Ministry of Health. Sudan.

Mohamed, G.K. 2000. Management of revolving drug fund: experience of Khartoum state-Sudan. M.Sc. Thesis Dissertation. The School of Pharmacy, University of Bradford: UK.

Narasimhan, V, Brown, H, Pablos-Mendez, A, Adams, O, Dussault, G, Elzinga G. 2004. Responding to the global human resources crisis. *The Lancet* 363:1469-1472.

Omer, A.M. 1994. Socio-Cultural aspects of water supply and sanitation in Sudan. *NETWAS*, Vol.2, No.4, Nairobi: Kenya.

Quick JD, Schulze, JD, Ashiru, DA, Khela, MK, Evans, DF, Patel, R and Parsons, GE. 1997. Managing drug supply: the selection, procurement, distribution and use of pharmaceuticals (2nd edition). West Hardford, CT, Kumarian Press.

Shepherd, J. 1995. Findings of training need analysis for qualified nurse practioners. *Journal of Advancing Nursing* 22:66-71.

Show, P.R., and Griffin, C.C. 1995. Financing health care in Sub-Saharan Africa through user fees and insurance. The World Bank. Washington D.C: USA.

The Act 2001. 2001. Pharmacy poisons and medical devices act. Ministry of Health (MOH). Khartoum: Sudan.

Van Damme W., Meessen B., and Von Schreeb J. 2001. Sotnikum New Deal. The first year better income for health staff: better service to the population. Cambodia: Medicines Sans Frontiers; Antwerp: Institute of Tropical Medicine; Phnom Penh: National Institute of Public Health; Brussels: AEDES; Cambodia: *UNICEF*. http://www.msf.be/fr/pdf/cambodia.pdf.

Velasquez G, Vergara, C, Mardones, C, Reyes, J, Tapia, RA, Quina, F, Atala, E. 1998. Health reform and drug financing. *Health Economic and Drugs*. DAP series, No. 6. WHO/DAP/98.3.

WHO. 2004. The World Medicines Situation. WHO/EDM/PAR/2004.5. World Health Organisation (WHO): Geneva: Switzerland.

Chapter 7

SUSTAINABLE ENERGY DEVELOPMENT AND ENVIRONMENT

ABSTRACT

People are relying upon oil for primary energy and this for a few more decades. Other conventional sources may be more enduring, but are not without serious disadvantages. The renewable energy resources are particularly suited for the provision of rural power supplies and a major advantage is that equipment such as flat plate solar driers, wind machines, etc., can be constructed using local resources and without the advantage results from the feasibility of local maintenance and the general encouragement such local manufacture gives to the buildup of small-scale rural based industry. This chapter comprises a comprehensive review of energy sources, the environment and sustainable development. It includes the renewable energy technologies, energy efficiency systems, energy conservation scenarios, energy savings in greenhouses environment and other mitigation measures necessary to reduce climate change. This study gives some examples of small-scale energy converters, nevertheless it should be noted that small conventional, i.e., engines are currently the major source of power in rural areas and will continue to be so for a long time to come. There is a need for some further development to suit local conditions, to minimise spares holdings, to maximise interchangeability both of engine parts and of the engine application. Emphasis should be placed on full local manufacture. It is concluded that renewable environmentally friendly energy must be encouraged, promoted, implemented and demonstrated by full-scale plant (device) especially for use in remote rural areas.

Keywords: renewable energy technologies, energy efficiency, sustainable development, emissions, environment

NOMENCLATURE

a	annum
ha	hectares
l	litre
HFU	heat flor unit
MSW	municipal sewage waste

1. INTRODUCTION

Power from natural resources has always had great appeal. Coal is plentiful, though there is concern about despoliation in winning it and pollution in burning it. Nuclear power has been developed with remarkable timeliness, but is not universally welcomed, construction of the plant is energy-intensive and there is concern about the disposal of its long-lived active wastes. Barrels of oil, lumps of coal, even uranium come from nature but the possibilities of almost limitless power from the atmosphere and the oceans seem to have special attraction. The wind machine provided an early way of developing motive power. The massive increases in fuel prices over the last years have however, made any scheme not requiring fuel appear to be more attractive and to be worth reinvestigation. In considering the atmosphere and the oceans as energy sources the four main contenders are wind power, wave power, tidal and power from ocean thermal gradients. The sources to alleviate the energy situation in the world are sufficient to supply all foreseeable needs. Conservation of energy and rationing in some form will however have to be practised by most countries, to reduce oil imports and redress balance of payments positions. Meanwhile development and application of nuclear power and some of the traditional solar, wind and water energy alternatives must be set in hand to supplement what remains of the fossil fuels.

The encouragement of greater energy use is an essential component of development. In the short-term, it requires mechanisms to enable the rapid increase in energy/capita, while in the long-term it may require the use of energy efficiency without environmental and safety concerns. Such programmes should as far as possible be based on renewable energy resources.

Large-scale, conventional, power plant such as hydropower has an important part to play in development although it does not provide a complete solution. There is however an important complementary role for the greater use of small-scale, rural based, and power plants. Such plants can be employed to assist development since they can be made locally. Renewable resources are particularly suitable for providing the energy for such equipment and its use is also compatible with the long-term aims.

In compiling energy consumption data one can categorise usage according to a number of different schemes:

- Traditional sector- industrial, transportation, etc.
- End-use- space heating, process steam, etc.
- Final demand- total energy consumption related to automobiles, to food, etc.
- Energy source- oil, coal, etc.
- Energy form at point of use- electric drive, low temperature heat, etc.

2. RENEWABLE ENERGY POTENTIAL

The increased availability of reliable and efficient energy services stimulates new development alternatives (Omer, 2009a). This communication discusses the potential for such integrated systems in the stationary and portable power market in response to the critical need for a cleaner energy technology. Anticipated patterns of future energy use and consequent environmental impacts (acid precipitation, ozone depletion and the greenhouse effect or global warming) are comprehensively discussed in this approach. Throughout the theme several issues relating to renewable energies, environment and sustainable development are examined from both current and future perspectives. It is concluded that renewable environmentally friendly energy must be encouraged, promoted, implemented and demonstrated by full-scale plant (device) especially for use in remote rural areas. Globally, buildings are responsible for approximately 40% of the total world annual energy consumption. Most of this energy is for the provision of lighting, heating, cooling, and air conditioning. Increasing awareness of the environmental impact of CO_2, NO_x and CFCs emissions triggered a renewed interest in environmentally friendly cooling, and heating technologies. Under the 1997 Montreal Protocol, governments agreed to phase out chemicals used as refrigerants that have the potential to destroy stratospheric ozone. It was therefore considered desirable to reduce energy consumption and decrease the rate of depletion of world energy reserves and pollution of the environment. One way of reducing building energy consumption is to design buildings, which are more economical in their use of energy for heating, lighting, cooling, ventilation and hot water supply. Passive measures, particularly natural or hybrid ventilation rather than air-conditioning, can dramatically reduce primary energy consumption. However, exploitation of renewable energy in buildings and agricultural greenhouses can, also, significantly contribute towards reducing dependency on fossil fuels. Therefore, promoting innovative renewable applications and reinforcing the renewable energy technologies market will contribute to preservation of the ecosystem by reducing emissions at local and global levels. This will also contribute to the amelioration of environmental conditions by replacing conventional fuels with renewable energies that produce no air pollution or greenhouse gases.

There is strong scientific evidence that the average temperature of the earth's surface is rising. This is a result of the increased concentration of carbon dioxide and other GHGs in the atmosphere as released by burning fossil fuels. This global warming will eventually lead to substantial changes in the world's climate, which will, in turn, have a major impact on human life and the built environment. Therefore, effort has to be made to reduce fossil energy use and to promote green energies, particularly in the building sector. Energy use reductions can be achieved by minimising the energy demand, by rational energy use, by recovering heat and the use of more green energies. This study was a step towards achieving that goal. The adoption of green or sustainable approaches to the way in which society is run is seen as an important strategy in finding a solution to the energy problem. The key factors to reducing and controlling CO_2, which is the major contributor to global warming, are the use of alternative approaches to energy generation and the exploration of how these alternatives are used today and may be used in the future as green energy sources (Omer, 2009b). Even with modest assumptions about the availability of land, comprehensive fuel-wood farming programmes offer significant energy, economic and environmental benefits. These benefits would be dispersed in rural areas where they are greatly needed and can serve as linkages for further rural economic development. The nations as a whole would benefit from savings in foreign exchange, improved energy security, and socio-economic improvements. With a nine-fold increase in forest – plantation cover, a nation's resource base would be greatly improved. The international community would benefit from pollution reduction, climate mitigation, and the increased trading opportunities that arise from new income sources.

Table 1. Sources of energy

Energy source	Energy carrier	Energy end-use
Vegetation	Fuel-wood	Cooking Water heating Building materials Animal fodder preparation
Oil	Kerosene	Lighting Ignition fires
Dry cells	Dry cell batteries	Lighting Small appliances
Muscle power	Animal power	Transport Land preparation for farming Food preparation (threshing)
Muscle power	Human power	Transport Land preparation for farming Food preparation (threshing)

The non-technical issues, which have recently gained attention, include: (1) Environmental and ecological factors, e.g., carbon sequestration, reforestation and revegetation. (2) Renewables as a CO_2 neutral replacement for fossil fuels. (3) Greater recognition of the importance of renewable energy, particularly modern biomass energy carriers, at the policy and planning levels. (4) Greater recognition of the difficulties of gathering good and reliable renewable energy data, and efforts to improve it. (5) Studies on the detrimental health efforts of biomass energy particularly from traditional energy users. The renewable energy resources are particularly suited for the provision of rural power supplies and a major advantage is that equipment such as flat plate solar driers, wind machines, etc., can be constructed using local resources and without the advantage results from the feasibility of local maintenance and the general encouragement such local manufacture gives to the buildup of small-scale rural based industry. This study gives some examples of small-scale energy converters, nevertheless it should be noted that small conventional, i.e., engines are currently the major source of power in rural areas and will continue to be so for a long time to come. There is a need for some further development to suit local conditions, to minimise spares holdings, to maximise interchangeability both of engine parts and of the engine application. Emphasis should be placed on full local manufacture.

Table 2. Renewable applications

Systems	Applications
Water supply	Rain collection, purification, storage and recycling
Wastes disposal	Anaerobic digestion (CH_4)
Cooking	Methane
Food	Cultivate the 1 hectare plot and greenhouse for four people
Electrical demands	Wind generator
Space heating	Solar collectors
Water heating	Solar collectors and excess wind energy
Control system	Ultimately hardware
Building fabric	Integration of subsystems to cut costs

The renewable energy resources are particularly suited for the provision of rural power supplies and a major advantage is that equipment such as flat plate solar driers, wind machines, etc., can be constructed using local resources and without the high capital cost of more conventional equipment. Further advantage results from the feasibility of local maintenance and the general encouragement such local manufacture gives to the build up of small-scale rural based industry. Table 1 lists the energy sources available.

Currently the 'non-commercial' fuels wood, crop residues and animal dung are used in large amounts in the rural areas of developing countries, principally for heating and cooking; the method of use is highly inefficient. Table 2 presented some renewable applications.

Table 3 lists the most important of energy needs.

Considerations when selecting power plant include the following:

- Power level- whether continuous or discontinuous.
- Cost- initial cost, total running cost including fuel, maintenance and capital amortised over life.
- Complexity of operation.
- Maintenance and availability of spares.
- Life.
- Suitability for local manufacture.

Table 3. Energy needs in rural areas

Transport, e.g., small vehicles and boats
Agricultural machinery, e.g., two-wheeled tractors
Crop processing, e.g., milling
Water pumping
Small industries, e.g., workshop equipment
Electricity generation, e.g., hospitals and schools
Domestic, e.g., cooking, heating, and lighting
Water supply, e.g., rain collection, purification, and storage and recycling
Building fabric, e.g., integration of subsystems to cut costs
Wastes disposal, e.g., anaerobic digestion (CH_4)

Table 4 listed methods of energy conversion.

The household wastes, i.e., for family of four persons, could provide 280 kWh/yr of methane, but with the addition of vegetable wastes from 0.2 ha or wastes from 1 ha growing a complete diet, about 1500 kWh/yr may be obtained by anaerobic digestion (Omer, 2009c). The sludge from the digester may be returned to the land. In hotter climates, this could be used to set up a more productive cycle (Figure 1).

There is a need for greater attention to be devoted to this field in the development of new designs, the dissemination of information and the encouragement of its use. International and government bodies and independent organisations all have a role to play in renewable energy technologies.

Society and industry in Europe and elsewhere are increasingly dependent on the availability of electricity supply and on the efficient operation of electricity systems. In the European Union (EU), the average rate of growth of electricity demand has been about 1.8% per year since 1990 and is projected to be at least 1.5% yearly up to 2030 (Omer, 2009c).

Currently, distribution networks generally differ greatly from transmission networks, mainly in terms of role, structure (radial against meshed) and consequent planning and operation philosophies.

Table 4. Methods of energy conversion

Muscle power	Man, animals
Internal combustion engines	
Reciprocating	Petrol- spark ignition
	Diesel- compression ignition
	Humphrey water piston
Rotating	Gas turbines
Heat engines	
Vapour (Rankine)	
Reciprocating	Steam engine
Rotating	Steam turbine
Gas Stirling (Reciprocating)	Steam engine
Gas Brayton (Rotating)	Steam turbine
Electron gas	Thermionic, thermoelectric
Electromagnetic radiation	Photo devices
Hydraulic engines	Wheels, screws, buckets, turbines
Wind engines (wind machines)	Vertical axis, horizontal axis
Electrical/mechanical	Dynamo/alternator, motor

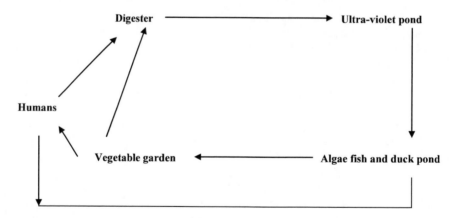

Figure 1. Biomass energy utilisation cycle.

3. ENERGY CONSUMPTION

Over the last decades, natural energy resources such as petroleum and coal have been consumed at high rates. The heavy reliances of the modern economy on these fuels are bound to end, due to their environmental impact, and the fact that conventional sources might eventually run out. The increasing price of oil and instabilities in the oil market led to search for energy substitutes.

In addition to the drain on resources, such an increase in consumption consequences, together with the increased hazards of pollution and the safety problems associated with a large nuclear fission programmes. This is a disturbing prospect. It would be equally unacceptable to suggest that the difference in energy between the developed and developing countries and prudent for the developed countries to move towards a way of life which, whilst maintaining or even increasing quality of life, and reduce significantly the energy consumption per capita. Such savings can be achieved in a number of ways:

- Improved efficiency of energy use, for example better thermal insulation, energy recovery, and total energy.
- Conservation of energy resources by design for long life and recycling rather than the short life throwaway product.
- Systematic replanning of our way of life, for example in the field of transport.

Energy ratio is defined as the ratio of energy content of the food product/ energy input to produce the food.

$$Er = Ec/Ei \qquad\qquad\qquad (1)$$

where Er is the energy ratio, Ec is the energy content of the food product, and Ei is the energy input to produce the food.

A review of the potential range of recyclables is presented in Table 5.

Currently the non-commercial fuelwood, crop residues and animal dung are used in large amounts in the rural areas of developing countries, principally for heating and cooking, the method of use is highly inefficient. As in the developed countries, the fossil fuels are currently of great importance in the developing countries. Geothermal and tidal energy are less important though, of course, will have local significance where conditions are suitable. Nuclear energy sources are included for completeness, but are not likely to make any effective contribution in the rural areas. Economic importance of environmental issue is increasing, and new technologies are expected to reduce pollution derived both from productive processes and products, with costs that are still unknown.

4. BIOGAS PRODUCTION

Biogas is a generic term for gases generated from the decomposition of organic material. As the material breaks down, methane (CH_4) is produced as shown in Figure 2. Sources that generate biogas are numerous and varied. These include landfill sites, wastewater treatment plants and anaerobic digesters (Omer, 2009d). Landfills and wastewater treatment plants emit biogas from decaying waste. To date, the waste industry has focused on controlling these emissions to our environment and in some cases, tapping this potential source of fuel to power gas turbines, thus generating electricity (Omer, 2009d). The primary components of landfill gas are methane (CH_4), carbon dioxide (CO_2), and nitrogen (N_2). The average concentration of methane is ~45%, CO_2 is ~36% and nitrogen is ~18% (Omer, and Yemen, 2001). Other components in the gas are oxygen (O_2), water vapour and trace amounts of a wide range of non-methane organic compounds (NMOCs). Landfill gas-to-cogeneration projects present a win-win-win situation. Emissions of particularly damaging pollutant are avoided, electricity is generated from a free fuel and heat is available for use locally.

Heat tariffs may include a number of components such as: a connection charge, a fixed charge and a variable energy charge. Also, consumers may be incentivised to lower the return temperature. Hence, it is difficult to generalise but the heat practice for any DH company no matter what the ownership structure can be highlighted as follows:

- To develop and maintain a development plan for the connection of new consumers.
- To evaluate the options for least cost production of heat.
- To implement the most competitive solutions by signing agreements with other companies or by implementing own investment projects.
- To monitor all internal costs and with the help of benchmarking, and improve the efficiency of the company.
- To maintain a good relationship with the consumer and deliver heat supply services at a sufficient quality.

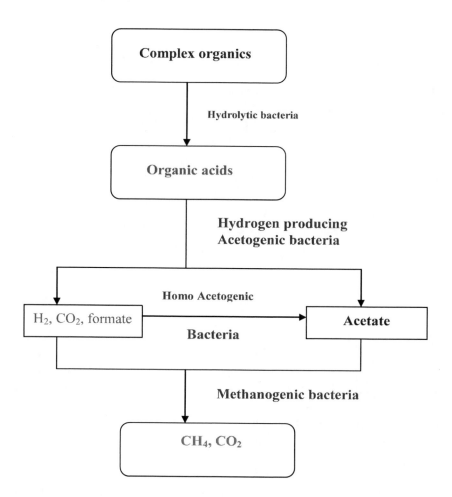

Figure 2. Biogas production process.

**Table 5. Summary of material recycling practices in the construction sector
(Robinson, 2007)**

Construction and demolition material	Recycling technology options	Recycling product
Asphalt	Cold recycling: heat generation; Minnesota process; parallel drum process; elongated drum; microwave asphalt recycling system; finfalt; surface regeneration	Recycling asphalt; asphalt aggregate
Brick	Burn to ash, crush into aggregate	Slime burn ash; filling material; hardcore
Concrete	Crush into aggregate	Recycling aggregate; cement replacement; protection of levee; backfilling; filter
Ferrous metal	Melt; reuse directly	Recycled steel scrap
Glass	Reuse directly; grind to powder; polishing; crush into aggregate; burn to ash	Recycled window unit; glass fibre; filling material; tile; paving block; asphalt; recycled aggregate; cement replacement; manmade soil
Masonry	Crush into aggregate; heat to 900°C to ash	Thermal insulating concrete; traditional clay
Non-ferrous metal	Melt	Recycled metal
Paper and cardboard	Purification	Recycled paper
Plastic	Convert to powder by cryogenic milling; clopping; crush into aggregate; burn to ash	Panel; recycled plastic; plastic lumber; recycled aggregate; landfill drainage; asphalt; manmade soil
Timber	Reuse directly; cut into aggregate; blast furnace deoxidisation; gasification or pyrolysis; chipping; moulding by pressurising timber chip under steam and water	Whole timber; furniture and kitchen utensils; lightweight recycled aggregate; source of energy; chemical production; wood-based panel; plastic lumber; geofibre; insulation board

Installing DH should be pursued to meet the objectives for improving the environment through the improvement of energy efficiency in the heating sector. At the same time DH can serve the consumer with a reasonable quality of heat at the lowest possible cost. The variety of possible solutions combined with the collaboration between individual companies, the district heating association, the suppliers and consultants can, as it has been in Denmark, be the way forward for developing DH in the United Kingdom. Three scales of the CHP which were largely implemented in the following chronological order: (1) Large-scale CHP in cities (>50 MWe), industrial and small-scale CHP. (2) Small (5 kWe – 5 MWe) and medium-scale (5-50 MWe).

5. WAVE POWER CONVERSION DEVICES

The patent literature is full of devices for extracting energy from waves, i.e., floats, ramps, and flaps, covering channels (Swift-Hook, et al., 1975). Small generators driven from air trapped by the rising and falling water in the chamber of a buoy are in use around the world (Swift-Hook, et al., 1975). Wave power is one possibility that has been selected. Figure 3 shows the many other aspects that will need to be covered. A wave power programme would make a significant contribution to energy resources within a relatively short-time and with existing technology.

Wave energy has also been in the news recently. There is about 140 megawatts per mile available round British coasts. It could make a useful contribution people needs in the UK. Although very large amounts of power are available in the waves, it is important to consider how much power can be extracted. A few years ago only a few percent efficiency had been achieved. Recently, however, several devices have been studied which have very high efficiencies. Some form of storage will be essential on a second-to-second and minute-to-minute basis to smooth the fluctuations of individual waves and wave's packets but storage from one day to the next will certainly not be economical. This is why provision must be made for adequate standby capacity.

The increased availability of reliable and efficient energy services stimulates new development alternatives. This study discusses the potential for such integrated systems in the stationary and portable power market in response to the critical need for a cleaner energy technology. Anticipated patterns of future energy use and consequent environmental impacts (acid precipitation, ozone depletion and the greenhouse effect or global warming) are comprehensively discussed in this theme. Throughout the theme several issues relating to renewable energies, environment and sustainable development are examined from both current and future perspectives. It is concluded that renewable environmentally friendly energy must be encouraged, promoted, implemented and demonstrated by full-scale plant (device) especially for use in remote rural areas. Globally, buildings are responsible for approximately 40% of the total world annual energy consumption. Most of this energy is for the provision of lighting, heating, cooling, and air conditioning. Increasing awareness of the environmental impact of CO_2, NO_x and CFCs emissions triggered a renewed interest in environmentally friendly cooling, and heating technologies. Under the 1997 Montreal Protocol, governments agreed to phase out chemicals used as refrigerants that have the potential to destroy stratospheric ozone. It was therefore considered desirable to reduce energy consumption and decrease the rate of depletion of world energy reserves and pollution of the environment. One way of reducing building energy consumption is to design buildings, which are more economical in their use of energy for heating, lighting, cooling, ventilation and hot water supply. Passive measures, particularly natural or hybrid ventilation rather than air-conditioning, can dramatically reduce primary energy consumption. However, exploitation of renewable energy in buildings and agricultural greenhouses can, also, significantly contribute towards reducing dependency on fossil fuels. Therefore, promoting innovative renewable applications and reinforcing the renewable energy market will contribute to preservation of the ecosystem by reducing emissions at local and global levels. This will also contribute to the amelioration of environmental conditions by replacing conventional fuels with renewable energies that produce no air pollution or greenhouse gases.

The provision of good indoor environmental quality while achieving energy and cost efficient operation of the heating, ventilating and air-conditioning (HVAC) plants in buildings represents a multi variant problem. The comfort of building occupants is dependent on many environmental parameters including air speed, temperature, relative humidity and quality in addition to lighting and noise. The overall objective is to provide a high level of building performance (BP), which can be defined as indoor environmental quality (IEQ), energy efficiency (EE) and cost efficiency (CE). During the past few decades has found many valuable uses for these complex chemical substances, manufacturing from them plastics, textiles, fertilisers and the various end products of the petrochemical industry. Each decade sees increasing uses for these products.

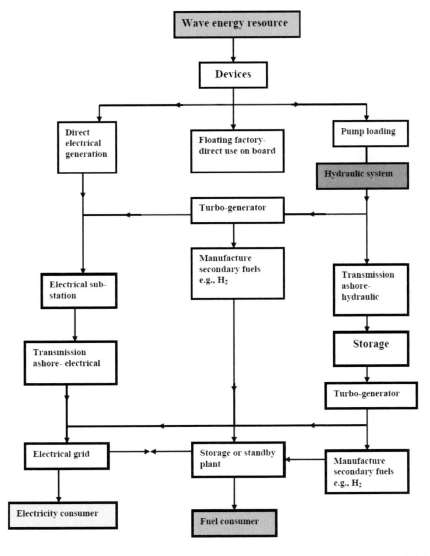

Figure 3. Possible systems for exploiting wave power, each element represents an essential link in the chain from sea waves to consumer.

6. ETHANOL PRODUCTION

Alternative fuels were defined as methanol, ethanol, natural gas, propane, hydrogen, coal-derived liquids, biological material and electricity production (Sims, 2007). The fuel pathways currently under development for alcohol fuels are shown in Figure 4.

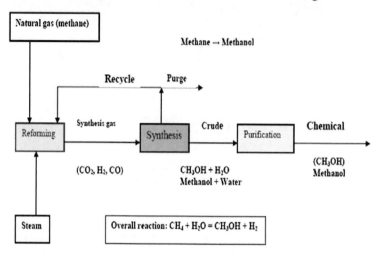

Figure 4. Schematic process flowsheet.

The production of agricultural biomass and its exploitation for energy purposes can contribute to alleviate several problems, such as the dependence on import of energy products, the production of food surpluses, the pollution provoked by the use of fossil fuels, the abandonment of land by farmers and the connected urbanisation. Biomass is not at the moment competitive with mineral oil, but, taking into account also indirect costs and giving a value to the aforementioned advantages, public authorities at national and international level can spur its production and use by incentives of different nature. In order to address the problem of inefficiency, research centres around the world have investigated the viability of converting the resource to a more useful form, namely solid briquettes and fuel gas (Sims, 2007) (Figure 5).

The main advantages are related to energy, agriculture and environment problems, are foreseeable both at regional level and at worldwide level and can be summarised as follows:

- Reduction of dependence on import of energy and related products.
- Reduction of environmental impact of energy production (greenhouse effect, air pollution, and waste degradation).
- Substitution of food crops and reduction of food surpluses and of related economic burdens, and utilisation of marginal lands and of set aside lands.
- Reduction of related socio-economic and environmental problems (soil erosion, urbanisation, landscape deterioration, etc.).
- Development of new know-how and production of technological innovation.

Biomass resources play a significant role in energy supply in all developing countries. Biomass resources should be divided into residues or dedicated resources, the latter including firewood and charcoal can also be produced from forest residues. Ozone (O_3) is a naturally occurring molecule that consists of three oxygen atoms held together by the bonding of the oxygen atoms to each other. The effects of the chlorofluorocarbons (CFCs) molecule can last for over a century.

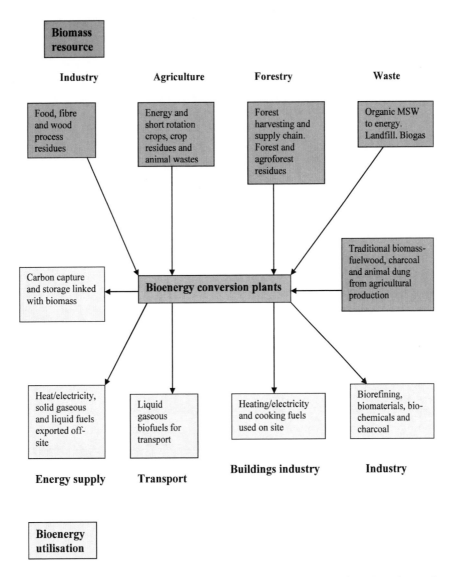

Figure 5. Biomass resources from several sources is converted into a range of products for use by transport, industry and building sectors (Sims, 2007).

This reaction is shown in Figure 6.

It is a common misconception that the reason for recycling old fridge is to recover the liquid from the cooling circuit at the back of the unit. The insulating foams used inside some fridges act as sinks of CFCs- the gases having been used as blowing agents to expand the foam during fridge manufacture. Although the use of ozone depleting chemicals in the foam in fridges has declined in the West, recyclers must consider which strategy to adopt to deal with the disposal problem they still present each year.

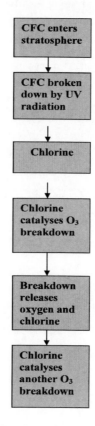

Figure 6. The process of ozone depletion (Trevor, 2007).

It is common practice to dispose of this waste wood in landfill where it slowly degraded and takes up valuable void space. This wood is a good source of energy and is an alternative to energy crops. Agricultural wastes are abundantly available globally and can be converted to energy and useful chemicals by a number of microorganisms. The success of promoting any technology depends on careful planning, management, implementation, training and monitoring. Main features of gasification project are:

- Networking and institutional development/strengthening.
- Promotion and extension.
- Construction of demonstration projects.
- Research and development; and training and monitoring.

7. BIOMASS CHP

Combined heat and power (CHP) installations are quite common in greenhouses, which grow high-energy, input crops (e.g., salad vegetables, pot plants, etc.). Scientific assumptions for a short-term energy strategy suggest that the most economically efficient way to replace the thermal plants is to modernise existing power plants to increase their energy efficiency and to improve their environmental performance. However, utilisation of wind power and the conversion of gas-fired CHP plants to biomass would significantly reduce the dependence on imported fossil fuels. Although a lack of generating capacity is forecasted in the long-term, utilisation of the existing renewable energy potential and the huge possibilities for increasing energy efficiency are sufficient to meet future energy demands in the short-term.

A total shift towards a sustainable energy system is a complex and long process, but is one that can be achieved within a period of about 20 years. Implementation will require initial investment, long-term national strategies and action plans. However, the changes will have a number of benefits including a more stable energy supply than at present, and major improvement in the environmental performance of the energy sector, and certain social benefits. A national vision (Omer, 2009d) used a methodology and calculations based on computer modelling that utilised:

- Data from existing governmental programmes.
- Potential renewable energy sources and energy efficiency improvements.
- Assumptions for future economy growth.
- Information from studies and surveys on the recent situation in the energy sector.

In addition to realising the economic potential identified by the National Energy Savings Programme, a long-term effort leading to a 3% reduction in specific electricity demand per year after 2020 is proposed. This will require further improvements in building codes, and continued information on energy efficiency.

Industry's use of fossil fuels has been blamed for our warming climate. When coal, gas and oil are burnt, they release harmful gases, which trap heat in the atmosphere and cause global warming. However, there has been an ongoing debate on this subject, as scientists have struggled to distinguish between changes, which are human induced, and those, which could be put down to natural climate variability. Industrialised countries have the highest emission levels, and must shoulder the greatest responsibility for global warming.

The environmental Non Governmental Organisations (NGOs) are urging the government to adopt sustainable development of the energy sector by:

- Diversifying of primary energy sources to increase the contribution of renewable and local energy resources in the total energy balance.
- Implementing measures for energy efficiency increase at the demand side and in the energy transformation sector.

The price of natural gas is set by a number of market and regulatory factors that include supply and demand balance and market fundamentals, weather, pipeline availability and deliverability, storage inventory, new supply sources, prices of other energy alternatives and regulatory issues and uncertainty. Classic management approaches to risk are well documented and used in many industries. This includes the following four broad approaches to risk:

- Avoidance includes not performing an activity that could carry risk. Avoidance may seem the answer to all risks, but avoiding risks also means losing out on potential gain.
- Mitigation/reduction involves methods that reduce the severity of potential loss.
- Retention/acceptance involves accepting the loss when it occurs. Risk retention is a viable strategy for small risks. All risks that are not avoided or transferred are retained by default.
- Transfer means causing another party to accept the risk, typically by contract.

Methane is a primary constituent of landfill gas (LFG) and a potent greenhouse gas (GHG) when released into the atmosphere. Globally, landfills are the third largest anthropogenic emission source, accounting for about 13% of methane emissions or over 818 million tones of carbon dioxide equivalent (MMTCO$_2$e) (Brain, and Mark 2007) as shown in Figures 7-9.

8. GEOTHERMAL ENERGY

Geothermal steam has been used in volcanic regions in many countries to generate electricity. The use of geothermal energy involves the extraction of heat from rocks in the outer part of the earth. It is relatively unusual for the rocks to be sufficiently hot at shallow depth for this to be economically attractive. Virtually all the areas of present geothermal interest are concentrated along the margins of the major tectonic plates, which form the surface of the earth. The forced or natural circulation of water through permeable hot rock conventionally extracts heat.

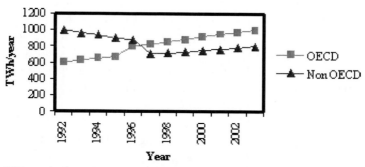

Figure 7. Global CHP trends from 1992-2003 (IEA, 2007).

1 Food, 2 Textile, 3 Pulp & paper, 4 Chemicals, 5 Refining, 6 Minerals, 7 Primary metals, and 8 others

The energy conservation scenarios include rational use of energy policies in all economy sectors and use of combined heat and power systems, which are able to add to energy savings from the autonomous power plants. Electricity from renewable energy sources is by definition the environmental green product. Hence, a renewable energy certificate system is an essential basis for all policy systems, independent of the renewable energy support scheme. It is, therefore, important that all parties involved support the renewable energy certificate system in place.

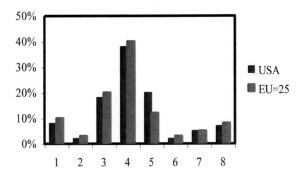

Figure 8. Distribution of industrial CHP capacity in the EU and USA (IEA, 2007).

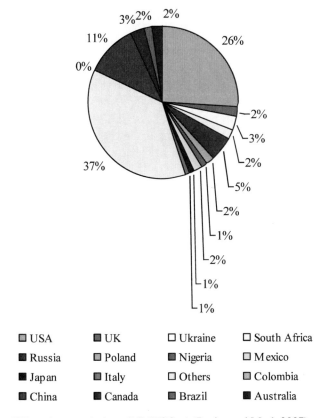

Figure 9. World landfill methane emissions ($MMTCO_2e$) (Brain, and Mark 2007).

There are various practical difficulties and disadvantages associated with the use of geothermal power:

Transmission: geothermal power has to be used where it is found. In Iceland it has proved feasible to pipe hot water 20 km in insulated pipes but much shorter distances are preferred.

Environmental problems: these are somewhat variable and are usually not great. Perhaps the most serious is the disposal of warm high salinity water where it cannot be reinjected or purified. Dry steam plants tend to be very noisy and there is releases of small amounts of methane, hydrogen, nitrogen, amonia and hydrogen sulphide and of these the latter presents the main problem.

The geothermal fluid is often highly chemically corrosive or physically abrassive as the result of the entrained solid matter it carries. This may entail special plant design problems and unusually short operational lives for both the holes and the installations they serve.

Because the useful rate of heat extraction from a geothermal field is in nearly all cases much higher than the rate of conduction into the field from the underlying rocks, the mean temperatures of the field is likely to fall during exploitation. In some low rainfall areas there may also be a problem of fluid depletion. Ideally, as much as possible of the geothermal fluid should be reinjected into the field. However, this may involve the heavy capital costs of large condensation installations. Occasionally, the salinity of the fluid available for reinjection may be so high (as a result of concentration by boiling) that is unsuitable for reinjection into ground. Ocasionally, the impurities can be precipitated and used but this has not generally proved commercially attractive.

World capacity of geothermal energy is growing at a rate of 2.5% per year from a 2005 level of 28.3 GW (Rawlings, 1999). The GSHPs account for approximately 54% of this capacity almost all of it in the North America and Europe (Rawlings, 1999). The involvement of the UK is minimal with less than 0.04% of world capacity and yet is committed to substantial reduction in carbon emission beyond the 12.5% Kyoto obligation to be achieved by 2012. The GSHPs offer a significant potential for carbon reduction and it is therefore expected that the market for these systems will rise sharply in the UK in the immediate years ahead given to low capacity base at present.

There are numerous ways of harnessing low-grade heat from the ground for use as a heat pump source or air conditioning sink. For small applications (residences and small commercial buildings) horizontal ground loop heat exchangers buried typically at between 1 m and 1.8 m below the surface can be used provided that a significant availability of land surrounding the building can be exploited which tends to limit these applications to rural settings.

Heat generation within the earth is approximately 2700 GW, roughly an order of magnitude greater than the energy associated with the tides but about four orders less than that received by the earth from the sun (Oxburgh, 1975).

Temperature distributions within the earth depend on:

- The abundance and distribution of heat producing elements within the earth.
- The mean surface temperature (which is controlled by the ocean/atmosphere system).
- The thermal properties of the earth's interior and their lateral and radial variation.
- Any movements of fluid or solid rock materials occurring at rates of more than a few millimetres per year.

Of these four factors the first two are of less importance from the point of view of geothermal energy. Mean surface temperatures range between $0\text{-}30^\circ\text{C}$ and this variation has a small effect on the useable enthalpy of any flows of hot water. Although radiogenic heat production in rocks may vary by three orders of magnitude, there is much less variation from place to place in the integrated heat production with depth. The latter factors, however, are of great importance and show a wide range of variation. Their importance is clear from the relationship:

$$\beta = q/k \tag{1}$$

where:

β is the thermal gradient for a steady state ($^{\circ}$C/km), q is the heat flux (10^{-6} cal cm^{-2} sec^{-1}) and k is the thermal conductivity (cal cm^{-1} sec^{-1} $^{\circ}$C^{-1}).

The first requirement of any potential geothermal source region is that β being large, i.e., that high rock temperatures occur at shallow depth. Beta will be large if either q is large or k is small or both. By comparison with most everyday materials, rocks are poor conductors of heat and values of conductivity may vary from 2×10^{-3} to 10^{-2} cal cm^{-1} sec^{-1} $^{\circ}$C^{-1}. The mean surface heat flux from the earth is about 1.5 heat flow units (1 HFU = 10^{-6} cal cm^{-2} sec^{-1}) (Oxburgh, 1975). Rocks are also very slow respond to any temperature change to which they are exposed, i.e., they have a low thermal diffusivity:

$$K = k/\rho C_p \tag{2}$$

where:

K is thermal diffusivity; ρ and C_p are density and specific heat respectively.

These values are simple intended to give a general idea of the normal range of geothermal parameters (Table 6). In volcanic regions, in particular, both q and β can vary considerably and the upper values given are somewhat nominal.

Table 6. Values of geothermal parameters

Parameter	Lower	Average	Upper
q (HFU)	0.8	1.5	3.0 (non volcanic) $\approx$100 (volcanic)
k =cal cm^{-2} sec^{-1} $^{\circ}$C^{-1}	2×10^{-3}	6×10^{-3}	12×10^{-3}
β =$^{\circ}$C/km	8	20	60 (non volcanic) $\approx$300 (volcanic)

9. LANDFILL GAS

Landfill gas (LFG) is currently extracted at over 1200 landfills worldwide for a variety of energy purposes (Table 7), such as:

- Creating pipeline quality gas or an alternative fuel for vehicles.
- Processing the LFG to make it available as an alternative fuel to local industrial or commercial customers.
- Generation of electricity with engines, turbines, micro-turbines and other emerging technologies.

In terms of solid waste management policy, many NGOs have changed drastically in the past ten years from a mass production and mass consumption society to 'material-cycle society' (Abdeen, 2008). In addition to national legislation, municipalities are legally obliged to develop a plan for handling the municipal solid waste (MSW) generated in administrative areas. Such plans contain:

- Estimates of future waste volume.
- Measures to reduce waste.
- Measures to encourage source separation.
- A framework for solid waste disposal and the construction and management of solid waste management facilities.

Landfilling is in the least referred tier of the hierarchy of waste management options: waste minimisation, reuse and recycling, incineration with energy recovery, and optimised final disposal. The key elements are as follows: construction impacts, atmospheric emissions, noise, water quality, landscape, visual impacts, socio economics, ecological impacts, traffic, solid waste disposal and cultural heritage.

Table 7. Types of the LFG implemented recently worldwide

Landfill caps	Electricity generation	Fuel production
		❑ Medium BTU gas
❑ Soil caps	❑ Reciprocating	❑ High BTU gas
❑ Clay caps	engines	❑ Liquefied methane
❑ Geo-membrane	❑ Combustion	
caps	turbines	
	❑ Micro-turbines	**Thermal generation**
	❑ Steam turbines	
LFG destruction	❑ Fuel cells	❑ Boilers
		❑ Kilns
❑ Flares		❑ Greenhouse heaters

10. ENERGY EFFICIENCY

Energy efficiency is the most cost-effective way of cutting carbon dioxide emissions and improvements to households and businesses. It can also have many other additional social, economic and health benefits, such as warmer and healthier homes, lower fuel bills and company running costs and, indirectly, jobs. Britain wastes 20 per cent of its fossil fuel and electricity use. This implies that it would be cost-effective to cut £10 billion a year off the collective fuel bill and reduce CO_2 emissions by some 120 million tons. Yet, due to lack of good information and advice on energy saving, along with the capital to finance energy efficiency improvements, this huge potential for reducing energy demand is not being realised. Traditionally, energy utilities have been essentially fuel providers and the industry has pursued profits from increased volume of sales. Institutional and market arrangements have favoured energy consumption rather than conservation. However, energy is at the centre of the sustainable development paradigm as few activities affect the environment as much as the continually increasing use of energy. Most of the used energy depends on finite resources, such as coal, oil, gas and uranium. In addition, more than three quarters of the world's consumption of these fuels is used, often inefficiently, by only one quarter of the world's population. Without even addressing these inequities or the precious, finite nature of these resources, the scale of environmental damage will force the reduction of the usage of these fuels long before they run out.

Throughout the energy generation process there are impacts on the environment on local, national and international levels, from opencast mining and oil exploration to emissions of the potent greenhouse gas carbon dioxide in ever increasing concentration. Recently, the world's leading climate scientists reached an agreement that human activities, such as burning fossil fuels for energy and transport, are causing the world's temperature to rise. The Intergovernmental Panel on Climate Change has concluded that "the balance of evidence suggests a discernible human influence on global climate". It predicts a rate of warming greater than anyone had seen in the last 10,000 years, in other words, throughout human history. The exact impact of climate change is difficult to predict and will vary regionally. It could, however, include sea level rise, disrupted agriculture and food supplies and the possibility of more freak weather events such as hurricanes and droughts. Indeed, people already are waking up to the financial and social, as well as the environmental, risks of unsustainable energy generation methods that represent the costs of the impacts of climate change, acid rain and oil spills. The insurance industry, for example, concerned about the billion dollar costs of hurricanes and floods, has joined sides with environmentalists to lobby for greenhouse gas emissions reduction. Friends of the earth are campaigning for a more sustainable energy policy, guided by the principal of environmental protection and with the objectives of sound natural resource management and long-term energy security. The key priorities of such an energy policy must be to reduce fossil fuel use, move away from nuclear power, improve the efficiency with which energy is used and increase the amount of energy obtainable from sustainable and renewable energy sources. Efficient energy use has never been more crucial than it is today, particularly with the prospect of the imminent introduction of the climate change levy (CCL). Establishing an energy use action plan is the essential foundation to the elimination of energy waste. A logical starting point is to carry out an energy audit that enables the assessment of the energy use and determine what actions to take. The actions are best categorised by splitting measures into the following three general groups:

(1) High priority/low cost:

These are normally measures, which require minimal investment and can be implemented quickly. The followings are some examples of such measures:

- Good housekeeping, monitoring energy use and targeting waste-fuel practices.
- Adjusting controls to match requirements.
- Improved greenhouse space utilisation.
- Small capital item time switches, thermostats, etc.
- Carrying out minor maintenance and repairs.
- Staff education and training.
- Ensuring that energy is being purchased through the most suitable tariff or contract arrangements.

(2) Medium priority/medium cost:

Measures, which, although involve little or no design, involve greater expenditure and can take longer to implement. Examples of such measures are listed below:

- New or replacement controls.
- Greenhouse component alteration, e.g., insulation, sealing glass joints, etc.
- Alternative equipment components, e.g., energy efficient lamps in light fittings, etc.

(3) Long term/high cost:

These measures require detailed study and design. They can be best represented by the followings:
- Replacing or upgrading of plant and equipment.
- Fundamental redesign of systems, e.g., CHP installations.

This process can often be a complex experience and therefore the most cost-effective approach is to employ an energy specialist to help.

11. POLICY RECOMMENDATIONS FOR A SUSTAINABLE ENERGY FUTURE

Sustainability is regarded as a major consideration for both urban and rural development. People have been exploiting the natural resources with no consideration to the effects, both short-term (environmental) and long-term (resources crunch). It is also felt that knowledge and technology have not been used effectively in utilising energy resources. Energy is the vital input for economic and social development of any country. Its sustainability is an important factor to be considered. The urban areas depend, to a large extent, on commercial energy sources.

The rural areas use non-commercial sources like firewood and agricultural wastes. With the present day trends for improving the quality of life and sustenance of mankind, and environmental issues are considered highly important. In this context, the term energy loss has no significant technical meaning. Instead, the exergy loss has to be considered, as destruction of exergy is possible. Hence, exergy loss minimisation will help in sustainability.

The development of a renewable energy in a country depends on many factors. Those important to success are listed below:

(1) Motivation of the Population

The population should be motivated towards awareness of high environmental issues and rational use of energy in order to reduce cost. Subsidy programme should be implemented as incentives to install biomass energy plants. In addition, image campaigns to raise awareness of renewable energy technology.

(2) Technical Product Development

To achieve technical development of biomass energy technologies the following should be addressed:

- Increasing the longevity and reliability of renewable energy technology.
- Adapting renewable energy technology to household technology (hot water supply).
- Integration of renewable energy technology in heating technology.
- Integration of renewable energy technology in architecture, e.g., in the roof or façade.
- Development of new applications, e.g., solar cooling.
- Cost reduction.

(3) Distribution and Sales

Commercialisation of biomass energy technology requires:

- Inclusion of renewable energy technology in the product range of heating trades at all levels of the distribution process (wholesale, retail, etc.).
- Building distribution nets for renewable energy technology.
- Training of personnel in distribution and sales.
- Training of field sales force.

(4) Consumer Consultation and Installation

To encourage all sectors of the population to participate in adoption of biomass energy technologies, the following has to be realised:

- Acceptance by craftspeople, and marketing by them.
- Technical training of craftspeople, initial and follow-up training programmes.
- Sales training for craftspeople.
- Information material to be made available to craftspeople for consumer consultation.

(5) Projecting and Planning

Successful application of biomass technologies also require:

- Acceptance by decision makers in the building sector (architects, house technology planners, etc.).
- Integration of renewable energy technology in training.
- Demonstration projects/architecture competitions.
- Biomass energy project developers should prepare to participate in the carbon market by:
  - Ensuring that renewable energy projects comply with Kyoto Protocol requirements.
  - Quantifying the expected avoided emissions.

- Registering the project with the required offices.
- Contractually allocating the right to this revenue stream.

Other ecological measures employed on the development include:

- Simplified building details.
- Reduced number of materials.
- Materials that can be recycled or reused.
- Materials easily maintained and repaired.
- Materials that do not have a bad influence on the indoor climate (i.e., non-toxic).
- Local cleaning of grey water.
- Collecting and use of rainwater for outdoor purposes and park elements.
- Building volumes designed to give maximum access to neighbouring park areas.
- All apartments have visual access to both backyard and park.

(6) Energy Saving Measures

The following energy saving measures should also be considered:

- Building integrated solar PV system.
- Day-lighting.
- Ecological insulation materials.
- Natural/hybrid ventilation.
- Passive cooling, and passive solar heating.
- Solar heating of domestic hot water.
- Utilisation of rainwater for flushing.

Improving access for rural and urban low-income areas in developing countries must be through energy efficiency and renewable energies. Sustainable energy is a prerequisite for development. Energy-based living standards in developing countries, however, are clearly below standards in developed countries. Low levels of access to affordable and environmentally sound energy in both rural and urban low-income areas are therefore a predominant issue in developing countries. In recent years many programmes for development aid or technical assistance have been focusing on improving access to sustainable energy, and many of them with impressive results.

Apart from success stories, however, experience also shows that positive appraisals of many projects evaporate after completion and vanishing of the implementation expert team. Altogether, the diffusion of sustainable technologies such as energy efficiency and renewable energies for cooking, heating, lighting, electrical appliances and building insulation in developing countries has been slow.

Energy efficiency and renewable energy programmes could be more sustainable and pilot studies more effective and pulse releasing if the entire policy and implementation process was considered and redesigned from the outset. New financing and implementation processes are needed, which allow reallocating financial resources and thus enabling countries themselves to achieve a sustainable energy infrastructure. The links between the energy policy framework, financing and implementation of renewable energy and energy efficiency projects have to be strengthened and capacity building efforts are required. Energy resources are needed for societal development. Their sustainable development requires a supply of energy resources that are sustainably available at a reasonable cost and can cause no negative societal impacts.

12. ENVIRONMENTAL ASPECTS OF ENERGY CONVERSION AND USE

Environment has no precise limits because it is in fact a part of everything. Indeed, environment is, as anyone probably already knows, not only flowers blossoming or birds singing in the spring, or a lake surrounded by beautiful mountains. It is also human settlements, the places where people live, work, rest, the quality of the food they eat, the noise or silence of the street they live in. Environment is not only the fact that our cars consume a good deal of energy and pollute the air, but also, that we often need them to go to work and for holidays.

Obviously man uses energy just as plants, bacteria, mushrooms, bees, fish and rats do. Man largely uses solar energy- food, hydropower, wood- and thus participates harmoniously in the natural flow of energy through the environment. But man also uses oil, gas, coal and nuclear power. By using such sources of energy, man is thus modifying his environment.

The atmospheric emissions of fossil fuelled installations are mosty aldehydes, carbon monoxide, nitrogen oxides, sulpher oxides and particles (i.e., ash) as well as carbon dioxide. Table 8 shows estimates include not only the releases occuring at the power plant itself but also cover fuel extraction and treatment, as well as the storage of wastes and the area of land required for operations. Table 9 shows energy consumption in different regions of the world.

13. GREENHOUSES ENVIRONMENT

Greenhouse cultivation is one of the most absorbing and rewarding forms of gardening for anyone who enjoys growing plants. The enthusiastic gardener can adapt the greenhouse climate to suit a particular group of plants, or raise flowers, fruit and vegetables out of their natural season. The greenhouse can also be used as an essential garden tool, enabling the keen amateur to expand the scope of plants grown in the garden, as well as save money by raising their own plants and vegetables.

There was a decline in large private greenhouses during the two world wars due to a shortage of materials for their construction and fuel to heat them. However, in the 1950s mass-produced, small greenhouses became widely available at affordable prices and were used mainly for raising plants (John, 1993). Also, in recent years, the popularity of conservatories attached to the house has soared. Modern double-glazing panels can provide as much insulation as a brick wall to create a comfortable living space, as well as provide an ideal environment in which to grow and display tender plants.

The comfort in a greenhouse depends on many environmental parameters. These include temperature, relative humidity, air quality and lighting. Although greenhouse and conservatory originally both meant a place to house or conserve greens (variegated hollies, cirrus, myrtles and oleanders).

A greenhouse today implies a place in which plants are raised while conservatory usually describes a glazed room where plants may or may not play a significant role. Indeed, a greenhouse can be used for so many different purposes. It is, therefore, difficult to decide how to group the information about the plants that can be grown inside it.

Table 8. Annual greenhouse emissions from different sources of power plants

Primary source of energy	Emissions (x 10³ metric tons)		Waste (x 10³ metric tons)	Area (km²)
	Atmosphere	Water		
Coal	380	7-41	60-3000	120
Oil	70-160	3-6	negligible	70-84
Gas	24	1	-	84
Nuclear	6	21	2600	77

Table 9. Energy consumption in different continents

Region	Population (millions)	Energy (Watt/m²)
Africa	820	0.54
Asia	3780	2.74
Central America	180	1.44
North America	335	0.34
South America	475	0.52
Western Europe	445	2.24
Eastern Europe	130	2.57
Oceania	35	0.08
Russia	330	0.29

Throughout the world urban areas have increased in size during recent decades. About 50% of the world's population and approximately 76% in the more developed countries are urban dwellers (UN, 2001). Even though there is an evidence to suggest that in many 'advanced' industrialised countries there has been a reversal in the rural-to-urban shift of populations, virtually all population growth expected between 2000 and 2030 will be concentrated in urban areas of the world. With an expected annual growth of 1.8%, the world's urban population will double in 38 years (UN, 2001). This represents a serious contributing to the potential problem of maintaining the required food supply. Inappropriate land use and management, often driven by intensification resulting from high population pressure and market forces, is also a threat to food availability for domestic, livestock and wildlife use. Conversion to cropland and urban-industrial establishments is threatening their integrity. Improved productivity of peri-urban agriculture can, therefore, make a very large contribution to meeting food security needs of cities as well as providing income to the peri-urban farmers. Hence, greenhouses agriculture can become an engine of pro-poor 'trickle-up' growth because of the synergistic effects of agricultural growth such as (UN, 2001):

- Increased productivity increases wealth.
- Intensification by small farmers raises the demand for wage labour more than by larger farmers.
- Intensification drives rural non-farm enterprise and employment.
- Alleviation of rural and peri-urban poverty is likely to have a knock-on decrease of urban poverty.

Despite arguments for continued large-scale collective schemes there is now an increasingly compelling argument in favour of individual technologies for the development of controlled greenhouses. The main points constituting this argument are summarised by (UN, 2001) as follows:

- Individual technologies enable the poorest of the poor to engage in intensified agricultural production and to reduce their vulnerability.
- Development is encouraged where it is needed most and reaches many more poor households more quickly and at a lower cost.
- Farmer-controlled greenhouses enable farmers to avoid the difficulties of joint management.

Such development brings the following challenges (UN, 2001):

- The need to provide farmers with ready access to these individual technologies, repair services and technical assistance.
- Access to markets with worthwhile commodity prices, so that sufficient profitability is realised.
- This type of technology could be a solution to food security problems. For example, in greenhouses, advances in biotechnology like the genetic engineering, tissue culture and market-aided selection have the potential to be applied for raising yields, reducing pesticide excesses and increasing the nutrient value of basic foods.

However, the overall goal is to improve the cities in accordance with the Brundtland Report (WCED, 1987) and the investigation into how urban green could be protected. Indeed, greenhouses can improve the urban environment in multitude of ways. They shape the character of the town and its neighbourhoods, provide places for outdoor recreation, and have important environmental functions such as mitigating the heat island effect, reduce surface water runoff, and creating habitats for wildlife. Following analysis of social, cultural and ecological values of urban green, six criteria in order to evaluate the role of green urban in towns and cities were prescribed (WCED, 1987). These are as follows:

- Recreation - everyday life and public health.
- Maintenance of biodiversity - preserving diversity within species, between species, ecosystems, and of landscape types in the surrounding countryside.
- City structure - as an important element of urban structure and urban life.
- Cultural identity - enhancing awareness of the history of the city and its cultural traditions.

- Environmental quality of the urban sites - improvement of the local climate, air quality and noise reduction.
- Biological solutions to technical problems in urban areas - establishing close links between technical infrastructure and green-spaces of a city.

The main reasons why it is vital for greenhouses planners and designers to develop a better understanding of greenhouses in high-density housing can be summarised as follows (WCED, 1987):

- Pressures to return to a higher density form of housing.
- The requirement to provide more sustainable food.
- The urgent need to regenerate the existing, and often decaying, houses built in the higher density, high-rise form, much of which is now suffering from technical problems.

The connection between technical change, economic policies and the environment is of primary importance as observed by most governments in developing countries, whose attempts to attain food self-sufficiency have led them to take the measures that provide incentives for adoption of the Green Revolution Technology (Herath, 1985). Since, the Green Revolution Technologies were introduced in many countries actively supported by irrigation development, subsidised credit, fertiliser programmes, and self-sufficiency was found to be not economically efficient and often adopted for political reasons creating excessive damage to natural resources. Also, many developing countries governments provided direct assistance to farmers to adopt soil conservation measures. They found that high costs of establishment and maintenance and the loss of land to hedgerows are the major constraints to adoption (Herath, 1985).

The soil erosion problem in developing countries reveals that a dynamic view of the problem is necessary to ensure that the important elements of the problem are understood for any remedial measures to be undertaken. The policy environment has, in the past, encouraged unsustainable use of land (Herath, 1985). In many regions, government policies such as provision of credit facilities, subsidies, price support for certain crops, subsidies for erosion control and tariff protection, have exacerbated the erosion problem. This is because technological approaches to control soil erosion have often been promoted to the exclusion of other effective approaches. However, adoption of conservation measures and the return to conservation depend on the specific agro-ecological conditions, the technologies used and the prices of inputs and outputs of production.

13.1. Types of Greenhouses

Choosing a greenhouse and setting it up are important, and often expensive, steps to take. Greenhouses are either freestanding or lean-to, that is, built against an existing wall. A freestanding greenhouse can be placed in the open, and, hence, take advantage of receiving the full sun throughout the day. It is, therefore, suitable for a wide range of plants. However, its main disadvantage when compared to a lean-to type is that more heat is lost through its larger surface area. This is mainly why lean-to greenhouses have long been used in the walled gardens of large country houses to grow Lapageria rosea and other plants requiring cool, constant temperature, such as half-hardly ferns. However, generally, good ventilation and shading in the spring and summer to prevent overheating are essential for any greenhouse. The high daytime temperatures will warm the back wall, which acts as a heat battery, releasing its accumulated heat at night. Therefore, plants in a greenhouse with this orientation will need the most attention, as they will dry out rapidly.

Also, greenhouses vary considerably in their shapes and internal dimensions. Traditional greenhouses have straight sides, which allow the maximum use of internal space, and are ideal for climbers (Herath, 1985). On the other hand, greenhouses with sloping sides have the advantage of allowing the greatest penetration of sunlight, even during winter (Herath, 1985). The low winter sun striking the glass at 90°C lets in the maximum amount of light. Where the sun strikes the glass at a greater or lesser angle, a proportion of the light is reflected away from greenhouse. Sloping sides, also, offer less wind resistance than straight sides and therefore, less likely to be damaged during windy weather. This type of greenhouse is most suitable for short winter crops, such as early spring lettuce, and flowering annuals from seed, which do not require much headroom.

A typical greenhouse is shown schematically in Figure 10. However, there are several designs of greenhouses, based on dimensions, orientation and function. The following three options are the most widely used:

- A ready-made design.
- A designed, which is constructed from a number of prefabricated modules.
- A bespoke design.

Of these, the prefabricated ready-made design, which is utilised to fit the site, is the cheapest greenhouses and gives flexibility. It is, also, the most popular option (Herath, 1985).

Specific examples of commercially available designs are numerous. Dutch light greenhouses, for example, have large panes of glass, which cast little shade on the plants inside. They are simple to erect, consisting of frames bolted together, which are supported on a steel framework for all but the smallest models. They are easy to move and extra sections can be added on to them, a useful attraction (Herath, 1985). Curvilinear greenhouses, on the other hand, are designed primarily to let in the maximum amount of light throughout the year by presenting at least one side perpendicular to the sun. This attractive style of greenhouse tends to be expensive because of the number of different angles, which require more engineering (Herath, 1985).

Likewise, the uneven span greenhouses are designed for maximum light transmission on one side. These are generally taller than traditional greenhouses, making them suitable for tall, early season crops, such as cucumbers (Herath, 1985). Also, the polygonal greenhouses are designed more as garden features than as practical growing houses, and consequently, are expensive. Their internal space is somewhat limited and on smaller models over-heading can be a problem because of their small roof ventilations. They are suitable for growing smaller pot plants, such as pelargoniums and cacti (Herath, 1985). Another example is the solar greenhouses.

These are designed primarily for areas with very cold winters and poor winter light. They take the form of lean-to structures facing the sun, are well insulated to conserve heat and are sometimes partially sunk into the ground. They are suitable for winter vegetable crops and early-sown bedding plants, such as begonias and pelargoniums (Herath, 1985). Mini lean-to greenhouses are suitable for small gardens where space is limited. They can, also, be used to create a separate environment within larger greenhouses. The space inside is large enough to grow two tomato or melon plants in growing bags, or can install shelves to provide a multi-layered growing environment, ideal for many small potted plants and raising summer bedding plants (Herath, 1985).

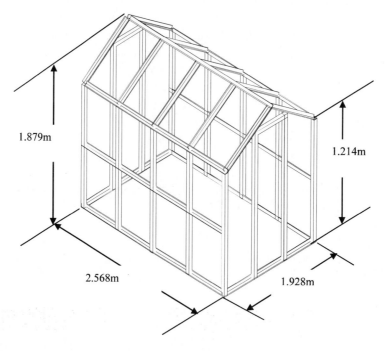

Figure 10. Greenhouse and base with horticultural glass.

13.2. Construction Materials

Different materials are used for the different parts. However, wood and aluminium are the two most popular materials used for small greenhouses. Steel is used for larger structures and UPVC for conservatories (Jonathon, 1991).

13.3. Ground Radiation

Reflection of sunrays is mostly used for concentrating them onto reactors of solar power plants. Enhancing the insolation for other purposes has, so far, scarcely been used. Several years ago, application of this principle for increasing the ground irradiance in greenhouses, glass covered extensions in buildings, and for illuminating northward facing walls of buildings was proposed (Achard and Gicqquel, 1986). Application of reflection of sun's rays was motivated by the fact that ground illuminance/irradiance from direct sunlight is of very low intensity in winter months, even when skies are clear, due to the low incident angle of incoming radiation during most of the day. This is even more pronounced at greater latitudes. As can be seen in Figure 11, which depicts a sunbeam split into its vertical and horizontal components, nearly all of the radiation passes through a greenhouse during most of the day.

Large-scale, conventional, power plant such as hydropower has an important part to play in development. It does not, however, provide a complete solution. There is an important complementary role for the greater use of small-scale, rural based and power plant (device). Such plant can be used to assist development since it can be made locally using local resources, and enabling a rapid built-up in total equipment to be made without a corresponding and unacceptably large demand on central funds. Renewable resources are particularly suitable for providing the energy for such equipment and its use is also compatible with the long-term aims. It is possible with relatively simple flat plate solar collectors (Figure 12) to provide warmed water and enable some space heating for homes and offices, which is particularly useful when the buildings are well insulated and thermal capacity sufficient for the carryover of energy from day to night is arranged.

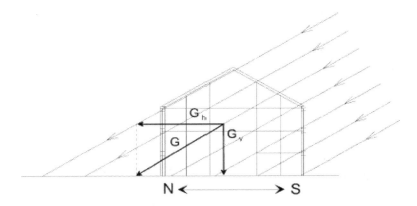

Figure 11. Relative horizontal and vertical components of solar radiation.

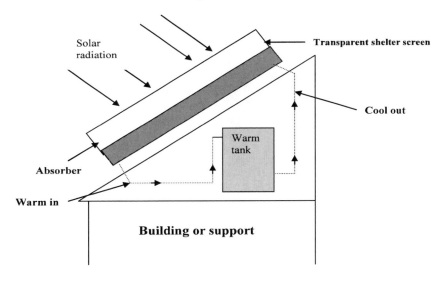

Figure 12. Solar heaters for hot water.

13.4. Greenhouse Environment

It has been known for long time now that urban centres have mean temperatures higher than their less developed surroundings. The urban heat increases the average and peak air temperatures, which, in turn, affect the demand for heating and cooling. Higher temperatures can be beneficial in the heating season, lowering fuel use, but they exacerbate the energy demand for cooling in summer time. In temperate climates neither heating nor cooling may dominate the fuel use in a building, and the balance of the effect of the heat is less. The solar gains, however, would affect the energy consumption. Therefore, lower or higher percentage of glazing, or incorporating of shading devices might affect the balance between annual heating and cooling load. As the provision of cooling is expensive with higher environmental cost, ways of using innovative alternative systems like mop fans will be appreciated (Figures 13-15). Indeed, considerable research activities have been devoted to the development of alternative methods of refrigeration and air-conditioning. The mop fan is a novel air-cleaning device that fulfils the functions of de-dusting of gas streams, and removal of gaseous contaminations from gas streams and gas circulation (Erlich, 1991). Hence, the mop fan seems particularly suitable for applications in industrial, agricultural and commercial buildings and greenhouses.

Figure 13. Mop fan systems.

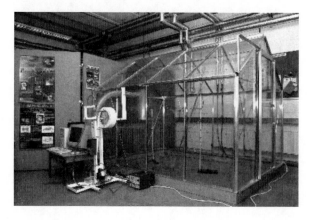

Figure 14. Mop fan in greenhouse.

Indoor conditions are usually fixed by comfort conditions, with air temperatures ranging from 15°C to 27°C, and relative humidities ranging from 50% to 70% (Abdeen, 2008). The system coefficient of performance (COP) is defined as the ration between the cooling effect in the greenhouse and the total amount of air input to the mop fan. Hence,

COP = cooling delivered/air input to the mop fan   (3)

Therefore, system performance (COP) varies with indoor and outdoor conditions. A lower ambient temperature and a lower ambient relative humidity lead to a higher COP. This means that the system will be, in principle, more efficient in colder and drier climates. The effect of indoor (greenhouse) conditions and outdoor (ambient) conditions (temperature and relative humidity) on system performance is illustrated in Figure 16.

Figure 15. Plants in greenhouse.

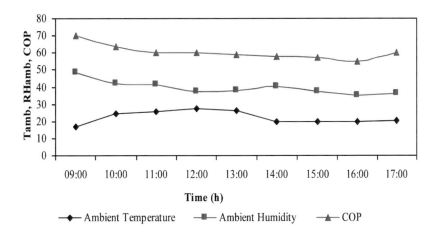

Figure 16. Ambient temperature, relative humidity and COP.

CONCLUSION

There is strong scientific evidence that the average temperature of the earth's surface is rising. This is a result of the increased concentration of carbon dioxide and other GHGs in the atmosphere as released by burning fossil fuels. This global warming will eventually lead to substantial changes in the world's climate, which will, in turn, have a major impact on human life and the built environment. Therefore, effort has to be made to reduce fossil energy use and to promote green energies, particularly in the building sector. Energy use reductions can be achieved by minimising the energy demand, by rational energy use, by recovering heat and the use of more green energies. This study was a step towards achieving that goal. The adoption of green or sustainable approaches to the way in which society is run is seen as an important strategy in finding a solution to the energy problem. The key factors to reducing and controlling CO_2, which is the major contributor to global warming, are the use of alternative approaches to energy generation and the exploration of how these alternatives are used today and may be used in the future as green energy sources. Even with modest assumptions about the availability of land, comprehensive fuel-wood farming programmes offer significant energy, economic and environmental benefits. These benefits would be dispersed in rural areas where they are greatly needed and can serve as linkages for further rural economic development. The nations as a whole would benefit from savings in foreign exchange, improved energy security, and socio-economic improvements. With a nine-fold increase in forest – plantation cover, a nation's resource base would be greatly improved. The international community would benefit from pollution reduction, climate mitigation, and the increased trading opportunities that arise from new income sources. The non-technical issues, which have recently gained attention, include: (1) Environmental and ecological factors, e.g., carbon sequestration, reforestation and revegetation. (2) Renewables as a CO_2 neutral replacement for fossil fuels. (3) Greater recognition of the importance of renewable energy, particularly modern biomass energy carriers, at the policy and planning levels. (4) Greater recognition of the difficulties of gathering good and reliable renewable energy data, and efforts to improve it. (5) Studies on the detrimental health efforts of biomass energy particularly from traditional energy users. Two of the most essential natural resources for all life on the earth and for man's survival are sunlight and water. Sunlight is the driving force behind many of the renewable energy technologies. The worldwide potential for utilising this resource, both directly by means of the solar technologies and indirectly by means of biofuels, wind and hydro technologies is vast. During the last decade interest has been refocused on renewable energy sources due to the increasing prices and fore-seeable exhaustion of presently used commercial energy sources. Plants, like human beings, need tender loving care in the form of optimum settings of light, sunshine, nourishment, and water. Hence, the control of sunlight, air humidity and temperatures in greenhouses are the key to successful greenhouse gardening. The mop fan is a simple and novel air humidifier; which is capable of removing particulate and gaseous pollutants while providing ventilation. It is a device ideally suited to greenhouse applications, which require robustness, low cost, minimum maintenance and high efficiency. A device meeting these requirements is not yet available to the farming community. Hence, implementing mop fans aides sustainable development through using a clean, environmentally friendly device that decreases load in the greenhouse and reduces energy consumption.

REFERENCES

Abdeen, MO. 2008. Chapter 10: Development of integrated bioenergy for improvement of quality of life of poor people in developing countries, In: *Energy in Europe: Economics, Policy and Strategy- IB*, Editors: Flip L. Magnusson and Oscar W. Bengtsson, 2008 NOVA Science Publishers, Inc., 341-373, New York, USA.

Achard, P; Gicqquel, R. 1986. European passive solar handbook. Brussels: Commission of the European Communities.

Bos, E; My, T; Vu, E; Bulatao R. 1994. World population projection: 1994-95. Edition, published for the World Bank by the John Hopkins University Press. Baltimore and London.

Brain, G; Mark, S. 2007. Garbage in, energy out: landfill gas opportunities for CHP projects. *Cogeneration and On-Site Power*, 8 (5), 37-45.

Erlich, P. 1991. *Forward Facing up to Climate Change, in Global Climate Change and Life on Earth*. R.C. Wyman (Ed), Chapman and Hall, London.

Herath, G. 1985. The green revolution in Asia: productivity, employment and the role of policies. *Oxford Agrarian Studies*, *14*, 52-71.

International Energy Agency (IEA). 2007. Indicators for Industrial Energy Efficiency and CO_2 Emissions: A Technology Perspective.

Jonathon, E. 1991. Greenhouse gardening. The Crowood Press Ltd. UK.

John, W. 1993. *The glasshouse garden.* The Royal Horticultural Society Collection. UK.

Omer, AM; Yemen, D. 2001. Biogas an appropriate technology. *Proceedings of the 7th Arab International Solar Energy Conference*, P.417, Sharjah, UAE, 19-22 February 2001.

Omer, AM. 2009a. Environmental and socio-economic aspect of possible development in renewable energy use, In: *Proceedings of the 4th International Symposium on Environment,* Athens, Greece, 21-24 May 2009.

Omer, AM. 2009b. Energy use, environment and sustainable development, In: *Proceedings of the 3rd International Conference on Sustainable Energy and Environmental Protection (SEEP 2009),* Paper No.1011, Dublin, Republic of Ireland, 12-15 August 2009.

Omer, AM. 2009c. Energy use and environmental: impacts: a general review, *Journal of Renewable and Sustainable Energy*, Vol.1, No.053101, 1-29, United State of America, September 2009.

Omer, AM. 2009d. Chapter 3: Energy use, environment and sustainable development, *In: Environmental Cost Management,* Editors: Randi Taylor Mancuso, 2009 NOVA Science Publishers, Inc., 129-166, New York, USA.

Oxburgh, ER. 1975. *Geothermal energy.* Aspects of Energy Conversion. 385-403.

Rawlings, RHD. 1999. *Technical Note TN 18/99 – Ground Source Heat Pumps: A Technology Review.* Bracknell. The Building Services Research and Information Association.

Robinson, G. 2007. Changes in construction waste management. *Waste Management World,* 43-49. May-June 2007.

Sms, RH. 2007. Not too late: IPCC identifies renewable energy as a key measure to limit climate change. *Renewable Energy World,* 10 (4): 31-39.

Swift-Hook, DT; et al. 1975. Characteristics of a rocking wave power devices. Nature, 254: 504.

Trevor, T. 2007. Fridge recycling: bringing agents in from the cold. *Waste Management World*, 5, 43-47.

United Nations (UN). 2001. *World Urbanisation Prospect: The 1999 Revision.* New York. The United Nations Population Division.

World Commission on Environment and Development (WCED). 1987. *Our common future.* New York. Oxford University Press.

Chapter8

THE ABSENCE OF THE CURRICULUM IN SCIENTIFIC RESEARCH IN DEVELOPING COUNTRIES

ABSTRACT

A booming economy, high population, land-locked locations, vast area, remote separated and poorly accessible rural areas, large reserves of oil, excellent sunshine, large mining sector and cattle farming on a large-scale, are factors which are most influential to the total water scene in Africa. This communication reviews the development of scientific and technological research highlights the efforts made sporadic in this area. In this study I try to put some solutions and recommendations that help the advancement of scientific research to resolve issues in the Sudanese society.

Despite the obstacles, the movement of scientific research did not stop completely, because a number of researchers still believe in the inevitability of continued scientific research to benefit the maximum of what is available (and the efforts of individual) to attain the objectives of development, prosperity and keep pace with scientific development. Research has become a pedestal to build a modern state in today's world, and became the backbone for all plans developed nations and even developing countries. And enter the world in the era of World Trade Organisation (WTO) and intellectual property Guanyin and the demands of globalisation for the next century that followed, we are in need to change. We should employ scientific research to address the backlog of cases over the years such as the issue of poverty and human capacity development and exploitation of natural resources of the country and the fight against desertification and settling of scientific technologies for the stability of the pastoral communities, and others.

Keywords: Sudan, education, technical scientific research, sustainable development

INTRODUCTION

Research has become a pedestal to build a modern state in today's world, and became the backbone for all plans developed nations and even developing countries. And enter the world in the era of World Trade Organisation (WTO) and intellectual property Guanyin and the demands of globalisation for the next century that followed, we are in need to change. We should employ scientific research to address the backlog of cases over the years such as the issue of poverty and human capacity development and exploitation of natural resources of the country and the fight against desertification and settling of scientific technologies for the stability of the pastoral communities, and others (Faisal, 1993). It must employ scientific research to address the backlog of cases over the years such as the issue of poverty and human capacity development and exploitation of natural resources of the country and the fight against desertification and settling of scientific technologies for the stability of the pastoral communities, and others. The ramifications in the fields of science and technological abounded to the point where it became impossible to take in all its aspects and its subsidiaries and became to be the selection of research needed by the various communities and the selection of educational curricula have efficacious in both the vertical (specialised) or horizontal (destruction). These thoughts came together and a hint of shame and sometimes pour sometimes without reference to party or institution. It is about frameworks general extrapolated to infer and devise solutions and treatments. I hope those who care about the subject of research and funding in the current international situation, which stripped the developing nations of the features that I found of Southeast Asia and other developing their country's' tradition of innovations and products to developed countries.

The financial accountability is also easier and more transparent. 'Global Change' consists of the linked and interacting phenomena of rapid, modern and widespread change in land cover and land use; atmospheric composition; climate; biological diversity; economic organisation; population size; distribution and consumption patterns; and trade patterns. Together these factors pose a great challenge to human development. Global change research is a large, interdisciplinary and worldwide effort to find solutions to these challenges, in order that human development may be sustained and equitable. Africa is particularly vulnerable to many of the negative consequences of global change. Every aspect of the water, energy, health, agricultural and biodiversity agenda is impacted in some significant way.

Many hundreds of Africa-based researchers are already engaged full or part-time on global change research. In addition, there are similar numbers of researchers based outside Africa, focused on global change research relating to Africa. A rich set of overlapping networks links global change researchers in Africa to each other and to the global community. These networks should form the basis of future coordination efforts. There are key constraints to the development of a larger, more productive and better distributed (Karim, 1997). Global change research community in Africa is the lack of adequate dedicated, accessible and stable funding sources to support this research area. Secondary constraints of insufficient highly skilled human resources, and in-continent access to certain technologies, could be resolved within a decade if funding were available. Specific recommendations include:

- Strengthen the existing networks rather than creating completely new sets, supplementing them where necessary. An overarching structure that caters for human sciences with a more development focus as well as for biophysical sciences with a longer-term focus is suggested.
- Establish centres of excellence in aspects of global change research in south, east, central, west and North Africa, using existing concentrations of researchers in most cases.
- Focus international collaborative research campaigns on large, complex key uncertainty areas in Africa, including (for example, and among others) the processes

and model characterisation of tropical and subtropical rainfall; climate land-surface feedbacks; hydrological cycle impacts in the presence of elevated CO_2; impacts and adaptation of biodiversity; responses to changes in vector-borne diseases and the emergence of novel diseases.

- Increase the availability of small to medium sized competitive research grants for African researchers in global change.

BACKGROUND

Sudan deals with science and technology since a long time, especially in the most important aspects of research and development. It did not identify specific policies or plans for science, technology and scientific research in Sudan in the post-independence period in spite of direct interest and activity, which is done by scientists and experts such as the Sudanese Arquette Conference for the overall development (Muhammad, 1998). The first composition of structural and institutional framework for science technology and scientific research establishment of the National Council for Research in 1970 as a government responsible for policy, planning and programming has been included under its five boards specialising in agricultural research and scientific and technological research, medical research, renewable energy research and the economic and social research. And be under the umbrella of some national committees in the environment, population and genetic engineering. It also established some of the centres and units to conduct scientific research in disciplines that are not available to universities and other research centres and the most important dimension of the Remote Sensing Centre and the Centre for Documentation and Information.

The new world order and the globalisation agenda of the twenty-first century are controlled by the revolution and the Sultan of information and amazing and successive developments in science communications, electronics, biotechnology, genetic engineering and computer science. All these challenges require the use of scientific research in order to survive competition and does not have a will of its strength does not have a decision in his or her liberty in all its dimensions. Sudan has the latest scientific advances by the end of the twentieth century new phenomena, including the authority of informatics and the authority of the international media and social change and systems in a sophisticated modern technology, production and management. It is more complicated systems using modern technologies and the accelerated globalisation of information and knowledge become the methods of scientific research of the basics and requirements of sustainable development of any country seeks to true independence. Because the people have the information and improve their use will have an advantage in the selection of appropriate technologies and in reducing the cost of production and raising rates and competition in foreign markets and meet the challenges of globalisation in all fields (Hashim, 1996).

All of this, the experience accumulated around us and our needs of actual and our well-defined should be our starting point to develop a strategy applied scientific research to which we aspire, otherwise we will apply to us the rule which says those who seek not what we want, and who want not what we need, which we need cannot access it. This communication dealt with the subject of scientific research in the Sudan, also highlights the efforts made sporadic in this area. And try to put some proposals and recommendations that help the advancement of scientific research in order to achieve its objectives and even going in the right direction, which can achieve sustainable development and in line with the requirements of globalisation.

The National Research Council in 1981 the first plan for science and technology and scientific research in the form of a programme known as ((the programme of building a science of modern)) in between 1981 and 1990 included the sub-programmes to exploit science and technology and to encourage scientific research and building science cities. In 1989 established the Ministry of Higher Education and Scientific Research and became a ministry responsible for setting policies of science and technology, scientific research and has developed a new management structure of the National Centre for Research and issued law in 1991 with the aim of the centre to conduct scientific and applied research. The entrusted the National Centre for Research to conduct scientific and applied research for economic and social development within the framework of state policy and under its auspices, and in coordination and integration with the institutions of scientific research in Sudan menu such as:

- Centre for research and industrial consultancy.
- Agricultural research corporation (and forests).
- The research of livestock.
- The geological and mining research.

RESEARCH DEVELOPMENT

At a time when the pace of the western world strides in the field of scientific research until they reached the maximum range in recent years, the country has been developing, including Sudan in the early stages, its features are not clear yet. It may be appropriate to begin the study for reasons including:

- Can take a closer look and comprehensive review in this area during the past decades.
- Defend economic, social and technical, which is moving towards wider horizons the better and the worst.
- Technological machines, which jumped from steam to nuclear reactors and equipment to lightning satellites.
- Globalisation and the consequence of openness are difficult to control it.

That is why the world's post-year 2000 would be governed, including the decisions taken today, and decisions taken by the views at the top of the state bodies in the development of policies, strategies and plans. Those decisions should be taken along the lines of solid base of knowledge and skill. Knowledge and skill that needed to research and application, research and application need for political and professional experience and knowledge and practice.

Modern scientific research began in Sudan and establishing a laboratory chemical in the second year of last century (Welcome Chemical Laboratory) and cotton research stations, followed by several research centres, all linked to areas of scientific and technological and related ministries and various government departments. After independence, turned institutions and scientific schools and the College of Gordon to form the University of Khartoum and obtained scientific research has attached great importance especially after the increase in graduates from the University of Khartoum and open the opportunity for many of them to engage in scientific research, this was a breakthrough for the development of scientific research in various levels and the response to a number of issues and try to address the problems associated with its evolution (Esmat, 1997a).

SCIENTIFIC RESEARCH IN SUDAN

The scientific research in the broad meaning is to harness science and technology for the benefit of society. We have met all platforms scientific (meetings, seminars, conferences, workshops, etc.), whether regional or global, which discussed the major challenges that science and technology has become the way real disposable in today's world to achieve economic and social development and that the difference between the developed and underdeveloped countries is due primarily to the difference between them in the scientific and technical capabilities. No way for these countries to achieve economic and social development only through science and technology and scientific research. Among the most important factor in increasing production and development is the scientific and technological progress and the ratio of the strong relationship between economic and social development and scientific and technological progress. Known to Sudan to deal with scientific research with the beginning of last century has been established plant (Welcome) in 1902 to take control of Epidemiology and parasites. In 1904 established the first station of the Agricultural Research. The established the Kitchener Medical School 1924 and then the School of Veterinary and Agriculture in 1938 and the interests of geology. The objective of this research in the colonial era to provide simple needs in the fields of agricultural, health and mining, but linked, in particular raw materials and make it suitable for export was also associated with imported goods in order to make it fit with local needs. Expansion of scientific research in the fields of agriculture and veterinary medicine, forestry and fish thanks to these colleges and professors to become a stand-alone there is a College of Graduate Studies at the University of Khartoum.

Also, as the correlation of universities and research institutions of these sectors is one of the important indicators in determining the level of scientific, technical and can be measured by the number of contracts related to research and development in the public or private-sector and also offers advice and other technical services for the benefit of society.

ABOUT THE HISTORY OF SCIENTIFIC RESEARCH IN SUDAN

Scientific research aims to find solutions to the problems facing the human to follow the scientific methods of the organisation based on a systematic consideration of sound. This concept began scientific research in the modern sense Sudan with the beginning of last century, and the research efforts, individually and in limited areas of improved goods and raw materials as well as in some of the veterinary services and quality control. The unit veterinary quarantines the first nucleus for scientific research in Sudan, where the user defined the British administration in 1900 in the interest of military veterinary time and paid attention to study some common diseases in the bovine animal plague and prevention.

In 1902 established a laboratory (Welcome), where the active and clear in the fight against diseases and insects harmful to humans and animals and plants, has contributed to the laboratory to discover some of the endemic diseases, especially schistosomiasis in Sudan. In 1904 established a separate unit for agricultural research included in 1912 to the benefit of knowledge of Sudan and had been cooperating closely with the laboratory (Welcome) and in 1919 turned into agricultural research to the Department of Agriculture and confined their efforts in the period to monitor the quality of the crops and protect them from pests. In 1934 opened the Kitchener Medical School and later joined to the rest of the sections of Gordon Memorial College to form the (Khartoum University College in 1951). The 1938 milestone in the history of scientific research in Sudan, where was established the School of Veterinary Medicine and School of Agriculture and have evolved later to the colleges of Agricultural Sciences and Veterinary Science and the graduates of those schools are leading the Sudanese first in the field of scientific research in agriculture, veterinary science and was almost an impact in enriching scientific research in Sudan qualified cadres and the development of curricula and research methods (Esmat, 1997b). Not only research activity in Sudan on the human and veterinary medicine and agriculture, but efforts have been made in multiple areas of research in science and engineering in different branches in industry, food, construction, geology, mining and areas of economics, sociology and the humanities, etc.

The owner of those official efforts by non-governmental scientific associations in various fields and made some magazines and scientific journals limited distribution and the country began to hear about conferences, seminars and scientific seminars. The independence of Sudan at the beginning of 1956 witnessed the beginning of a new stage in quality and quantity of scientific research and approaches and decisions of the College of Khartoum university marked the birth of the University of Khartoum, which recognised the nuances of scientific world and provide for scientific research potential of the largest and expanded opportunities for government missions and non-governmental organisations to universities and colleges outside the country, which shares in the country to provide scientific expertise in the various branches of modern knowledge. The spread scientific research centres, especially in agriculture and livestock in various parts of the country to reach areas of production and animal communities and expanded the work of general survey of the potential of the country and its natural resources. A closer cooperation between Sudan and the friendly countries and bodies and international organisations are very important. Technical innovation and employment are to serve the Sudanese community, in collaboration with universities and institutions of Higher Education and Scientific research, and the other country.

STRUCTURE MANAGEMENT AND COORDINATION OF SCIENCE AND TECHNOLOGY AND SCIENTIFIC RESEARCH

The country continued the scientific research during the period that followed independence until the 1970 year of transition the great scientific where a National Research

Council as an independent sponsor scientific research in the country under the care directly to the president and the president of the National Council for Research in the rank of minister and participate in meetings of the Council of Ministers and led the National Council Research march to scientific research by universities. Sudan announced in early 1982 that the era of the eighties would be the start-up phase in building a modern science. In 1991 the centre was established the National Centre for Research (NCR) as a substitute for independent, multidisciplinary research and development concern in the field of applied developmental and under the auspices of the Ministry of Higher Education and Scientific Research (Ahmed, 1998).

SECTOR OF HIGHER EDUCATION AND SCIENTIFIC RESEARCH

This sector is supervised by the terms of the development of policies, strategies and plans, National Council for Higher Education and Scientific Research under the auspices of head of state and is chaired by the Minister of Higher Education and Scientific Research, representing the universities and the NCR heads of their boards and directives members in addition to the membership of institutions of higher education colleges and private (Ahli). Established the National Council for Higher Education and Scientific Research of the Research Committee as one of its standing committees and entrusted with the task of coordination between universities, higher institutes and the NCR.

RESEARCH AND DEVELOPMENT SECTOR MINISTRIES

This sector consists of institutions for research and development to follow the federal ministries. The largest of these institutions, the Agricultural Research Organisation, which consists of agricultural research centres and the Food Research and affiliated with the Ministry of Agriculture and Natural Resources, in the same form belongs to the Livestock Research of the Ministry of Livestock, and Medical Laboratory of the Ministry of Health, the Centre for Research and Industrial Consultancy of the Ministry of Industry, Research Institute of Hydraulic of the Ministry of Irrigation, and the Institute of Geological research of the Ministry of Energy and Mining.

THE SUPREME COUNCIL FOR ENVIRONMENT
AND NATURAL RESOURCES

Established under the chairmanship of the president of the Republic and the membership of the Minister of Agriculture and Livestock Minister and the Ministers of Higher Education and Scientific Research, health, industry, irrigation and finance. The Board shall leave policies that are interested in the environment, natural resources and the coordination between governmental and private institutions in their respective fields. Based on the above are policy development and coordination and evaluation in the field of science and technology at the moment on several levels in different sectors. The National Council for Higher Education and Scientific Research, policy development and coordination between universities and the NCR on science and technology through the Standing Committee on Research. The policy development and coordination between national research institutions are at the level of the Supreme Council for Environment and Natural Resources and the Federal Cabinet.

QUALIFIED PERSONNEL IN THE FIELD OF SCIENCE AND TECHNOLOGY AND SCIENTIFIC RESEARCH

Classified as developing nations when compared to the rest of the world it is lagging behind scientifically percentage of scientists is very low, and also spending in science and technology does not affect the rate of 1% of the GNP set by UNESCO as a minimum for the development of science and technology. There are many indications and are known to help assess the degree of development of science and technology, including:

- Number of qualified cadres in the field of research and development in every million people.
- Percentage of expenditure on research and development of the gross national product.

Number of scientific personnel and technical ability in the thousands in every million people, it is the duty of developing countries to develop research and development in the structures of local scientific even expanding the scientific base and are used and increase their effectiveness and strive for the developed countries and benefit from the model of scientific as most of the scientific personnel and technical assistance to these developed countries working in industry in the area of quality control and production development. In the field of research and development, the labour force and high-trained staff are a great wealth for the success of the application of science and technology in moving the wheel of development. According to UNESCO standards require developing countries with a per capita GNP between USA $500-1000 to 6000 specialised scientific and engineer in every million people that 10% of whom specialise in the field of scientific research and development. Also need this number of qualified personnel to a large number of intermediate cadres of technicians and skilled workers.

SUDAN TRENDS IN SCIENCE AND TECHNOLOGY AND SCIENTIFIC RESEARCH

Sudan are evaluating the effectiveness of teachers and researchers in universities and research institutions in the light of the results of their research and their experience in addition to their participation and contribution in the productive sectors, including industrial, agricultural and services sector. Also, as the correlation of universities and research institutions of these sectors is one of the important indicators in determining the level of scientific, technical and can be measured by the number of contracts related to research and development in the public or private-sector and also offers advice and other technical services for the benefit of society.

SCOPE OF PROCEDURE

Sudan has taken in recent years a number of positive steps towards supporting science and technology and its applications in the field of formal commitment and policies are the following steps:

- A comprehensive national strategy included an approved plan for development of science and technology and its applications. This is the first formal commitment in the modern history of Sudan.
- To create the National Centre for Research in 1991 as an independent, multidisciplinary research and development concerned and aimed to strengthen the scientific capacity in the field of applied and developmental.
- Formation of the Supreme Council for Environment and Natural Resources in order to support, coordination and balance environmental and sustainable development.
- Horizontal and vertical expansion in the training of scientific personnel and professional establishment of regional universities and specialised institutes and increase the proportion of the number of students of higher education. As the expansion of education above college and branched specialties.
- Restructuring of the agricultural research for independence and support staff and financial resources and linking them to the application farm in the major agricultural enterprises. And take advantage of technological packages for the establishment of crops and increasing the vegetable production, and were separated bodies dealing with livestock research and development in livestock enormous.
- And encourage scientific publishing and media support scientific publications and programmes at different levels of information in audio-visual media tools and governmental interest in public awareness in science and technology.
- Establishment of the council to develop the manufacture of Sudan under the auspices of Centre for Research and Industrial Consultancy, Ministry of Industry.

OUTSIDE

Sudan has been a great interest in regional and international cooperation in science and technology to provide what is available has the potential of scientific and material assistance to the brotherly and friendly countries as Sudan committed themselves to the agreements and protocols, regional and global within this framework:

- Sudan signed in 1992 the Vienna Convention and Montreal Protocol on Substances that Deplete the Ozone Layer.

- Sudan signed in 1992 the Convention on International Centre for Biotechnology and Genetic Engineering and established a national focal point of the National Centre for Research.
- Sudan has signed conventions on biodiversity and global climate during the International Conference of the United Nations Environment and Development in Rio de Janeiro in 1992.
- Sudan adopts many of the activities of the Federation of Arab Scientific Research Councils.
- Sudan participated actively in the programmes of regional and international organisations.

There is still a need for greater effort in different sectors to create an enabling environment adapted to the increased production of science and technology and its applications to bring about growth and development through:

- An interest in developing the capacities of institutes and research units in universities and scientific centres to provide the funding and training opportunities within and outside Sudan and supplement laboratory equipment, libraries and documentation centres and information.
- Care for the environment and the provision of scientific supplies necessary for researchers and technicians, professionals and motivate them to stability in Sudan and to increase their production and technological research and to encourage and honour of or excelled in them.
- Involvement of scientists and modern technology (Altqanya) priorities in the formulation and application of science and technology on scientific grounds.
- Working to upgrade the basic science curriculum in higher education with interest in teaching mathematics and modern science such as genetic engineering and computer science.
- Work on doubling the number of researchers, engineers and technicians with stimulate outstanding students to join the institutes of education and training and in the training of the intermediate.

To find the appropriate formulas for linking scientific research areas of consulting, production and so on:

- Insurance needs of the citizens in the basic food, clothing, medicine and energy through productive projects attractive to investors so as to ensure self-sufficiency.
- Self-production complexes typical for rural development and balanced regional cities and to alleviate the bottlenecks. These complexes will provide opportunities for the transfer of research results to the fields and factories.
- Vertical development is done through modernisation of management and means of production and services in various sectors and the use of computer data and takes advantage of genetic engineering and the introduction of sprinkler and drip irrigation and other inputs from modern technology.

- Double centres and computer networks and arrived in the commercial and political centres in Sudan, computer information networks that are created the National Council of the Computer.
- Establishment of laboratories and high technology and advanced to the development and consolidation of technological rules to advancing development, such as microelectronics laboratories and laboratories of biotechnology and genetic engineering laboratories and information and solar power, atomic energy and science of the desert and remote sensing and medical science.
- And/or attract scientists and professional staff, technical and technological requirements and provide better service and equipment specific to the progressive work and a better working environment.
- Qualification and training and improving production in order to focus on human development.

It is clear that universities set up distributed to various states of Sudan and the following must be considered:

(1) Confirm the identity of the nation and establish it through the curriculum which are approved by the university and applied.
(2) Conduct scientific and applied research related to the different needs of society and renewable energy in order to service and upgrade.
(3) Interest in desert ecology, medicine, land and industry in the context of interest in the development of Sudan in general.
(4) Attention to issues of human development, thought and religious values.
(5) Concern for the environment of Sudan in general and the state concerned, especially the environment and the rehabilitation staff is able to upgrade and resolve jurisdictional issues relating to the environment.
(6) Technical innovation and employment to serve the Sudanese community, in collaboration with universities and institutions of Higher Education and Scientific Research, and the other country.
(7) Interaction with the citizen understanding of rural problems and recognition of the knowledge and experience, and work with them to develop according to his needs and values.
(8) To prepare students and give them the vacation of science.

From here it is clear that scientific research theory and practice-oriented and is the mainstay of development plans in the country, the first stage of each development project, or a step forward in the world today based mainly on new technologies. However, the concept of scientific research and unfortunately may like a large mixing in Sudan since the prevailing concept of scientific research is the work leading to the new discoveries and inventions of the things that were not known before. And this became the prevailing sense of politicians, planners and executives, as well as ordinary citizens, resulting in the full dimension of scientific research on the movement planning in the country, and do all the development projects both big and small, without that preceded or followed by scientific research. As in the side of development projects across all the previous era, we have witnessed the collapse of a large number of agricultural projects, especially those established by the private-sector, for simple reasons for the disqualification of soil or poverty, or lack of validity of the climate, or after the area of marketing, or certain types of insects in the region concerned and others. In the industry it is even worse and more bitter, and the whole Sudanese society is still bemoans the textile project Gadow, and a project for manufacturing fertilisers, and the project of producing yeast and others. Even the existing plants, it works because the specifications of low cards do not match temperature in Sudan (Arab - British Trade, 1998a).

THE GOALS OF SCIENTIFIC RESEARCH IN HIGHER EDUCATION INSTITUTIONS

Derived from the Sudanese old universities reputation and fame and reputation locally, regionally and globally from academic and research excellence. Prestigious universities have sought to regulate scientific research to devote our specialised departments dealing with the affairs of scientific research under the direct supervision of the Departments of University. It is the most important functions of the Department of Scientific Research is the supervision, coordination, and funding for outstanding scientific research. To hide that the cross-fertilisation between scientific research centres, which is the main tributaries of research activity in universities has become an absolute necessity in the pursuit of excellence and integrated work for scientific research, basic and applied universities, as both are complementary to one another because of the organic unity that exists between Higher Education and Scientific Research, education includes research activities in all its aspects and specialisations, includes research activities in all its aspects and specialisations, and comes through the provision of scientific research results in basic and applied. But is not limited scientific research on science and technology, but knocking all areas of social sciences and human and economic, etc. Since the homes are prestigious universities has particular expertise harnessed their potential and efficiency of research faculty members to conduct applied research of the character, which contribute to achieving development goals, by linking research and the urgent need for the development of society and help solve problems that impede development (Educational Statistics, 1991).

AREAS OF SCIENTIFIC RESEARCH INSTITUTIONS OF HIGHER EDUCATION

The link between higher education institutions are preoccupied with the country necessitated the commitment engagement in all areas of research available, which greatly helped in securing the rise reflected an integrated development and prosperity and the prosperity and development of comprehensive and the most important areas of research:

(1) Medical research and health.
(2) Research and industrial engineering.
(3) Research livestock and wildlife.
(4) Agricultural researches.

(5) Research forestry and natural resources.
(6) Research geological, earth sciences and mining.
(7) Water research.
(8) Energy research and renewable energy.
(9) Economic research and development.
(10) Research for strategic studies.
(11) Research sovereign, political and judicial.
(12) Research, educational, social and intellectual.
(13) Security research.

Tasks of scientific research centres specialised in:

(1) There are centres of radiation of modern technologies such as remote sensing, biotechnology and space science, renewable energies and others.
(2) There are centres of training for staff in all institutions in order to settle these technologies in the country.
(3) To be advisers to the state of the scientific studies necessary preceded any development project in order to avoid confusion and loss, and planning random.
(4) To ensure that resettlement technologies in the global agriculture, industry and animal husbandry, and others.
(5) That the necessary studies in order to achieve international standards in all national products and national wealth in order to be competitive globally.
(6) That works to detect the exploitation of natural resources existing in the country, such as plants medical and natural fertilisers and pesticides, plant, solar and others.
(7) To spread the spirit of innovation, invention, disclosure, encourages and seeks to achieve scientific leadership.
(8) To establish research projects of integrated human development and fighting poverty.

PROBLEMS OF EVALUATING SCIENTIFIC RESEARCH

Scientific research in Sudan is concentrated in three or four ministries which are the Ministry of Higher Education and Scientific Research and led by the National Centre for Research and Universities, Ministry of Agriculture and Natural Resources Development and the Agricultural Research Corporation, and the Ministry of Livestock and the Body Livestock Research, Ministry of Industry and the Centre for Research and Industrial Consultancy. There are small units in some other ministries such as the geological research, and the National Department of Energy, and General Management of Forests (UNESCO, 1990).

- An escalating costs of research materials necessary laboratory and field and logistical.
- Migration researchers outside the country and sometimes internal, non-research sites to get an income guarantee.
- Lack of funding for general research, a very small proportion of national income.
- Imbalance ratios between scientific research and technical assistant in all disciplines.

- Social and economic conditions that adversely affect the researcher and the lack of appropriate financial incentives and lack of access to opportunities to participate in conferences and meetings of global and regional.
- Quality of research and its incompatibility with the application directly.
- Coordination between producers of research and its users.
- Failure to provide supplies to apply the results of scientific research in terms of management and human resources, and financing frameworks.
- Busy bodies in the productivity boost production, and incentives and lucrative bonuses.
- Weak contribution of research organisations in solving the actual bottlenecks in production lines, including punishment of confidence between the parties.
- Non-priority-setting researches properly respond to urgent cases and with tangible economic returns.
- Poor management of scientific research.

DIFFICULTIES THAT HINDER THE COURSE OF THE DEVELOPMENT OF RESEARCH AND DEVELOPMENT SYSTEM

The following difficulties are discussed:

Coordination in the Development of Research and Development Programmes

Lack of coordination in the development of research and development policies and development between the national plan and research programmes in universities, research centres and lack of familiarity with the problems and needs of the industrial sector and agriculture and due to the lack of authority, which sponsors scientific research and then entrusted with the accounting research directions.

Coordination between Universities and Research Centres

Not to encourage researchers in universities on the use of laboratories, research centres and the results of previous research affiliate, no facilities and special benefits for the exchange of information and experience between them but no interaction is limited to each and everyone in his own world.

Coordination between the Productive Sector and the Service, and Scientific Research

Solve the problems of the factories away from the centres of scientific research, whether in universities or research institutions and often the solution will cut the import of new or foreign assistance. The results of the research not concern the productive sector to lack of confidence and accommodating the extent of its success and realistic cost.

INTERNAL DIFFICULTIES IN RESEARCH INSTITUTIONS

There are several internal constraints impeding research institutions to interact with the productive and service sectors, such as not qualified for the latest Altqanya in sufficient numbers to carry out maintenance and operations research under the supervision of senior staff. These intermediate cadres need to be a comprehensive training and there is a lack of familiarity with the needs of these sectors and how to market services to them. There is no sense of the role of media failure and the absence of a mechanism to activate the association and coordination and integration between research institutions and sectors of the recipient.

FINANCING OF RESEARCH AND DEVELOPMENT

The agreement in the research and development is one of the direct indicators to assess the scientific status of any country; there is a high correlation between investment in research and development and economic growth. Developed countries maintain the progress and prosperity through the availability of credits in the field of research and development, while developing countries in contrast to the belief that the interest in investment in science and technology does not come benefits urgent. In the time that developed countries spend high percentages (between 2 to 3 %) of their gross national product in research and development, we find that the majority of developing countries does not exceed more than 2%-3% and 0.2%-0.5% of their gross national product in research and development (Scientific Statistics Department, 1998).

HUMAN RESOURCE

The human element is a powerful tool in bringing about economic and social development through the use of other factors of production, and the human at the same time the target development. Dealing with human development as the development of population characteristics, abilities and organically linked with the overall development of a society where rights and the means or purpose. The population of Sudan, about thirty million people (Statistics 1996) and estimated population growth rate of about 8.2% per year and notes that the increase of population in urban areas of 5.7% per year, while increasing the rural population rate of 5.1% per annum and is due mainly to migration from the countryside to cities. It is also noted that a high proportion of the population under the age of 15 years (it follows that the existence of the proportion of approved high) and the population density is about 10 inhabitants per square kilometre and up in agricultural areas populated to 390 and there is variation in the distribution of population between different states (ALECSO, 1987).

The number of economically active population is about 48% of the total population and contribute to economic for males is about 60% and women is about 4.6%, which is very low and worthy of review and processing and the contribution of women in rural areas is higher than in urban areas and areas rich in the top of the poor. By the urbanisation process and continuous improvement of health services, the youngest age bracket increases for the composition of the population. The estimated labour force (15-64 years) at about 54% of the total economically active population who are estimated to be about 5.8 million of whom 1.2 million urban and 4.6 million rural. One of the problems that must be addressed, increase-mounting losses in the various stages of education, has the unemployment rate in 1993 about 17% while it was 5.10% in 1983. And to promote population characteristics and development of human resources and development skills of the workforce to raise the efficiency of production to achieve the development goals desired, while maintaining the stability, entitled to the problems of population and workforce awareness and mobilisation and organisation, guidance and information to improve the characteristics of population while maintaining the highest rates of population growth commensurate with the geographical expanse and objectives strategic addition to the employment potential of the population and higher rates of growth while preserving natural resources and ensure the continuation of its bid for ways to raise awareness and promote the means of production and legislation (World Development Report, 1990). The workforce may be seeking to address issues of illiteracy and reduction of waste education and technical training and vocational skills development and linked to appropriate technologies and the local environment and the needs of the labour market. Vary the contribution of each of these systems in the GDP and in the importance of economic and living conditions depend Sudanese economy heavily on irrigated agriculture and rain-fed agriculture mechanism and the two together contribute about 80% of food production (maize) and estimated its contribution of about 50% of the value of agricultural exports has recently contributions to the growing livestock and gum Arabic in the Outbox (United Nations, 1986). Successive governments have continued to exercise parental role in this sector in terms of providing the requirements of foreign exchange for inputs and operational requirements and the obstacles that have emerged have accompanied the influential sector in its overall performance (Ministry of Labour, 1996).

STRATEGIC OBJECTIVES BASED ON THREE AXES

Three procedures required accessing the completeness of the goals of the observed and the strategy includes three main components:

1. Formulation of objectives.
2. Planning work programmes to reach the objectives.
3. Ensure that the capabilities required.

Management strategy must be seen as an ongoing process and it does relate to the goals and plans, and monitor the risks and restraints, and support aids, and the decline of priorities, and believe in possibilities, and the sequence of implementation, and the results. Even a strategy of scientific research does must believe:

1. Maturity of the strategy and the plans.
2. Safety priorities.
3. Integrated management process (technically, administratively and financially).
4. The stability of scientific research institutions structurally and functionally.

Types of research can be summed up as follows:

1. Research strategy.
2. Applied research.
3. Technical research (field technology) research.
4. Research methods (methodology matters).

STAGES OF DEVELOPMENT OF THE CAPACITY OF RESEARCH INSTITUTIONS

Studies agree that there are three key stages to reach this goal and every stage of which needs to periodic review in order to achieve its goal. And these stages can be summarised as follows:

1. At the consensus of opinion.
2. Building capacities.
3. Reinforcement of structural and functional (consolidation).

SCIENTIFIC RESEARCHES AND THE CHALLENGE OF CIVILISATION

Believes much of the so-called third world (including Sudan) that the transfer of technology in its final form represents an ideal solution to the problems of poverty and underdevelopment are experiencing, and an indispensable tool in narrowing the differences of civilisation between them and the developed countries, the owner of this technology and then adopt to monitor huge amounts of their income low in order to achieve this goal at a time are neglected these countries. Scientific research, both basic and applied largely ignored and clearly foreshadows this meagre allocations upon the absence of a plan and conscious of actual research priorities, methods, and operates intellectual capacity through both scientific researches mentioned it affects each in other in a constant cycle of movement and interaction, leading to community-building and face the problems of scientific solutions which take into account the specificity of the community and the uniqueness of its components. One manifestation of lack of attention to scientific research as well as the weakness of the preparation of national human resources trained and capable of understanding the techniques of the times and working with them for its development or at least bear in mind in order to achieve optimal use of them, reduces mistakes, and vanished with its negative effects on social and cultural life (Working Paper of the Comprehensive National Strategy, 1991; ILO 1986; Abdeen, 1995a; Abdeen, 1997; Abdeen, 1998; National Research Council, 1970a; and 1981b; Ahmed, 1984; Yacoub, 1970; Eisa, 1997).

Technicians, technology, information revolution, the Internet are all echoes of the scientific research and development has reached the world for hundreds of years in research and experience. And for the world technology, we find that pre-empt access to knowledge that ensure comfort and superiority to others, especially to maintain national security. We find that some developed countries have given the efforts of the scientific and intellectual interest in each of their children and ridiculed their all kinds of support. And others learned the importance of the human mind to migrate minds and marshalled for the service of science and technology even possessed the world. It has become clear that the intellectual capital, which came by the information revolution, is a source of wealth, a new capital.

Scientific research is the examination and investigation of the fact that the orderly and systematic follow scientific methods. Since we are in the era of globalisation and privatisation has received the importance of scientific research in the definition of globalisation as the entry because of the evolution of the information revolution and technology in the process of the development of civilisation. Development is the activation of natural resources and the human-cantered human mind in its search for how to use and activated. And development stage does not come only by the accumulation of quantitative and qualitative research and ongoing efforts. Relationship is still scientific research institutional development the subject of controversy, this despite that the relationship between them cannot be separated. Developmental planning is putting the priorities and needs of the scientific research programme while feeding the proper research institutions planning information and the correct data, which would work to increase production and reduce costs and make the upcoming changes. It is sad to know that the development of research is an investment in itself, where studies showed that the average rate of return of scientific research than 1000% of expenditure on it. Although this ratio is differ according to different states in Sudan. But this is just the opposite (Saeid, 1998; Abdullah; 1997a; Selah; 1998; Hassan, 1996; Eisa, 1998; Abdeen, 2001; Abdeen, 1995b; Mohammed, 1994; Abdeen, 1999; Arab Ministerial Conference on Environment and Development, environmental problems in agriculture and long-term use of natural resources in the Arab world, 1991; World Resource Institute, 1994; Arab - British Trade, 1998b; Arab - British Trade, 1998c).

CONCEPTUAL FRAMEWORK

First, the Cultural Fabric of Social Considerations

Begin to say that the experiences of forty years of development efforts that shows growing modern industry is not subject to the elements of traditional production, which also says the comprehensive national strategy for only a tenuous link does not carry an abundance of raw materials and energy resources or human density flows or foreign aid. Growing modern industry depends on cultural fabric of social, a product of more than a political, social, cultural spiritual and the fruits of social mobilisation. Court, unity and national vision inclusive and effective, is the result of the victory on the crises of national and confusion, and waste and conflict, which is the result of finding programmes that cruel to national priorities and resource allocation and the provisions of the settings and promotion of manpower. Even with modest assumptions about the availability of land, comprehensive fuel-wood farming programmes offer significant energy, economic and environmental benefits.

Second, Macro-economic Policies

The growing modern industry requires the provision of appropriate macro-economic environment, which manufactured and rooted macro-economic policies effects directly and indirectly. That macro-economic policy is appropriate to require Shell industrial development programmes and obstruction of public and industrial performance, and industrial investment.

Third, National Constitutional Changes

It is necessary to fit the industrial development policies and regulations of the Department of Industrial Development with the political and constitutional changes and major social changes that are organised home. Hijacking on the meanings and values of autonomy (the terms of reference state and the competence of local government) is exceeded on the principles of democratic governance and the principles of decentralisation and basic services. Needs to be a better means of economic management and capacity-building performance in the climate market economies, and then fit with the dramatic changes in the global economic climate.

Fourth, Changes in Economic Life and the Global Wave of Globalisation

This is characterised by a climate of international economic life depletion constraints of official international aid organisations and to link international inferior pressures and constraints. And the national industry will face new challenges in the post-General Agreement on Tariffs and Trade (GATT), not only the challenges of competition in foreign markets but in the challenges against competitors in the market within the national itself. There is no way an expression of that cruel to reach high levels of economic efficiency and quality, support and development. The advantages are of natural systems, the relative economic efficiency and high quality industrial dynamic, and that the keys to the future are a science and technology, information technology and knowledge-based industry. Otherwise the industry will not get from agricultural products and endemic but rates of subsistence and go value added and employment opportunities and employment and gain skills to employers industries. The industry is oil refining and electricity production and transfer of mineral ores and agricultural products to high-value multi-purpose. At this stage of our economic development, it is necessary to link the agricultural development to industrial development and a tight linkage to the full coordination between the arbitrator agricultural programmes and industrial programmes. The world is at the stage of knowledge-based economy and information technology in the pattern and type of industrial products and a stunning development in the patterns of industrial goods, the traditional and the transition to industrial products, new goods and products to the intellectual capacity (Brain Power Products) this economy is based on the miracle of conversion brainpower to material products (Abdullah, 1998b).

Economy of the future requires a workforce with high skills and superior capabilities of mind and fullness of the spirit of innovation and initiative. Economy of the future requires a new culture and new modes of behaviour and affiliation unwavering to the values of harmony and unity and team spirit and requires a firm relationship between membership, production and the forces of science and knowledge and powers of the soul and technology. We have to deal with the changes of contemporary systems, the division of labour (Division of Labour) in the international economy and the resettlement of large groups of industries to train industrial industries like truck, automotive, industrial equipment, steel mills, fertiliser and cement and traditional textiles and leather tanning, etc.

Fifth, Schools (Mzhabiyat) Industrial Development

We should beware of liquid industrial development strategy in any Mzhabiyat or schools of thought or any rules M_khash theory (Cook Book Rule) and be careful of the transfer and application of the theoretical basis and ignore the real situation in the country and the actual conditions. And the development of the necessary tools to address the actual situations of national industry and solve problems based or expected in the national industry.

Sixth, Congenital Natural Economic

On the natural moral economic technical homeland outline of a strategy of industrial development in the country, the strategy for the development of real resources of the country and to protect and enrich the education advantages enjoyed by the country's crops and wealth dynamic industrial technology and the highest levels of economic value added and economic efficiency and innovation. And a summary of the matter is that the foundation stone of the industrial strategy is the treatment of industrial processes or manufacturing of materials and primary resources of the country (such as availability of cotton, oil seeds, leather, corn, vegetables and fruit) that requires the full coordination and harmonisation of agricultural and industrial policies. Should adopt a strategy of industrial development to address the real ills and build on the diagnosis and deep analysis, realistic and critical ills following:

(1) Small industrial processes (shallow), and low added value.
(2) Greater reliance on wills inputs.
(3) Imbalance and the weakness of the front and rear linkages with other sectors of the economy.
(4) Imbalance and the weakness of the interdependence of the industrial sector still rolling.
(5) Lack of productive diversification, and low use of industrial by-products.
(6) The inability to integrate in the global market.

INTERNATIONAL AGREEMENTS

(1) United Nations Convention to combat desertification, particularly in Africa, signed in 1995 one hundred and five countries and thus become legally binding. Whale Convention forty articles and four annexes and the aim are to combat desertification and mitigate the effects of drought.

(2) World Federation for the Conservation of Nature and the headquarters of the Union in Gland, Switzerland and the Union's regional and continental offices and national site in 35 in various parts of the world, the number of member states 133 and the number of government institutions and NGOs. Must be accompanied by training local management based on the use of this technology, and this must be a programmer to choose the local people sperm delivery of the project they must participate in the process of technology selection, design, construction and delivery and to participate in a simple run under the supervision of experts.

(3) The need to comply with base balance between the sources of energy, environment and promote cooperation in the field of renewable energy and focus on the applications of biomass energy for rural development and the expansion of interest in them. And dissemination and promotion of solar technologies for the introduction of new technologies is not harmful to the environment.

(4) The rationalisation of consumption of firewood and coal, and thus to preserve the wealth of forest and environmental impacts, and optimum utilisation of agricultural residues and animal heat and raise the value.

(5) Awareness of the importance of native landscaping, and encourage farming planted forests and shelter belts and the introduction of fast-growing tree species and improve care of trees and forest management and the fight against destructive factors unjust random, and the rationalisation of consumption and improved household stoves next to improve manufacturing techniques.

(6) The need for organisational units and lead the process of preserving the environment and achieve the objectives of development of natural resources in the cooperation and coordination.

(7) There is a need for fundamental change in energy systems to bring them into line with sustainable development.

And the need to change dictated by social and economic issues, environmental and security situation in the account with the following:

a) To promote universal access to modern energy.

b) Building local capacity.

c) Establishment and maintenance of fair rules of the game (by removing the permanent subsidies and make energy prices reflect the external costs (such as social and environmental costs)).

d) To single out the roles of stakeholders (environmentalists, consumers, current and potential, etc.) not belonging to the private-sector [28-39].

e) The entry of the regular formation of the new generation of technologies that are used to cleaner fossil fuels, and renewable sources and efficiency improvements.

STRATEGIC OBJECTIVES FOR THE ENVIRONMENT IN SUDAN

(1) Preservation of the environment and development and prevent disasters.
(2) Stop environmental degradation and reconstruction.
(3) To maintain the balance and stability of environmental components (systems ecology).
(4) Development agencies working in the environmental field.
(5) To develop relations with other countries and international institutions and organisations in the environmental field.
(6) Development of the balance of wildlife and exploitation in accordance with sustainable development.

POLICY

(1) To follow a sound approach in the rational exploitation of natural resources.
(2) Achieve sustainable development in pace with global efforts to protect the environment and natural resources.
(3) Status of a comprehensive plan for scientific research in the fields of the environment.
(4) Issuance of environmental legislation for each collector and the fundamental principles of public policy for the protection of the environment.
(5) Reconstruction of the southern Sudan environment affected by the war and reconstruction of areas affected by drought and environmental degradation in coordination with the relevant authorities.
(6) Sudan to fulfill commitments to international conventions and organisations in the area.
(7) Attention to cadres and the creation of specialised training at home and abroad.
(8) To encourage the voluntary associations and organisations in the environmental field.
(9) Allocation of new natural reserves and the promotion of new projects zoos states.
(10)Include the environment in the curriculum.
(11)Rational exploitation of resources, the environment and promotion of environmental awareness to all levels of coordination.
(12)To provide potential accommodation and complementary services of transport and communication.

HIGHER EDUCATION IN AGRICULTURE

(1) An expansion of a higher agricultural education system to include all food production from farm to market.
(2) Preparation of higher agricultural education programmes on the basis of teaching, research and extension.
(3) Re-structure of the curriculum so as to provide knowledge, skills and information technology including the use of computer and sustainable resources.

(4) Creating and strengthening links between universities (Colleges of Agriculture and Veterinary, etc.), and national research centres.
(5) Provision of training opportunities to raise the capabilities and efficiency of the faculty members through seminars, meetings and programmes.
(6) Support cooperation and distribution of information among educational institutions and other devices.
(7) Attention to sustainability of institutions of higher education because they suffer from a continued decline in funding and the erosion and degradation of the environment research and the inability to keep the faculty members in their positions.
(8) Consideration of the establishment of centres of excellence for postgraduate studies in some Arab countries.

The main objectives of agriculture as set out in comprehensive national strategy on:

(1) Food security.
(2) Sustainable agricultural development.
(3) Of the increase and diversity in crop and animal production.
(4) Raising the efficiency of resource use.
(5) Increasing productivity by using modern technologies to focus on small farmers and investment by the private-sector and the interest in the role of women in development.
(6) Integrated rural development and balanced.

The proportion of researchers working in the field of agricultural engineering:

• Agricultural Research 9.26%
• University of Khartoum 6.34%
• University of Gezira 5.15%
• Sudan University 5.11%
• Sesame Centres 5.11%

Organisational structures have remained constant with the change of project objectives and farming systems and policy changes are being reviewed in the structures and I think that there is a need for closer monitoring of the implementation of the new agricultural policy which requires re-formulation of structures.

(1) High cost of funding and limitations on the financing requirements of production and post-harvest operations, including marketing.
(2) Rising tax burden on the irrigated sector, particularly on cotton with the ease of what happened remains an urgent need for further easing the tax burden on agriculture.
(3) The escalation of production costs due to inflation.
(4) The escalation of the cost of administrative expenses.

Many developing countries have little incentive to use wind energy technologies, to reduce their emissions despite the fact that the most rapid growth in CO_2 emissions is in the developing world.

CONCLUSION AND RECOMMENDATIONS

The attention to scientific research has become the duty of the state-public and private-sector and it is difficult to enter the twenty-first century, the century of globalisation and planetary who became illiteracy is computer illiterate without proficiency in scientific research and technology development resources and technology. And developing countries face many problems in the transfer of technology or in the tradition of industry after the WTO and to ensure intellectual property rights related to Trade Related Issue of Intellectual Property Right (TRIPR) and the court of tendencies and not but to rely on themselves or the establishment of centres for joint research with the developing to be produced from an original effort, the thought of their children and technology adapted to the level of development. Of the richest on the discretion of the state of the importance of research and serious political commitment towards the creation of a climate and create a mechanism to develop, update and adopt a policy, e.g., a sound and priorities are on a scientific basis and adequate funding. The research is the way to provide scientific information systematically to develop programmes. The activities effects on sectors where it is applied research results and therefore on society and the state. After it is vital that we come to be vital for the development of technological and scientific research can be its recommendation as follows:

1. Development of strategic planning for the scientific research according to the desired available and can be leading, and setting priorities. Favourites including development issues and serve the community, and detailed analysis of the research carried out and the periodic review.
2. Coordination between the centres of scientific research in the education sector and government and private agencies in the country.
3. Research unit and the private-sector through the Sudanese Employers Union and coordination with the relevant authorities.
4. Coordination and integration, communication, and cooperation between scientific research bodies, universities, consulting, production locally, regionally and globally.
5. Financing of scientific research (laboratories, and field logistic), input of scientific research, scientific instruments, certain references, magazines, specialised libraries, scientific publishing, and attending conferences.
6. To identify a percentage of national income for scientific research and the involvement of the private-sector and the federation of employers in financing scientific research, and the imposition of fees on all productive sectors for the benefit of scientific research.
7. Support the translation and localisation to keep abreast of developments in scientific research.
8. Benefit from the results of scientific research in the development and promotion of production and services, industry and trade, according to development plans.

9. Focus on scientific research on rural development and regional balanced and sustainable.

10. Commitment of government and private agencies to provide information and data to researchers and establish a database in all science disciplines and double centres and computer networks.

11. Public awareness of the importance of scientific research in the dissemination of knowledge and the evolution of reality through the media of audio-visual and print media.

12. Improve the situation and living conditions of researchers and stop the migration of research centres, universities and others.

REFERENCES

Abdeen MO (1995a). *Rainfall patterns in Sudan,* NETWAS, Vol. 2, No. 7, Nairobi, Kenya.

Abdeen MO (1995b). *Water resources in Sudan,* NETWAS, Vol.2, No.8, Nairobi, Kenya, December, 1995.

Abdeen MO (1997). Compilation and evaluation of solar and wind energy resources in Sudan, *Renewable Energy Journal,* Vol. 12, No. 1, pp. 39 - 69, Reading, UK, September, 1997.

Abdeen MO (1998). Horizons of using wind energy and establishing wind stations in Sudan, *Dirasat Journal,* Vol. 25, No. 3, pp. 545 - 552, Jordan.

Abdeen MO (1999). *Energy and development and the environment in Sudan, Journal of Jordanian engineer,* No. 66, Year 34, pp. 78-80, Amman, Jordan, February 1999.

Abdeen MO (2001). *Water development in Sudan,* NETWAS, Vol.7, No.7, Nairobi, Kenya, September 2001.

Abdullah AA (1997a). *Agriculture and the challenges of globalisation and the revolution in Information Technology.* Ministry of Finance and Economics, December 1997.

Abdullah AA (1998b). *Agricultural engineering research, The Arab Organisation for Agricultural Development.* Khartoum, March 1998.

Ahmed AR (1984). *Carriages the Quran key scientific research,* Khartoum, in September 1984.

Ahmed J (1998). *The absence of the curriculum in scientific research,* Centre for Political and Strategic Studies. Al-Ahram, Cairo: Egypt.

ALECSO (1987). *The Arab situation in science and technology and its environment,* Report submitted to the Arab science strategy Committee.

Arab - British Trade. (1998a). Vol. 5, No. 10, June 1998.

Arab - British Trade. (1998b). Vol. 5, No. 11, August 1998.

Arab - British Trade. (1998c). Vol. 5, No. 12, December 1998.

Arab Ministerial Conference on Environment and Development, environmental problems in agriculture and long-term use of natural resources in the Arab world, Cairo, 1991.

Educational Statistics (1991). Ministry of Education, Sudan.

Eisa TI (1997). Energy and technology to achieve sustainable rural development. *The National Symposium to create the Sudanese economy to keep pace with globalisation and the revolution in Information Technology,* Khartoum, 3 to 4 December 1997.

Eisa TI (1998). *Sudan's energy and climate change, the Supreme Council for Environment and Natural Resources*. Khartoum, June 1998.

Esmat AR (1997a). Scientific research and the role of employment in the Arab world, *The Accompanying Scientific Conference for the Thirtieth Session of the Boards of the Union of Arab Universities,* Sana'a: Yemen.

Esmat AR (1997b). The problem facing the process of scientific research in Arab universities and means to overcome them, *A Symposium of Scientific Research in Arab Universities,* The Goals and Problems, the Union of Arab Scientific Research Councils. Baghdad: Iraq.

Faisal TO (1993). *Science and technology,* National Research Centre. Khartoum: Sudan.

Hashim MH (1996). The role of scientific research in the development and economic planning, *The Second Economic Conference,* Khartoum: Sudan.

Hassan OA (1996). The forests of Sudan can meet the growing demand for energy. *The Second Economic Conference,* Khartoum, October 1996.

Karim MS (1997). *Scientific research in institutions of Higher Education and Scientific Research, the development of the university environment,* Khartoum: Sudan.

ILO (1986). Employment and Economic Reform: Towards a strategy for the Sudan.

Ministry of Labour (1996*).Yearly Statistical Book.* Sudan.

Mohammed HH (1994). Projections of demand for the use of firewood in brick ambushes, *a special study prepared for the Food and Agriculture Organisation of the draft forestry development project within the limited consumption of forest products in Sudan.* Khartoum.

Muhammad AR (1998). The role of scientific research in new universities in achieving comprehensive national strategy, *A Symposium Role in the Development of Scientific Research.* Khartoum: Sudan.

National Research Council, (1970a). *The journal issued by the public relations office of the National Research Council, the first issue, the first year,* Khartoum, Sudan, October 1970.

National Research Council (1981b). *Scientific research facilities in the Sudan,* Khartoum, 1981 - 1990.

Saeid AB, (1998). *Development of water resources development and irrigation uses of irrigated agriculture in Sudan,* the Arab Organisation for Agricultural Development. Khartoum, March 1998.

Selah MN (1998). *The environmental impact of fossil fuel products.* Khartoum, March 1998.

Scientific Statistics Department (1998). Ministry of Higher Education and Scientific Research, Sudan.

UNESCO (1990). *Statistical Year Book.*

United Nations (1986). Reports on Statistics.

Working Paper of the Comprehensive National Strategy, Conference for the Decade of the 1991 - 2000, Khartoum, Sudan, 1991.

World Development Report (1990). The World Bank.

World Resource Institute, *World resources: A guide to the global environment, people and the environment.* 1994.

Yacoub QA (1970). Scientific research and development of Sudan. *Printing House University of Khartoum,* July 12, 1970.

Chapter 9

NON-CONVENTIONAL ENERGY SYSTEMS AND ENVIRONMENTAL POLLUTION CONTROL

ABSTRACT

The massive increases in fuel prices over the last years have however, made any scheme not requiring fuel appear to be more attractive and to be worth reinvestigation. In considering the atmosphere and the oceans as energy sources the four main contenders are wind power, wave power, tidal and power from ocean thermal gradients. The renewable energy resources are particularly suited for the provision of rural power supplies and a major advantage is that equipment such as flat plate solar driers, wind machines, etc., can be constructed using local resources and without the advantage results from the feasibility of local maintenance and the general encouragement such local manufacture gives to the buildup of small-scale rural based industry. The key factors to reducing and controlling CO_2, which is the major contributor to global warming, are the use of alternative approaches to energy generation and the exploration of how these alternatives are used today and may be used in the future as green energy sources. Even with modest assumptions about the availability of land, comprehensive fuel-wood farming programmes offer significant energy, economic and environmental benefits. These benefits would be dispersed in rural areas where they are greatly needed and can serve as linkages for further rural economic development. Self-renewing resources such as wind, sun, plants and heat from the earth can provide clean abundant energy through the development of renewable technologies. Virtually all regions of the world have renewable resources of one type or another. Research and development investments in the past 25 years in renewable technologies development has lead to important advances in performance and resulting cost effectiveness. Renewable resources currently account for about 9%-10% of the energy consumed in the world; most of this is from hydropower and traditional biomass sources. Wind, solar, biomass and geothermal technologies are cost effective today in an increasing number of markets and are making important steps to broader commercialisation. The present situation is best characterised as one of very rapid growth for wind and solar technologies and of significant promise for biomass and geothermal technologies. Each of the renewable energy technologies is in a different stage of research, development and commercialisation and all have differences in current and future expected costs, current industrial base, resource availability and potential impact on energy supply. This article discusses the potential for such integrated systems in the stationary and portable power market in response to the critical need for a cleaner energy technology. Anticipated patterns of future energy use and consequent environmental impacts (acid precipitation, ozone depletion and the greenhouse effect or global warming) are comprehensively discussed in this paper. Throughout the theme several issues relating to renewable energies, environment and sustainable development are examined from both current and future perspectives.

Keywords: energy, renewable energy technologies, energy efficiency, environment, sustainable development, global warming, emissions.

1. INTRODUCTION

Energy has been a vital input into the economic and social development. However, one third of the world population, living in developing and threshold countries, has no access to electricity. These people mostly live in remote and rural areas with low population density, lacking even the basic infrastructure. Accordingly, utility grid extension is not a cost-effective option and sometimes technically not feasible. Therefore, it is imperative to look for sustainable (i.e., cost-effective, environmentally benign and reliable) sources of energy for the development of these regions. Using locally available renewable energy sources (especially solar irradiation that is characterised by a sufficient availability on a daily basis), which are of high potential in most of these regions offers a strategic solution for their techno-economic development. From the point of view of technology, the design of system technology that meets electrification requirements and fulfils, if necessary, the requirements of integration into alternative current (AC) supply grids, has to be considered.

The modernisation of the system components and their power ranges which allow easy expandability of the supply structure, the standardisation of interfaces and the hybridisation by integration of different energy converters in order to increase the power availability, represent the most important measures from the point of view of system technology. Moreover, the use of renewable energy sources is essentially made easier if the existing reliable AC- technical standards of construction and extension of conventional electricity supply systems are adopted. Therefore, incompatibility cannot be taken as a reason to reduce the dissemination of renewable systems.

In the early years of photovoltaic (PV) history stand-alone as well as grid-connected PV-systems had been built as individual items or unique masterpieces. Until recent years the realisation of PV-systems still was characterised by monolithic system concepts resulting in a costly design and engineering process. Consequently a large number of different PV system components (e.g., inverter) were developed each tailored for the use in the dedicated application with its specific parameters (e.g., input voltage/current).

The exploitation of the energetic potential (solar and wind) for the production of electricity proves to be an adequate solution in isolated regions where the extension of the grid network would be a financial constraint. The use of wind as alternative energy source is increasing and research and development about this clean and unlimited resource is being carried out on various levels.

Likewise, energy savings from the avoidance of air conditioning can be very substantial. Whilst day-lighting strategies need to be integrated with artificial lighting systems in order to become beneficial in terms of energy use, reductions in overall energy consumption levels by employment of a sustained programme of energy consumption strategies and measures would have considerable benefits within the buildings sector. The perception often given however is that rigorous energy conservation as an end in itself imposes a style on building design resulting in a restricted aesthetic solution. It would perhaps be better to support a climate sensitive design approach, which encompassed some elements of the pure conservation strategy together with strategies, which work with the local ambient conditions making use of energy technology systems, such as solar energy, where feasible. In practice, low energy environments are achieved through a combination of measures that include:

- The application of environmental regulations and policy.
- The application of environmental science and best practice.
- Mathematical modelling and simulation.
- Environmental design and engineering.
- Construction and commissioning.
- Management and modifications of environments in use.

In summary, achieving low energy building requires comprehensive strategy that covers; not only building designs, but also considers the environment around them in an integral manner. Major elements for implementing such a strategy are as follows:

1.1. Efficiency Use of Energy

- Climate responsiveness of buildings.
- Good urban planning and architectural design.
- Good housekeeping and design practices.
- Passive design and natural ventilation.
- Use landscape as a means of thermal control.
- Energy efficiency lighting.
- Energy efficiency air conditioning.
- Energy efficiency household and office appliances.
- Heat pumps and energy recovery equipment.
- Combined cooling systems.
- Fuel cells development.

1.2. Utilise Renewable Energy

- Photovoltaics.
- Wind energy.
- Small hydros.
- Waste-to-energy.
- Landfill gas.
- Biomass energy and biofuels.

1.3. Reduce Transport Energy

- Reduce the need to travel.
- Reduce the level of car reliance.
- Promote walking and cycling.
- Use efficient public mass transport.
- Alternative sources of energy and fuels.

1.4. Increase Awareness

- Promote awareness and education.
- Encourage good practices and environmentally sound technologies.
- Overcome institutional and economic barriers.
- Stimulate energy efficiency and renewable energy markets.

2. ENERGY FROM WASTE

Measures to maximise the use of high-efficiency generation plants and on-site renewable energy resources are important for raising the overall level of energy efficiency. The world's view of waste has changed dramatically in recent years and it is now seen as a source to feed the ever-growing demand for energy (Figure 1). The road from the initial concept to the production of the first kilowatt of power is long and has many challenges, not least the need for adequate funding. Scientific research evidence, public awareness and increased levels of participation in environmental campaigning have led to governments' worldwide implementing regulations and legislation. Examples include:

- EU landfill diversion directive.
- Recycling targets.
- Climate change regulations.

The demand for nuclear power generation, wind farms, solar power and so on is now unstoppable and has created a whole new market, though each has its own challenges (Figure 2). The waste collection, transfer and landfill disposal business comprise a mature, slow-growth industry. Economic drivers to developing the waste and renewable energy sector have included:

- Waste disposal and landfill gate fees/landfill tax.
- Penalties/avoidance schemes (e.g., landfill allowance schemes and fines, carbon trading).
- Energy prices.
- Investments subsidies.

When considering the demand and opportunity in today's marketplace these points are prevalent:

- The demand for renewable energy is not going to go away.
- The public feeling is that governments across the world are responsible.
- The pressure caused by diminishing fossil fuel supplies is increasing.
- Investment funds are increasingly available from traditional sources.
- The needs for new technologies that can deliver carbon reduction and waste reduction outcomes are increasingly bankable, which opens up the market for all.

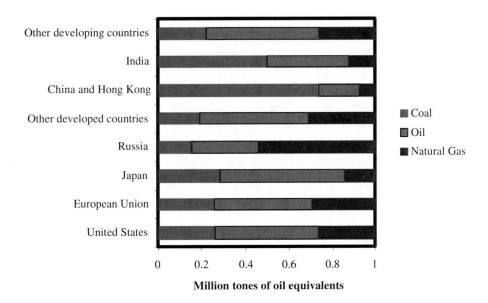

Figure 1. Global fossil fuel consumption.

Financial institutions across most global markets are gearing themselves up for the environmental revolution. Within the waste to renewable energy sector, history has shown a hesitancy to invest in projects not supported by four things:

- Adequate independent technology due diligence.
- Security of waste input and power off-take contracts.
- A site with planning permission.
- A reference plant, preferably at scale.

Reviewing the evolution of municipal sewage waste (MSW) management in general, waste collection has tended to progress from incomplete collection through to complete collection and finally to collection with separation into different waste streams. In turn, waste treatment has progressed from ad-hoc decentralised disposal to a strategy more dependent on controlled treatment and disposal, including the use of sanitary landfilling accompanied by waste reduction strategies. In developed countries, this evolution has taken place over a period of about 30-40 years. The standard of liner design and construction standard are summarised in Table 1.

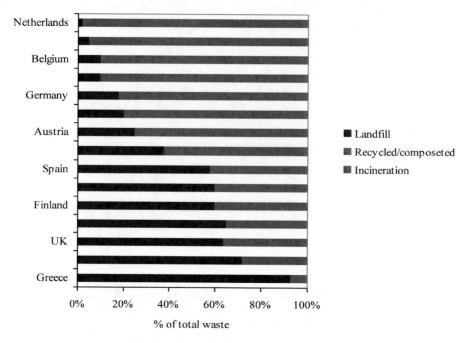

Figure 2. Municipal waste management in the European Union.

Table 1. Comparison of basic requirements for bottom liners in MSW sanitary landfills

Liner system requirement	Leachate drainage layer	Geomembrane line	Compacted clay layer
US EPA Standard (40CFR258)	$K > 1 \times 10^{-4}$ ms^{-3} Thickness 0.3 m	Thickness ≥ 0.75 mm Recommended 1.5 mm HDPE	$K \leq 1 \times 10^{-6}$ ms^{-3} Thickness ≈ 60 cm
EU Landfill Directive (1999/31/Dec)	Thickness 0.5 m	Not specified. Yet liner thickness should be 100 cm $K \leq 1 \times 10^{-6}$ ms^{-3}	With HDPE liner, thickness of clay layer > 50 cm
German Standard (TASI 1993)	$K > 1 \times 10^{-3}$ ms^{-3} Thickness 0.3 m	Thickness ≥ 2.5 mm HDPE	$K \leq 5 \times 10^{-10}$ ms^{-3} Thickness 3x25 cm
Chinese Standard (CJJ 113-2007)	$K > 1 \times 10^{-3}$ ms^{-3} Thickness >0.3 m	Thickness ≥ 1.5 mm HDPE	$K \leq 1 \times 10^{-6}$ ms^{-3} Thickness 75 cm

One of the negative results of growing prosperity worldwide has been an increase in waste generation from year to year. In response, policy-makers and researchers are examining how best to decouple waste growth and economic growth. In both developed and developing countries sanitary landfill sites can be operated in such a way that danger to residents and the environment, from Leachate, odours, fire and explosion is almost entirely eliminated. Table 2 summarised different parameters in waste compaction. Waste professionals use cross wrap machinery for its reliability and efficiency in storage and transport of waste materials.

2.1. Waste Shredding

With the demand for faster and more efficient recycling technologies showing no signs of abating, the market for faster, more efficient shredding equipment is of course on the up. To the man on the street the term shredding most likely brings to mind the transformation of business documents, bank and credit card statements into a bird's nest of paper- a practice now relies on worldwide to prevent fraudsters accessing the personal financial data and sensitive information. It means big business, as shredding of waste is common practice across almost all areas of the waste industry. Far from focusing simply on paper, shredding is a disposal technique for everything from agriculture to household waste and electrical to industrial waste.

The overall trend in today's market tends to be 'shred first and sort later'. Shredding of waste material as a precursor to sorting is useful for two reasons. It reduces the size of the waste, allowing for greater ease of transportation, but perhaps more importantly – at a time when recycling as much materials as effectively as possible is paramount- it allows for more effective sorting afterwards. And logically, effective sorting equals greater opportunity for recycling (Figure 3).

Table 2. The different parameters in waste compaction

Refuse	Item size Organic components Inert substances Slurry
Application technique	Thin layer operation Face operation Pushing distance
Compaction machine	Compactor Operating weight Wheel design
Types of waste disposal site	Pit type waste disposal site Raised refused disposal site Height of the refuse disposal site Refuse load
Weather conditions	Precipitation Temperature

While every customer on the lookout for a shredder is interested in efficiency, one thing that will also attract a potential buyer is energy efficiency. With the large environmental challenges facing the world today, waste industry professionals are increasingly aware of the need to make sure their business are as kind to the environment as possible. The operation of large-scale equipment such as shredding machines naturally uses a large amount of power and machines which can run effectively on a lesser amount have an advantage over their competitors.

Even with modest assumptions about the availability of land, comprehensive fuel-wood farming programmes offer significant energy, economic and environmental benefits. These benefits would be dispersed in rural areas where they are greatly needed and can serve as linkages for further rural economic development. The nations, as a whole would benefit from savings in foreign exchange, improved energy security, and socio-economic improvements. With a nine-fold increase in forest – plantation cover, the nation's resource base would be greatly improved. The international community would benefit from pollution reduction, climate mitigation, and the increased trading opportunities that arise from new income sources. The aim of any modern biomass energy systems must be:

- To maximise yields with minimum inputs.
- Utilisation and selection of adequate plant materials and processes.
- Optimum use of land, water, and fertiliser.
- Create an adequate infrastructure and strong R & D base.

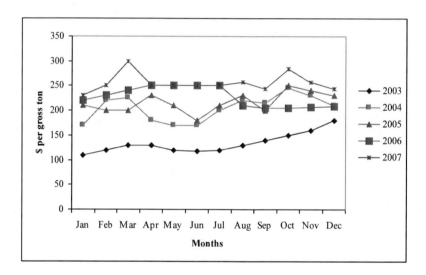

Figure 3. Monthly averages paper scraps.

2.2. Biomass CHP

Combined heat and power (CHP) installations are quite common in greenhouses, which grow high-energy, input crops (e.g., salad vegetables, pot plants, etc.). Scientific assumptions for a short-term energy strategy suggest, that the most economically efficient way to replace the thermal plants is to modernise existing power plants to increase their energy efficiency and to improve their environmental performance. However, utilisation of wind power and the conversion of gas-fired CHP plants to biomass would significantly reduce the dependence on imported fossil fuels. Although a lack of generating capacity is forecasted in the long-term, utilisation of the existing renewable energy potential and the huge possibilities for increasing energy efficiency are sufficient to meet future energy demands in the short-term.

A total shift towards a sustainable energy system is a complex and long process, but is one that can be achieved within a period of about 20 years. Implementation will require initial investment, long-term national strategies and action plans. However, the changes will have a number of benefits including: a more stable energy supply than at present and an improvement in the environmental performance of the energy sector, and certain social benefits. A vision used a methodology and calculations based on computer modelling that utilised:

- Data from existing governmental programmes.
- Potential renewable energy sources and energy efficiency improvements.
- Assumptions for future economy growth.
- Information from studies and surveys on the recent situation in the energy sector.

In addition to realising the economic potential identified by the National Energy Savings Programme, a long-term effort leading to a 3% reduction in specific electricity demand per year after 2020 is proposed. This will require: further improvements in building codes, and continued information on energy efficiency. The environmental Non Governmental Organisations (NGOs) are urging the government to adopt sustainable development of the energy sector by:

- Diversifying of primary energy sources to increase the contribution of renewable and local energy resources in the total energy balance.
- Implementing measures for energy efficiency increase at the demand side and in the energy generation sector.

Methane is a primary constituent of landfill gas (LFG) and a potent greenhouse gas (GHG) when released into the atmosphere. Globally, landfills are the third largest anthropogenic emission source, accounting for about 13% of methane emissions or over 818 million tones of carbon dioxide equivalent (MMTCO$_2$e) (Brain, and Mark, 2007) as shown in Figure 4.

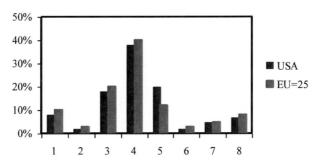

1 Food, 2 Textile, 3 Pulp & paper, 4 Chemicals, 5 Refining, 6 Minerals, 7 Primary metals, and 8 Others.

Figure 4. Distribution of industrial CHP capacity in the EU and USA.

The price of natural gas is set by a number of market and regulatory factors that include:
Supply and demand balance and market fundamentals, weather, pipeline availability and deliverability, storage inventory, new supply sources, prices of other energy alternatives and regulatory issues and uncertainty.

Classic management approaches to risk are well documented and used in many industries. This includes the following four broad approaches to risk:

- Avoidance includes not performing an activity that could carry risk. Avoidance may seem the answer to all risks, but avoiding risks also means losing out on potential gain.
- Mitigation/reduction involves methods that reduce the severity of potential loss.
- Retention/acceptance involves accepting the loss when it occurs. Risk retention is a viable strategy for small risks. All risks that are not avoided or transferred are retained by default.
- Transfer means causing another party to accept the risk, typically by contract.

3. WIND ENERGY

There are numerous factors that influence the overall prospects for the wind industry, though in the end it is the economics that will be the deciding factor (Table 3). The most important issues identified:

- Assessment of previous patterns of market development in similar markets.
- Increased engagement of utilities and large energy companies.
- National energy plans and government support for renewable energy.
- Technical development.
- Growth in market and the present dynamics of the industry.
- Information about specific large projects.
- Assessment of wind resources and how they can be used.

Table 3. Market shares 2005-2007

Years		2005		2006		2007	
Manufacturer	Country	Supplied	Share	Supplied	Share	Supplied	Share
Vestas	Denmark	3186	27.6%	4239	28.2%	4503	22.8%
Ge Wind	US	2025	17.5%	2326	15.5%	3283	16.6%
Gamesa	Spain	1474	12.8%	2346	15.6%	3047	15.4%
Enercon	Germany	1640	14.2%	2316	15.4%	2769	14.0%
Suzton	India	700	6.1%	1157	7.7%	2082	10.5%
Siemens	Denmark	629	5.4%	1103	7.3%	1397	7.1%
Acciona	Spain	224	1.9%	426	2.8%	873	4.4%
Goldwind	China	132	1.1%	416	2.8%	830	4.2%
Nordex	Germany	298	2.6%	505	3.4%	676	3.4%
Sinovel	China	3	0.0%	75	0.5%	671	3.4%
Others		1032	8.9%	1094	7.3%	2076	10.5%
Total		11343	98%	16003	107%	22207	112%

Most of these factors are favourable for the industry at the moment. There is strong political support for wind energy, both as engineering and supply chain problems that have been associated with rapid growth in the past.

Economic projections are difficult at the best of times, when economies are relatively stable and a reference 'business as usual' case can be used. However, there are numerous signals that the world faces very turbulent economic conditions for a while - a credit crunch may make some project finance difficult and the shortage of raw materials could lead to supply chain difficulties. However, the rapidly escalating price of oil is focusing a lot of attention on the price of energy and the hedge of electricity supply without a fuel cost is likely to become increasingly attractive to many companies and utilities. At some stage, rising fuel costs could lead to demand for wind energy becoming almost infinite. The main factors expected to influence the continuing growth of the wind sector are:

- The economies of the transition states (Russia and Central Asia) will start to grow.
- Increasing energy demand in Asia and South America.
- Oil prices will continue to remain high as will demand for fossil fuels.
- Continuing competitiveness of wind with fossil fuels.
- Many countries may find they are well off their international CO_2 reduction commitments and need to install some new renewable capacity very quickly.
- Security of supply questions will continue to support wind power.
- Deregulated markets will remove excess conventional power capacity and new capacity is likely to be more expensive than wind.

While wind energy can still seem a small industry compared with conventional power generation, the achievement of 1% of world electricity generation is potentially significant. In individual markets such as Denmark, Germany and Spain reaching 1% has been a breakthrough figure, establishing a critical mass and being followed by further rapid growth in each year market. If the same pattern is seen with world wind energy demand and the industry continues to establish itself as a significant player in the energy sector and pushes on rapidly to 30% of world electricity demand and beyond, then the glass should be seen as half full.

Wind energy is one of the low investments high yielding sources of power generation. The future of wind energy is extremely bright and there is no doubt that in the renewable energy sector, wind power would play a predominant role in adding to the national grids clean and non-polluting energy in the coming years (Table 4).

In recent years, demand for the micro wind turbines, of the output below 1 kW, is on the increase as monuments and educational materials. Most of the micro wind turbine that has a diameter less than 1.0 m is low blade tip speed ratio type on the market, by the problem of the frequency, the safety and the blade noise. In these circumstances, it would be necessary to develop the system characteristics of micro wind turbines for the purpose of much higher performance in spite of the low Reynolds number regions. Wind power generation is characterised by its stochastic nature, whereby supply and demand, in small grid systems in particular, mostly do not match. The combination of wind power with a second complementary power generation and/or direct/indirect storage technology therefore has, in principle, considerable potential. Wind-diesel, wind-water desalination and wind power in combination with hydrogen production are all potential options that have been high on the international renewable energy agenda for several years. A small-scale wind-PV hybrid power generator system for dairy farm is shown in Figure 5, to verify the possibilities to apply a power generating system and heating source for dairy farm. It is possible to apply the system for power supply and heat source to melt snow and process fertiliser.

Table 4. Installed capacity per year

Year	Europe (MW)	World (MW)
Before 2000	9.413	13.954
2000	13.306	18.449
2001	17.812	24.927
2002	23.832	32.037
2003	29.301	40.301
2004	34.725	47.912
2005	40.897	59.320
2006	48.628	74.517
2007	57.136	94.593
2008	66.785	120.458
2009	78.514	151.753
2010	93.590	191.318

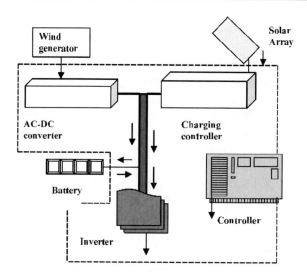

Figure 5. Wind-Photovoltaic hybrid generation systems.

4. SOLAR ENERGY

Global investment in renewables and energy efficiency now outpaces that for nuclear energy. Renewables also accounted for more than a fifth of new generation capacity built in 2007. The renewable energy and energy efficiency sectors seeing a level of commercial investment that most thought unattainable just a few years back. The risk factors pointing towards a 'bust' as identified and summed up in Figure 6.

Studies have been begun to estimate both the economic effects that climate change will have on global society as well as the costs of possible climate change mitigation and adaptation measures. Although the capacity to enact either a mitigation or adaptation strategy is based on country-specific conditions, technology and information availability, models have been used to calculate the approximate cost to stabilise atmospheric emissions at different levels.

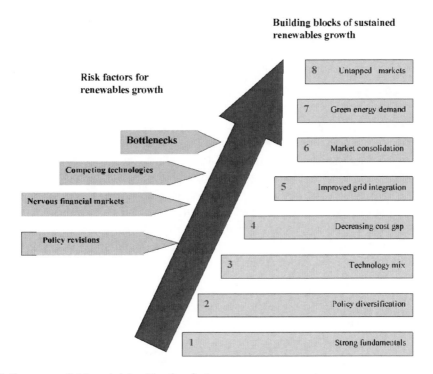

Figure 6. Summary of risk and risk mitigation factors.

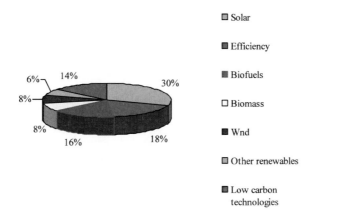

Figure 7. Investment in renewable and energy efficiency.

Wind power is far from the only clean energy sector on the rise and many of the technologies following in its tracks are much more decentralised, including roof-top systems like photovoltaics (PVs) or solar thermal and energy efficiency technologies on the demand side. Solar and energy efficiency were actually the two largest sectors in terms of venture capital investment, with solar bringing in 30% and efficiency 18% (Figure 7). Besides the high level of early stage investment, mostly focusted on new technology development, these two sectors also fared well on the public stock markets, ranking second and third after wind. Solar would have overtaken wind on the public markets.

The potential of electric power generation from incorporating PVs in buildings is enormous. If the electrical power demand of many countries is to be supplemented by the use of PV, it is deemed necessary to integrate such systems into many of the building faces. Many larger structures such as superstores, public buildings and most houses use mass produced tiles on their roofs. Such areas lend themselves useful in contributing to the energy used in the building or to export to the electrical grid when active roof tiles (PV-tiles) are introduced as a part of the roof structure. The integration of PVs within both domestic and commercial roof offers the largest potential market for PV especially in the developed world. Numerous national programmes are attempting to stimulate this market using standard PV modules as a roof element. However, roofs are not static, uniform structures and so are not ideally suited to modules which require precise, planar mounting structures. To date, attempts at producing a PV roof tile which accommodates current roof practice, in terms of both installation and aesthetics. The market for PV has historically been based on off-grid application where the relatively high cost of PV could be economically justified. In 2001 about 330 MWp of PV were produced and installed around the world and the growth rate of the industry is over 30% per year-100% per year in some countries with aggressive implementation schemes (Figure 8).

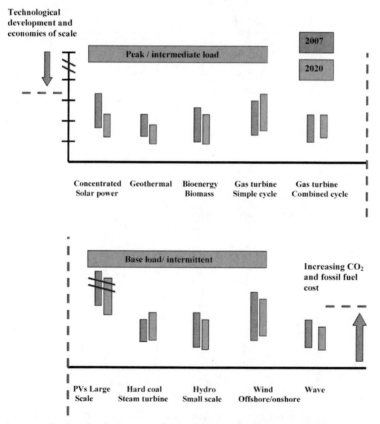

Figure 8. Cost increase for fossil primary energies and CO_2 emissions and continued technological development and economies of scale in renewables will improve competitiveness.

There have now been a number of successful large-scale programmes of systems deployed for basic power needs in rural households in developing and less developed countries. Remote applications servicing other applications such as telecommunications, cathodic protection, water pumping, etc., continue to grow as well. The PV market will continue to grow strongly for the next several years at least, driven by incentive programmes, cost reductions and greater market awareness. It is becoming evident, that the quality will be the key to the PV market (Figure 9). The European Commission's Altener Programme a Training Manual was developed by the Global Approval Programme for PVs (PV GAP) to help manufactures of PV products to introduce quality management in their production. The manual contain important up-dates of the PV GAP Manuals for PV manufactures, published by the World Bank in 1999, including alignment with the 2000 edition of ISO 9001. This revised training manual was also translated from English into French, German and Spanish.

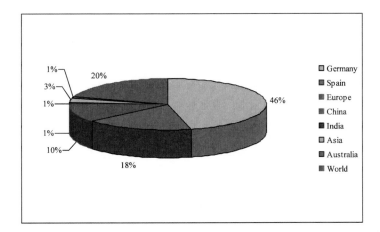

Figure 9. Differences in predicted PV market volumes worldwide until 2010.

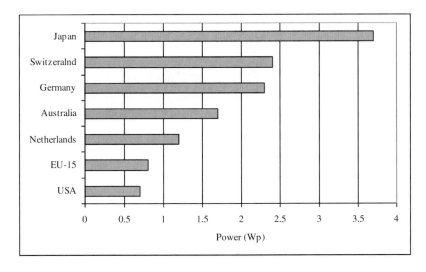

Figure 10. PV distributions for different countries per person.

Water pumping is one of PV modalities that are growing in rural areas, mainly in developing countries. Due to the importance for health and food production it may be regarded as one of the noblest solar PV uses in isolated areas (Figures 10-11). Reliability and autonomy or self-reliance is understood as main factors for this high growth rates.

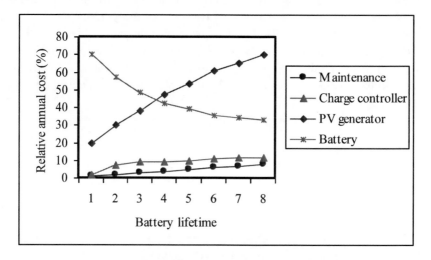

Figure 11. Relative annual costs of the components of small PV stand alone system with a lead-acid battery.

5. HYDROPOWER POTENTIAL

This section discusses various aspects of hydropower including: harnessing ocean energy, hydroelectric dams and micro hydropower systems. This section on hydropower explores the factors associated with utilising the actual potential of hydropower energy. The section covers all the technological details, along with issues and challenges faced during the utilisation of hydropower energy. Major projects, power plants, players in the industry, the major role of the United States in the global hydropower industry, and the various environmental benefits of using hydropower energy are all explored in depth in this section.

The growing worldwide demand for renewable energy projects is being driven by ever increasing global energy consumption and the availability of carbon and renewable energy credits. Renewable energy is entering a new phase with additional funding becoming available from governments, from socially responsible equity funds, and from public capital raisings.

Hydropower is the capture of the energy derived from moving water for some useful purpose. Prior to the widespread availability of commercial electric power, hydropower was used for irrigation, milling of grain, textile manufacture, and the operation of sawmills.

Hydropower produces essentially no carbon dioxide or other harmful emissions. In contrast to burning fossil fuels, this energy is not a significant contributor to global warming through production of CO_2.

Hydroelectric power can be far less expensive than the electricity generated from fossil fuel or nuclear energy. Areas with abundant hydroelectric power attract industry.

Environmental concerns about the effects of reservoirs may prohibit development of economic hydropower sources in some areas.

Hydropower currently accounts for approximately 20% of the world's electricity production, with about 650,000 MW installed and approximately 135,000 MW under construction or in the final planning stages. Notwithstanding this effort, there are large untapped resources on all continents, particularly in areas of the world that are likely to experience the greatest growth in power demand over the next century. It is estimated that only about a quarter of the economically exploitable water resources has been developed to date, leaving the potential for hydro to continue to play a large role in sustaining renewable global electricity production in the future.

Apart from a few countries with abundance, hydro power is normally applied to peak load demand because it can be readily stopped and started. Nevertheless, hydroelectric power is probably not a major option for the future of energy production in the developed nations, however, because most major sites within these nations are either already being exploited or are unavailable for other reasons, such as environmental considerations.

Future hydropower energy programmes must be put into practice in conjunction with sound policies that restrict the use of fossil fuels and natural resources and contribute to the reduction of emissions into the environment. Such a strategy should be based in a sound scientific basis, without ideology, politics or financial interests. It should be implemented on a worldwide basis and not limited to industrialised countries. To achieve this goal, existing hydropower energy options must be evaluated for implementation, new strategies must be formulated and new, innovative solutions have to be found.

All projects are required to have environmental impacts assessment conducted, covering all potential damage to the environment, mitigation and restoration, a reclamation plan including a resettlement programme for displaced residents, and the estimated implementation costs. All hydro projects are required to conduct an environmental impact study. Water is essential to industry for processes such as cooling, cleaning, diluting and sanitation. With increasingly stringent water abstraction limits, recent droughts and a growing interest in the environmental performance of businesses, there is a need for industry to reduce water use.

6. GASIFICATION

An important consideration for operators of wastewater treatment plants (WWTPs) is how to handle the disposal of the residual sludge in a reliable, sustainable, legal and economical way. This by-product of wastewater treatment contains abundant organic material, including many kinds of bacteria. It also contains heavy metals and its composition is generally unknown. The benefits of drying sludge can be seen in two main treatment options:

- Use of the dewatered sludge as a fertiliser or in fertiliser blends.
- Incineration with energy recovery.

Use as a fertiliser option takes advantage of the high organic content 40-70% of the dewatered sludge and its high levels of phosphorous and other nutrients. However, there are a number of concerns about this route including:

- The chemical composition of the sludge (e.g., heavy metals, hormones and other pharmaceutical residues).
- Pathogen risk (e.g., salmonella, Escherichia coli, prionic proteins, etc.).
- Potential accumulation of heavy metals and other chemicals in the soil.

Sludge can be applied as a fertiliser in three forms:

- Liquid sludge.
- Wet cake blended into compost.
- Dried granules.

Use as energy recovery option takes advantage of the energy available in the sludge's organic content. Drying the sludge reduces its water content, thus increasing its calorific value and making it easier to combust. It also reduces odours and improves handling, with lower transport and storage costs. Sludge from WWTPs is typically combusted in (Table 5):

- Cement kilns.
- Coal-fired power plants.
- Mono-incinerators, i.e., plants burning refuse-derived fuel or a single waste stream.
- Mixed waste incinerators, e.g., municipal waste incinerator.

Table 5. Energy recovery from WWTP sludge

Form of sludge and usage	Cement kiln	Coal-fired power plant	Mono-incinerator	Refuse incinerator
Dewatered sludge	No	Yes	Yes/No*	Yes
Fuel substitute	-	No	No	No
Thermal dry biosolids	Yes	Yes	Yes	Yes
Substitute for fuel	Yes	Yes	Yes/No	No
Substitute for minerals	Yes	No	No	No
Type of process	Residue-free process	Fuel substitution	Disposal	Disposal

*Depending on type of plant.

The recognised advantages of energy recovery from sludge include:

- The high calorific value (similar to lignite) of dewatered sludge.
- The use of dewatered sludge as a carbon dioxide (CO_2) neutral substitute for primary fuels such as oil, gas and coal.
- The use of dewatered sludge is a 'sink' for pollutants such as heavy metals, toxic organic compounds and pharmaceutical residues, thus offering a potential disposal route for these substances provided the combustion plant has adequate flue gas cleaning.

- The potential, under certain circumstances, to utilise the inorganic residue from sludge incineration (incinerator ash), such as in cement or gravel.

The demands placed on the drying system are therefore critical and include:

- High process stability.
- High mechanical reliability.
- High safety standards under all operating conditions.
- Compliance with environmental legislation such as emission limits.
- A product with properties suitable for a wide range of uses.

The energy efficiency formula takes into account the energy generated by the plant and puts it in relation to the calorific value of the municipal waste (Federal Ministry for the Environment, 2005). The energy introduced into the process from outside (such as fossil fuels or electricity) is subtracted. The energy efficiency can be improved by, for instance, reducing the input of fossil fuels. According to the European commission's formula, the energy efficiency for waste-to-energy (WTE) plants is calculated as follows:

$$\text{Energy efficiency} = E_p - (E_f + E_i) / 0.97 \ (E_W + E_f) \tag{1}$$

where:

E_p is the annual energy produced as heat or electricity in GJ/year. It is calculated with energy in the form of electricity multiplied by 2.6 and heat produced for commercial use multiplied by 1.1.

E_f is the annual energy input to the system in GJ/year from fuels contributing to the production of steam.

E_W is the annual energy in GJ/year contained in the treated waste calculated using the lowest net calorific value of the waste.

E_i is the annual energy imported in GJ/year, excluding E_W and E_f. For thermodynamic reasons, E_f must be deleted in the nominator of the equation as it is included twice- in the nominator and the denominator. An energy efficiency factor of 0.6, which has been proposed by commission, is too high for most existing plants. A threshold of 0.5 would be sufficient, with a further reduction of 0.1 for small plants and plants that produce electricity only due to a lack of demand for heat. A factor of 0.6 would disadvantage smaller plants as they generally need the same energy for operation as larger plants, but have a lower throughput. It should be noted, at this point, that the public tend to prefer smaller plants in order to reduce the distance waste is transported (Figures 12-13).

7. ENERGY RECOVERY

Mechanical-biological treatment (MBT) can be one option for improving the conservation of resources and energy in waste management systems. Mechanical-biological treatment enhances the conservation of embedded energy through recycling and allows potentially more efficient combustion or conversion of refuse-derived fuel (RDF). The MBT encompasses a wide range of technologies aiming to process solid waste by a mixture of mechanical and biological separation. It also enables metals and other dry recyclables to be recovered (Figure 14). There are five main types of the MBT process:

- Incorporating anaerobic digestion to generate biogas for electricity production. Anaerobic digestion also generates a digestate to be discharged or to be dewatered, producing a compost product.
- Producing an RDF product.
- Producing a compost product and/or a stabilised material for landfilling as well as a RDF product.
- Producing a compost product.
- Stabilising waste prior to landfill.

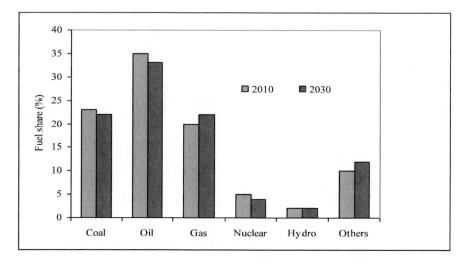

Figure 12. Outlook for world total primary energy supply. ('Others' includes combustible renewables and waste, geothermal, wind and tidal energy).

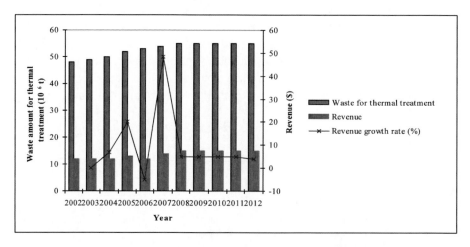

Figure 13. Volume shipments and revenue forecasts for the western European thermal waste treatment services.

Gas production in the gasification system is controlled in two ways:

- The system analyses the gas produced and proprietary software feeds instructions back to a control system, delivering constant management of operating variables.
- A surge and mixing tank blends the gas flow.

Social and environmental will benefit to the community from the utilisation of alternative energy and reduced fossil fuel consumption.

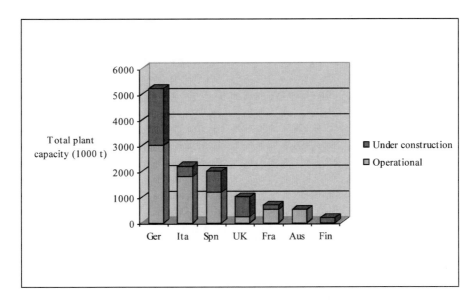

Figure 14. Capacity of plants in each country.

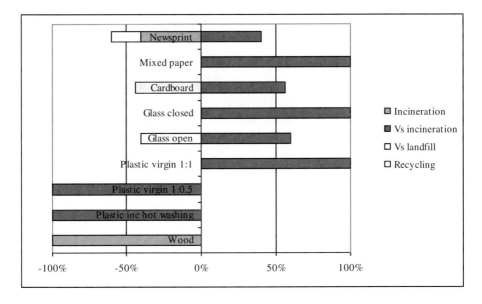

Figure 15. Performance of the life cycle analysis (LCA) on avoided greenhouse gas emissions.

Recycling is hugely beneficial, both from an economic and environmental perspective. The environment and economy will both benefit from an improved and more efficient reverse supply chain. This, in turn, would result in increased and optimised recycling of waste streams that reduce the depletion of scarce natural resources- water, oil and minerals- while at the same time protecting the environment by keeping these valuable materials in service. Figure 15 shows how the individual LCAs performed on avoided greenhouse gas emissions in CO_2 equivalent by material (WRAP, 2006). A cornerstone of recycling is effective waste collection and many local authorities are now looking to households and businesses to ensure this is in place. It is no longer sufficient to leave one or more bags of 'rubbish' at the end of a drive or back of a workplace.

Odorous emissions are a serious concern relating to biowaste treatment facilities and affect the likelihood of planning permission being granted particularly in urban areas. Research has shown odour is one of the prime concerns from urban residents where MBT plants are proposed. An often unrecognised aspect of the waste management field is the many examples of individuals giving back. A conventional biofilter operates by engineering the correct environment in a closed vessel, ensuring that the bacteria within it are cultivated to effectively biodegrade the compound in the air that passes through it. Recirculation of the airstreams improves the rate of degradation by the bacteria. This has several advantageous characteristics:

- A surface s structure that is excellent for supporting bacteria and has the ability to retain moisture in the event of temporary water failure.
- A self-supporting structure with good packing characteristics that ensures there is minimal pressure drop across the media bed (thus reducing running costs).
- The right conditions to encourage bacteria to metabolise more than grow, which minimises sludge production and hence associated disposal.

8. GEOTHERMAL HEAT

This study explores the factors associated with utilising the actual potential of geothermal energy. It also covers all the technological details, along with issues and challenges faced during the utilisation of geothermal energy. Major projects, power plants, players in the industry, the major role of the United States in the global geothermal industry, the active role of the US Department of Energy, and the various environmental benefits of using geothermal energy are all explored in-depth in this study.

Geothermal power is the use of geothermal heat to generate electricity. Geothermal comes from the Greek words geo, meaning earth, and therme, meaning heat. The utilisation of geothermal energy for the production of electricity dates back to the early part of the twentieth century. For 50 years the generation of electricity from geothermal energy was confined to Italy and interest in this technology was slow to spread elsewhere. In 1943 the use of geothermal hot water was pioneered in Iceland. Estimates of exploitable worldwide geothermal energy resources vary considerably.

The largest dry steam field in the world is the geysers, about 90 miles (145 km) north of San Francisco. The geysers began in 1960 which has 1360 MW of installed capacity and produces about 1000 MW net. Calpine Corporation now owns 19 of the 21 plants in the geysers and is currently the United States' largest producer of renewable geothermal energy. The other two plants are owned jointly by the Northern California Power Agency and Santa Clara Electric. Since the activities of one geothermal plant affects those nearby, the consolidation plant ownership at the geysers has been beneficial because the plants operate cooperatively instead of in their own short-term interest. The geysers are now recharged by injecting treated sewage effluent from the City of Santa Rosa and the Lake County sewage treatment plant. This sewage effluent used to be dumped into rivers and streams and is now piped to the geothermal field where it replenishes the steam produced for power generation.

Another major geothermal area is located in south central California, on the southeast side of the Salton Sea, near the cities of Niland and Calipatria, California. As of 2001, there were 15 geothermal plants producing electricity in the area. Cal Energy owns about half of them and the rest are owned by various companies. Combined the plants have a capacity of about 570 megawatts. Geothermal energy can be used as an efficient heat source in small end-use applications such as greenhouses, but the consumers have to be located close to the source of heat. Geothermal energy has a major environmental benefit because it offsets air pollution that would have been produced if fossil fuels were the energy source. Geothermal energy has a very minor impact on the soil - the few acres used look like a small light-industry building complex. Since the slightly cooler water is reinjected into the ground, there is only a minor impact, except if there is a natural geyser field closed by. The world's decentralised energy industry is right to look towards developing countries as potential markets for its wares. These countries have the opportunity to miss out the industrial development stage characterised by a single, centralised electricity transmission and distribution system, in favour of a hybrid of local grids liberally supplied by small-scale decentralised power generators.

9. GREENHOUSE GAS EMISSIONS

The typical challenge in the developing world is lack of everything, including electricity. When striving to feed the increasing power demand to enable social reform and industrial growth, local decision makers face the question of which route to take. Shall we copy the model of the rich countries, or could there be another, maybe better way? The need to reduce CO_2 emissions presents a new, additional challenge, difficult even for the richest of nations.

Conventional, centralised electricity networks are the norm in the developed world. However, the present energy infrastructure of the developed countries was mainly created during the monopolistic utility era of the past. The utilities had the power to decide what kind of capacity to construct and how to construct the grid. There was practically no competition. The main challenge for many utilities was to get construction permits for building new generation capacity- the construction itself was practically risk free as they could turn their cost structure into a solid power tariff. Capital was easily available and cost competitiveness and overall system cost optimisation were not the main concerns.

Today the market situation and rules have permanently changed. Progressive modern utilities have left the past behind and are striving towards a modern, competitive energy system. However, one remainder of the past, still to some extent maintaining the economies of scale thinking, is to look at and calculate power plant and grid investments separately, as if they had nothing, or very little, to do with each other. Installing flexible, high-efficiency peaking and grid stability generation capacity closer to the load pockets of cities and industrial areas would be a natural way to go, but it does not fit the model of old, large-scale utility thinking (Figure 16).

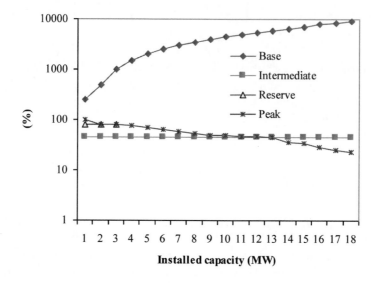

Figure 16. Load duration curve showing the base, intermediate, peaking load segments, and necessary contingency reserve.

A typical electricity system built by a rich country monopoly utility 'wasted' capital in the following ways:

- Large, steam-fired power plants, running on part load, were used for frequency and load control (unless hydro non-dispatchable wind condition, but the problem has been solved with the very strong grid, which can transmit the wind power over long distances to remote countries.
- Efficient peaking capacity was not constructed; instead, the so-called peaking plants were typically based on largest possible simple cycle industrial gas turbines located in critical points in grid. This capacity functional mainly as an emergency reserve at grid nodes and was hardly ever used as it has such poor heat rate and relatively long starting time.
- Excessive base-load capacity was constructed. This was possible as there was an ensured return on asset investments.
- The grid has been sized to transmit the full peak power from large remote power plants to the consumption centres in cities.

Utilities in the developing world face a major challenge in developing their electricity systems. The parameters to optimise at the same time are:

- Reliability of supply.
- System flexibility and preparedness for load growth.
- Economical competitiveness, i.e., cost effectiveness, elimination of 'wastes'.
- Access to power for the whole nation.
- Ensured access to fuels and fuel flexibility.

Combined heat and power (CHP) has been installed and used for many years as a highly efficient energy supply system to fulfil electricity and thermal energy needs in a range of applications around the world. As environmental issues, particularly those of climate change caused by greenhouse gases (GHGs), become as an effective means to reduce emissions is more pronounced than ever. However, despite CHP's significant contribution to the reduction of CO_2 emissions, calculation of its effects can result in preventing further dissemination of CHP. This is because, when an appropriate calculation method is used, there is a risk of under-estimating the effects of CHP and even, in some instances, estimating erroneously an increase in CO_2 emissions as a result of installing a CHP system. Proper estimation of the benefits of CHP will be vital in encouraging more businesses to install CHP systems.

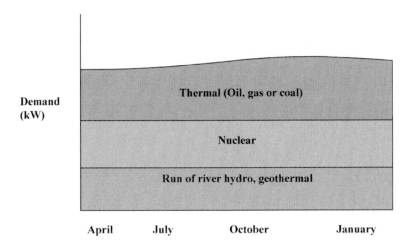

Figure 17. Image of annual load profile of power plants.

Considerable debate surrounds the calculation of CO_2 emissions as a result of the use of the two emission factors (Kyoto Protocol, 1997; and GHG Protocol, 2002):

- All fuels average emission factor (AEF) - used to calculate emissions from power consumption.
- Marginal emission factor (MEF) - used to estimate reduced amount of CO_2 emissions as a result of lower power consumption.

CO_2 emission by consumption of a fuel is calculated as follows:

Emission amount (kg CO_2) = emission factor of a fuel (kg CO_2/MJ, kWh, etc.) x energy consumed (MJ, kWh, etc.)   (2)

In the case of electricity, however, CO_2 is not emitted at the point of use. The power on the grid is usually generated by a mix of fuels – some CO_2 – emitting plants such as oil, coal and natural gas-based thermal power plants and non- CO_2 – emitting power sources such as nuclear, hydro and renewables. As a result, when calculating CO_2 emission by the use of grid power, the use of AEF or MEF is preferred. Therefore, CO_2 emission by an electricity user is estimated as follows:

CO_2 emission volume (kg CO_2) = AEF (kg CO_2/kWh) x power consumed (kWh)   (3)

CO_2 emission of grid electricity is not accounted for on the demand side, rather on the generation side. AEF is obtained by dividing the total CO_2 emission volume from thermal power plants connected to the grid by the total power supplied through the grid including those that are free of emissions, indicating average CO_2 emission per kWh of power on grid.

AEF (kg CO_2/kWh) = total CO_2 emissions from all thermal plants on grid (kg/ CO_2)/total power supplied to grid from all sources (kWh)   (4)

To calculate the amount of CO_2 emissions reduced because of lower power consumption, the emission factor for these plants must be multiplied by the reduced power consumption. The factor used is called the marginal emission factor or MEF. The reduced emission amount is obtained by the following calculation:

Emission reduction (kg CO_2) = MEF (kg CO_2/kWh) x electricity conserved (kWh)   (5)

10. ENERGY EFFICIENCY

Eventually renewable energies will dominate the world's energy supply system. There is no real alternative. Humankind cannot indefinitely continue to base its life on the consumption of finite energy resources. Today, the world's energy supply is largely based on fossil fuels and nuclear power. These sources of energy will not last forever and have proven to be contributors to our environmental problems. The environmental impacts of energy use are not new but they are increasingly well known; they range from deforestation to local and global pollution. In less than three centuries since the industrial revolution, humankind has already burned roughly half of the fossil fuels that accumulated under the earth's surface over hundreds of millions of years. Nuclear power is also based on a limited resource (uranium) and the use of nuclear power creates such incalculable risks that nuclear power plants cannot be insured.

Renewable sources of energy are an essential part of an overall strategy of sustainable development. They help reduce dependence of energy imports, thereby ensuring a sustainable supply. Furthermore renewable energy sources can help improve the competitiveness of industries over the long run and have a positive impact on regional development and employment. Renewable energy technologies are suitable for off-grid services, serving those in remote areas of the world without requiring expensive and complicated grid infrastructure. In 2007, the United States outlined plans to ease out of its foreign oil dependence through the use of renewable energy resources, and reduce gas usage by a full 20% in ten years through alternative fuels. Extending hope and opportunity depends on a stable supply of energy that keeps nation's economy running and environment clean.

Many wind/solar farms are located in remote areas served by low capacity radial distribution networks. These networks often have insufficient capacity to ship the power to demand centres when the turbines are generating at full capacity.

The public awareness of the depletion of the ozone layer has increased since 1970s. The ozone layer in the stratosphere protects life on earth against the ultraviolet radiation from the sun. Scientists and politicians argued for some years on the reasons for the reduced ozone layer, which was recognised especially over the Antarctic. The fact is that the ozone is depleted by the presence of chlorine. Further, very low temperatures as well as sunlight are required for having the process running. Vapour compression refrigeration systems using fluorocarbons, hydrocarbons or ammonia represent the established technology for household, commercial and industrial refrigeration and air-conditioning.

Understanding the earth and the processes that shape it is fundamental to the successful development and sustainable management of our planet. The evolution of the earth's crust over geological time has resulted in diverse, often beautiful, landscapes formed by earthquakes, oceans, fire and ice. These landscapes are also a source of mineral wealth and water, a base for engineering projects and a receptacle for the waste. Minerals are vital for manufacturing, construction, energy generation and agriculture. Some of the requirements can be met by increasing the use of recycled and renewable resources, but the need for new mineral resources continues. Minerals are important to maintaining the modern economy and lifestyle. Hence, everyone must make best use of these valuable assets whilst minimising the impact of their extraction on the environment. Global energy use will rise dramatically but world oil and gas production will eventually decline. Geological hazards account for huge loss of life and damage to property. Poorer countries often suffer the most due to inappropriate land-use planning and in the future climate change may worsen these effects. Clean drinking water is a basic human need. The environmental sustainability of water resources, especially when balancing ecological and human needs with economic growth, is of major concern to governments worldwide. Increases in the world's population, the growth of cities, industrial development and ever-increasing waste and pollution can increase pressure on the environment. Understanding the effects of climate change is a key issue for society and researches focus on how to predict future climate change events and how to mitigate them.

Two essential components for economic development are a soundly based knowledge of a country's natural resources, such as groundwater, minerals and energy, and an understanding of the geological environment. The latter includes the potential effects of earthquakes, volcanic eruptions, landslides and the forces of erosion and deposition on rivers and coastlines. Helping developing countries acquire this knowledge and apply it to promote economic growth, sustainable livelihoods and the protection of people. Sustainable land use involves protection of the natural environment (viable agriculture, forestry, water resources, and soil functions), the quality and character of the countryside and existing communities. Groundwater is an essential source of drinking water (70%). Although commonly regarded as pure, it is often vulnerable to pollution or may contain natural concentrations of elements that can adversely affect health. Population growth, intensive agriculture, urbanisation and higher standards of living have resulted in increasing, often unsustainable, demands on groundwater resources. These pressures and the likely effects of climate change mean that groundwater systems must be better understood so that they can be evaluated effectively, managed sustainably and protected securely.

District heating plays a very important role in transition economies. District heating systems in transition economies often face financial, technical or managerial problems largely created by an inadequate policy framework. These challenges include lack of customer focus, low efficiency, excess capacity, corruption and an uneven playing field. Figure 18 shows how these challenges are interrelated and create a vicious circle that undermines finances and competitiveness of district heating companies, jeopardising their long-term sustainability.

10.1. Energy Supply

The earth is believed to be close to a state of thermal equilibrium where the energy, which is received at the surface by solar radiation, is lost again at night and the much smaller amount of energy, which is generated by the decay of unstable isotopes of Uranium, Thorium and Potassium distributed within the earth is balanced by the small continuous heat flux from the earth's interior to the pecans and atmosphere.

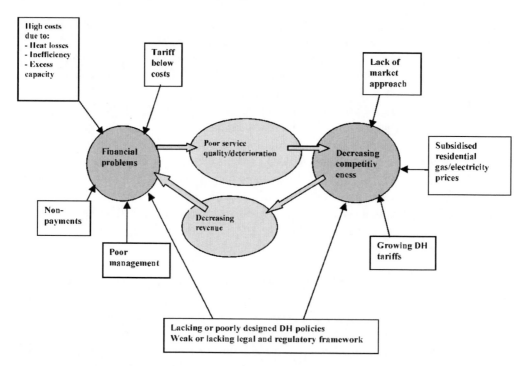

Figure 18. Key challenges for district heating systems in transition economies.

The use of geothermal energy involves the extraction of heat from rocks in the outer part of the earth. It is relatively unusual for the rocks to be sufficiently hot at shallow depth for this to be economically attractive. Virtually all the areas of present geothermal interest are concentrated along the margins of the major tectonic plates, which form the surface of the earth. Heat is conventionally extracted by the forced or natural circulation of water through permeable hot rock.

There are various practical difficulties and disadvantages associated with the use of geothermal power:

Transmission: geothermal power has to be used where it is found. In Iceland it has proved feasible to pipe hot water 20 km in insulated pipes but much shorter distances are preferred.

Environmental problems: these are somewhat variable and are usually not great. Perhaps the most serious is the disposal of warm high salinity water where it cannot be reinjected or purified. Dry steam plants tend to be very noisy and there is releases of small amounts of methane, hydrogen, nitrogen, amonia and hydrogen sulphide and of these the latter presents the main problem.

The geothermal fluid is often highly chemically corrosive or physically abrasive as the result of the entrained solid matter it carries. This may entail special plant design problems and unusually short operational lives for both the holes and the installations they serve.

Because the useful rate of heat extraction from a geothermal field is in nearly all cases much higher than the rate of conduction into the field from the underlying rocks, the mean temperatures of the field is likely to fall during exploitation. In some low rainfall areas there may also be a problem of fluid depletion. Ideally, as much as possible of the geothermal fluid should be reinjected into the field. However, this may involve the heavy capital costs of large condensation installations. Occasionally, the salinity of the fluid available for reinjection may be so high (as a result of concentration by boiling) that is unsuitable for reinjection into ground. Ocasionally, the impurities can be precipitated and used but this has not generally proved commercially attractive.

10.2. Refrigeration

The refrigeration industry has had to face environmental challenges. Regulation and behaviour related to the ozone layer differ from one country to another. Problems related to global warming are likely to cause similar problems in very different ways. These two kinds of problems may lead company managers to change/modernise equipment. These environmental factors must not hide other important development factors such as those related to hygiene, organoleptic quality, control methods and equipment packaging, etc. All equipment change and investment projects must take all of them into consideration. The more developed a country, is the more widely refrigeration is used. Food preservation is still the main use of refrigeration, followed by air conditioning, energy savings and transport (heat pump and liquefied gases), industrial processes and medicine. When consumers demand more and more fresh products, handling such products becomes increasingly difficult: cold stores, display cabinets, refrigerated trucks have to be much more effective. Difficulties are mostly related to:

- Very narrow ranges, in many cases, between temperatures involving microbial or chemical risk and temperatures that cause chilling injury or freezing.
- No or little overlapping between temperature ranges permitted for different products.
- No or little possibility for the consumer or the seller to evaluate the remaining life spans of the product or even to perceive any risk.

- Humidity control, vapour pressure differences have serious consequences, not only on the weight but also on the unit value of products.

The production and use of Chloro Fluoro Carbon (CFC) refrigerants have been phased out in developed countries. Hydro Chloro Fluoro Carbon (HCFC) refrigerants are now subject to regulation and are scheduled to be phased out in the twenty-first century. Recent developments in refrigeration technology and environment (climate) control have provided a better quality of life for mankind. As is known, refrigeration technology has played a key role in preserving the quality of food and eliminating waste (spoilage). The refrigerated transport of perishable products provides a critical link in the cold chain between the consumer and the producers, processors, shippers, distributors and retailers. The transport refrigeration industry incorporated mechanical vapour compression mechanisms into the first transport refrigeration units. Therefore, the physical, chemical and thermodynamic properties of these refrigerants had a major impact on the design of compressors and other refrigeration system components. Many refrigerants have excellent thermodynamic properties suitable for use in vapour compression refrigeration systems, but are limited by other properties such as toxicity, flammability and chemical stability within the refrigeration system.

Most district heating systems in former planned economies are less efficient and oversized. In other words, their supply infrastructure is larger than necessary to meet current demand. This problem can be exacerbated when they lose customers. Reforms, particularly for tariffs, are needed to improve finances and break the vicious circle of deteriorating competitiveness. Yet, policymakers need to design the reforms very carefully for them not to backfire and make matters worse, as illustrates in Figure 19.

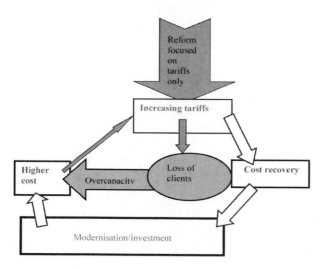

Figure 19. Unsustainable tariff growth.

10.3. Temperature Distributions

World capacity of geothermal energy is growing at a rate of 2.5% per year from a 2005 level of 28.3 GW (Kyoto Protocol, 1997). The GSHPs account for approximately 54% of this capacity almost all of it in North America and Europe (Lund, Freeston, and Boyd, 2005). The involvement of the UK is minimal with less than 0.04% of world capacity and yet is committed to substantial reduction in carbon emission beyond the 12.5% Kyoto obligation to be achieved by 2012. The GSHPs offer a significant potential for carbon reduction and it is therefore expected that the market for these systems will rise sharply in the UK in the immediate years ahead given to low capacity base at present. There are numerous ways of harnessing low-grade heat from the ground for use as a heat pump source or air conditioning sink. For small applications (residences and small commercial buildings) horizontal ground loop heat exchangers buried typically at between 1 m and 1.8 m below the surface can be used provided that a significant availability of land surrounding the building can be exploited which tends to limit these applications to rural settings. Horizontal ground loop heat exchangers can be used to circulate refrigerant (direct heat exchanger) or a water/antifreeze mixture (indirect heat exchange) and rely to some extent on solar input during summer for earth temperature recovery following winter heat extraction (Figures 20-21). For a more economical use of available land, a vertical ground loop heat exchange array or "borehole" array can be used typically involving a matrix of vertical borehole heat exchangers spaced at 5 m at depths of up to 180 m. high density plastic (typically high density polyethylene) tube of 20-40 mm nominal diameter is fed down to 100-150 mm diameter borehole to form a U-tube with the borehole subsequently grouted using a high conductivity hard-setting compound (usually bentonite). These and others methods of low grade-ground heat harvesting have been reviewed in details from the perspective of UK application by Rawlings (Rawlings, 1999).

Figure 20. A photograph of a laying ground-loop heat exchanger.

Heat generation within the earth is approximately 2700 GW, roughly an order of magnitude greater than the energy associated with the tides but about four orders less than that received by the earth from the sun (Oxburgh, 1975).

Temperature distributions within the earth depend on:

- The abundance and distribution of heat producing elements within the earth.
- The mean surface temperature (which is controlled by the ocean/atmosphere system).
- The thermal properties of the earth's interior and their lateral and radial variation.
- Any movements of fluid or solid rock materials occurring at rates of more than a few millimetres per year.

Figure 21. A photograph shows the earth loop vapour and liquid.

Of these four factors, the first two are of less importance from the point of view of geothermal energy. Mean surface temperatures range between 0-30°C and this variation has a small effect on the useable enthalpy of any flows of hot water. Although radiogenic heat production in rocks may vary by three orders of magnitude, there is much less variation from place to place in the integrated heat production with depth. The latter factors, however, are of great importance and show a wide range of variation. Their importance is clear from the relationship:

$$\beta = q/k \tag{6}$$

where:

β is the thermal gradient for a steady state (°C/km), q is the heat flux (10^{-6} cal cm^{-2} sec^{-1}) and k is the thermal conductivity (cal cm^{-1} sec^{-1} °C^{-1}).

The first requirement of any potential geothermal source region is that β being large, i.e., that high rock temperatures occur at shallow depth. Beta will be large if either q is large or k is small or both. By comparison with most everyday materials, rocks are poor conductors of heat and values of conductivity may vary from 2 x 10^{-3} to 10^{-2} cal cm^{-1} sec^{-1} °C^{-1}. The mean surface heat flux from the earth is about 1.5 heat flow units (1 HFU = 10^{-6} cal cm^{-2} sec^{-1}) (Oxburgh, 1975). Rocks are also very slow respond to any temperature change to which they are exposed, i.e., they have a low thermal diffusivity:

$$K = k/\rho C_p \qquad (7)$$

where:

K is thermal diffusivity; ρ and C_p are density and specific heat respectively.

These values are simple intended to give a general idea of the normal range of geothermal parameters (Table 6). In volcanic regions, in particular, both q and β can vary considerably and the upper values given are somewhat nominal.

In many geothermal areas natural heat exchangers exist in the form of large-scale, sub-surface water circulation systems, which commonly give rise to hot springs or geyser activity at the surface (Table 7).

Table 6. Values of geothermal parameters

Parameter	Lower	Average	Upper
q (HFU)	0.8	1.5	3.0 (non volcanic) $\approx$100 (volcanic)
k =cal cm^{-2} sec^{-1} $^{\circ}$C^{-1}	2×10^{-3}	6×10^{-3}	12×10^{-3}
β =$^{\circ}$C/km	8	20	60 (non volcanic) $\approx$300 (volcanic)

Table 7. Selected comparative cost data for geothermal energy

Geothermal field	Geothermal production	Local average, other fuel
Electricity, US mills/kWh		
Iceland	2.5-3.5	-
Italy	4.8-6.0	~7.5
Japan	4.6	~6.0
Mexico	4.1-4.9	~8.0
Russia	7.2	~10.0
USA	5.0	7.0
Space heating, US$/Gcal energy		
Iceland	4.0	6.7
Hungary	3.0	11.0
Refrigeration, US$/Gcal		
New Zealand	0.12	2.40
Drying diatomite, US$/ton		
Iceland	~2	~12

10.4. Thermodynamic Analysis of Refrigeration Cycles

Thermodynamics is the study of energy, its transformations and its relation to states of matter. A thermodynamic system is a region in space or a quantity of matter bounded by a closed surface. The surroundings include everything external to the system and the system is separated from the surroundings by the system boundaries. These boundaries can be movable or fixed, real or imaginary.

Potential energy (PE) is caused by attractive forces existing between molecules, or the elevation of system.

$$PE = mgz \tag{8}$$

where
m is the mass
g is local acceleration of gravity
z is the elevation above horizontal reference plane

[Net amount of energy added to system] = [Net increase of stored energy in system]   (9)

or
[Energy in] – [Energy out] = [Increase of stored energy in system]   (10)

Refrigeration cycles transfer thermal energy from a region of low temperature T_R to one of higher temperature. Usually the higher temperature heat sink is the ambient air or cooling water, at temperature T_o, the temperature of the surroundings.

Performance of a refrigeration cycle is usually described by a coefficient of performance (COP), defined as the benefit of the cycle (amount of heat removed) divided by the required energy input to operate the cycle:

COP = useful refrigerating effect / Net energy supplied from external sources   (11)

For a mechanical vapour compression system, the net energy supplied is usually in the form of work, mechanical or electrical and may include work to the compressor and fans or pumps. Thus,

COP = Qevap / Wnet   (12)

11. ENVIRONMENTAL CHALLENGE

The Montreal Protocol and its amendments (London 1990, Copenhagen 1992 and Vienna 1994) have given rise to the drawing up of regulation on the phase out the substances named Ozone Depleting Substances (ODS), including CFCs such as R12, R11 and HCFCs such as R22. Alternatives to CFCs include:

- Hydro Fluoro Carbons (HFCs) such as R134a and many mixtures.
- Naturally occurring refrigerants such as NH_3, hydrocarbons, water, air CO_2. Some of them are flammable and/or toxic or should be limited to relatively small niches.

- For other technologies such as absorption, solid sorption, etc., there are also some very interesting niches. However, refrigeration as a whole is not likely to shift there very soon because of the many other challenges it has to face.

The United Nations Framework Convention on Climate Change came into effect on March 21, 1994 and the Kyoto Protocol on December 10, 1997. Global warming is mostly caused by the rising percentage of various substances in the atmosphere, particularly CO_2, CH_4 and many others including CFCs, Hydro Chloro Fluoro Carbons (HCFCs) and HFCs. Refrigeration is involved because of:

- Refrigerants themselves, if they are released into the atmosphere, this is the 'direct effect'.
- CO_2 produced by energy consumption of the system throughout its lifespan, this is the 'indirect effect'.

The Fifth Framework Programme (FP5) is designed to ensure that European research efforts are translated more effectively into practical and visible results. The programme includes four thematic and three horizontal programmes.

Thematic programmes are:

- Quality of life and management of living resources.
- Users-friendly information society.
- Competitive and sustainable growth.
- Energy, environment and sustainable development.

Horizontal programmes are:

- Confirming the international role of community research.
- Promotion of innovation and encouragement of participation.
- Improving human research potential and the socio-economic knowledge base.

Research in agro-food industry, including refrigeration, has an important role, particularly under the key action "Food, Nutrition and Health".

11.1. Natural Disasters

The following is a detailed guide for communities and emergency operations team to develop and maintain a viable disaster management and recovery plan. In addition, it explains in details the concept of disaster management and various steps from planning to prepare against any disaster. It further highlights the role of government agencies and local authorities at the time of disaster as well as before and after it. It also discusses long and short-term goals for mitigation, planning and recovery from disaster along with initiatives that the US government has taken in recent years. It also, has a special focus on management of energy infrastructure during the time of disaster and presents a checklist for emergency response and recovery (Figure 22).

Every year, tornadoes, hurricanes and other natural disasters injure and kill thousands of people and damage billions of dollars worth of property in the United States. Most of the times, it is almost impossible to prevent the occurrence of these disasters and their damages. However, it is possible to reduce their impact by adopting suitable disaster management strategy. Disaster management is a systematic approach towards preparing for disaster before it happens and includes disaster response - emergency evacuation, quarantine, mass decontamination - as well as supporting and rebuilding society after natural disasters have occurred. Efficient disaster management relies on thorough integration of emergency plans at all levels of government and non-government involvement.

Disaster preparedness, emergency management and post disaster recovery is highly dependent on economic and social conditions local to the disaster. However, the basics steps for disaster management remain same in all scenarios. Preparedness is the first step to counter disaster, which involves developing plan of action and it, includes communication, chain of command development, proper maintenance and training of emergency services and development of emergency warning systems along with emergency shelters and evacuation plans. Next step is response, which includes mobilisation of the necessary emergency services such as fire fighters, police, and ambulance that may be supported by a number of secondary emergency services, such as specialist rescue teams. Recovery from disaster involves restoration of the affected area including destroyed property, re-employment and redevelopment of essential infrastructure. Mitigation efforts attempt at preventing hazards from developing into disasters or reducing the impact of disasters and it focuses on long-term measures for reducing or eliminating future risks.

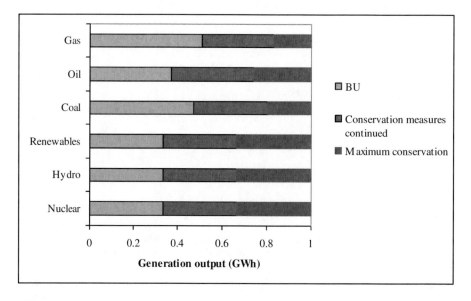

Figure 22. Energy conservation measures.

CONCLUSION

The following are concluded:

- Promoting innovation and efficient use of applicable renewable energy technologies.
- Identifying the most feasible and cost effective applications of renewable energy resources suitable for use.
- Highlighting the local, regional and global environmental benefits of renewable energy applications.
- Ensuring the renewable energy takes its proper place in the sustainable developments, supply and use of energy for greatest benefit of all, taking due account of research requirements, energy efficiency, conservation and cost criteria.
- Ensuring the financing of and institutional support for economic renewable energy projects.
- Encouraging education, research and training in renewable energy technology in the region.

12.1. Utilisation of Renewable Energy

Developing and implementing the use of renewable energy sources.

12.1.1. Policy and environment

- Efficiency, conservation, and policies.
- Renewable energy availability.
- Local and environmental concerns.
- Technology transfer.
- Financing requirements.
- Educational initiatives, legislative benchmarks.

12.1.2. Solar electrical technology

- PV technology manufacture, testing and certification.
- PV Stand-alone systems and components.
- PV for utility rural development and grid connection.
- PV Markets and commercialisation, financing schemes and national programmes.

12.1.3. Solar thermal technology

- Solar radiation prediction and analysis.
- Solar cooling, heating and rural applications.
- Solar thermal applications for power generation.
- Collector technology developments.

12.1.4. Solar and low energy architecture

- External Environment.
- Building, landscape design and urban communities and comfort productivity and health issues.
- Sustainable policy and social issues.
- Building simulation, material and design.
- Building refurbishment and integration of renewable energy; case studies.
- Internal environment.
- Thermal, ventilation and air movement.
- Lighting operation and control.

12.1.5. Wind energy technology and applications

- Small, micro-generation and hybrid systems.
- Machines and wind farms.
- Offshore wind power.
- Wind resources and environmental issues.
- Connection and integration.
- National and regional programmes.
- Economic and institutional issues.

12.1.6. Biomass conversion

- Heat and electricity generation.
- Energy crops and residues.
- Liquid fuels.
- Socio-economics, case studies, and environmental impacts.
- Gasification processes.

12.1.7. Fuel cells and hydrogen technology

- Fuel cells technology advances electricity generation.
- Fuel cells technology advances for transportation.
- Hydrogen production for fuel processing and systems.
- Proven commercialisation reports including economic and policy issues.

12.1.8. Marine/ocean energy

- Wave and tidal energy resources and their characterisation.
- Device modelling, testing and development.
- Device hydrodynamics, structural integrity and environmental analysis.
- Environmental impact assessment and standards.

- Balance of system - power take-off, sensors, controls and grid integration.
- Legislation, policy, finance and markets.
- Socio-economic assessment, education and training.

REFERENCES

Brain, G. & Mark, S. (2007). Garbage in, energy out: landfill gas opportunities for CHP projects. *Cogeneration and On-Site Power, 8 (5),* 37-45.

Federal Ministry for the Environment. (2005). Nature conservation and nuclear safety. Waste incineration- a potential danger. September 2005.

GHG Protocol. (2002). http://www.ghgprotocol.org/files/electricity_final.pdf

Kyoto Protocol (CDM). (1997). http://cdm.unfccc.int/methodologies/Tools/EB35_repan12_Tool_grid_emission.pdf

Lund, J. W., Freeston, D. H., and Boyd, T. L. (2005). Direct application of geothermal energy: 2005 Worldwide Review. *Geothermics*, 34, 691-727.

Oxburgh, E. R. (1975). Geothermal energy. *Aspects of Energy Conversion*, pp. 385-403.

Rawlings, R. H. D. (1999). Technical Note TN 18/99 – Ground Source Heat Pumps: A Technology Review. Bracknell. *The Building Services Research and Information Association*.

WRAP. (2006). Environmental benefits of recycling: an international review of life cycle comparisons for key materials in the UK recycling sector. 2006.

Chapter 10

ENERGY EFFICIENT SYSTEMS AND SUSTAINABLE DEVELOPMENT

ABSTRACT

People will have to rely upon mineral oil for primary energy and this will go on for a few more decades. Other conventional sources of energy may be more enduring, but are not without serious disadvantages. The renewable energy resources are particularly suited for the provision of rural energy supplies. A major advantage of using the renewable energy sources is that equipment such as flat plate solar driers, wind machines, etc., can be constructed using local resources and with the advantage of local maintenance which can encourage local manufacturing that can give a boost to the building of small-scale rural based industries. This study gives a comprehensive review of energy sources, the environment and sustainable development in Sudan. It reviews the renewable energy technologies, energy efficiency systems, energy conservation scenarios, energy savings in greenhouses environment and other mitigation measures necessary to reduce climate change. This communication gives some examples of small-scale energy converters, nevertheless it should be noted that small conventional, i.e., engines are currently the major source of power in rural areas and will continue to be so for a long time to come. There is a need for some further development to suit local conditions, to minimise spares holdings, to maximise interchangeability both of engine parts and of the engine application. Emphasis should be placed on full local manufacturing of some of the energy systems. It is concluded that renewable environmentally friendly energy must be encouraged, promoted, implemented and demonstrated by full-scale plant (device) especially for use in remote rural areas of many developing nations.

Keywords: Energy systems; consumption patterns, renewable technologies, environment; sustainable development, mitigation measures

MONOCULTURE

a	Greenhouse width (m)
b	Greenhouse length (m)
DR	Monthly average of diffuse radiation for a winter day (Wm^{-2})
G_T	Average ground irradiance in the presence of a reflecting wall (Wm^{-2})

$G_O^S(\varphi)$	Solar irradiance (radiation at a plane normal to incident solar rays) (Wm^{-2})
G_O^D	Diffuse (sky) irradiance (Wm^{-2})
G_G^D	Ground irradiance arising from diffuse radiation by direct insolation (Wm^{-2})
G_T^D	Total diffuse irradiance (Wm^{-2})
G_R^D	Reflected diffuse radiation (Wm^{-2})
G_T^S	Total ground irradiance from sunlight (Wm^{-2})
$G_{T,av}^S$	Average total ground irradiance from sunlight (Wm^{-2})
G_G^S	Ground irradiance from direct insolation (Wm^{-2})
G_R^S	Ground irradiance from the reflector (Wm^{-2})
H	Height of the reflecting wall (m)
P	Part of the ground not illuminated by sunlight
Q_T^S	Daily energy gain from solar radiation (Whm^{-2})
Q_G^S	Daily energy gain from ground irradiance (Whm^{-2})
R	Earth's radius (km)
ΔR	Thickness of the atmosphere (km)
S_V	Vertical width of the sunbeam falling onto a reflecting wall (m)
S_h	Horizontal width of the sunbeam falling onto a reflecting wall (m)
$S_ε$	Width of a beam of diffuse light arriving at an angle ε (degree)
S_O	Ground surface area of the greenhouse (m^2)
S	Ground surface area illuminated by reflection (m^2)
X_O	Distance from the reflecting wall traversed by reflected radiation (m)
z, y	Auxiliary variables
α	Inclination angle of the reflecting wall (degree)
α (φ)	Changing inclination of the reflecting wall (degree)
β, γ	Auxiliary angles (degree)
ε	Auxiliary angle (degree)
$η_D$	Reduction coefficient of diffuse ground radiation (%)
$η_S$	Enhancement coefficient of direct ground illumination by reflected sunlight (%)
$η_T$	Total enhancement coefficient (%)
$η_W$	Enhancement coefficient of average irradiance (%)
$η_{Wh}^S$	Energy enhancement coefficient (%)
φ	Incident angle of the sun (degree)
θ	Azimuth angle of the sun (degree)
$θ_C$	Critical azimuth angle at which half of the ground area is not receiving the reflected light (degree)
ρ	Reflectivity coefficient of the wall (%)
τ	Time of day (h)
τ	Transmission coefficient (km^{-1})

1. INTRODUCTION

This will also contribute to the amelioration of environmental conditions by replacing conventional fuels with renewable energies that produce no air pollution or greenhouse gases (during their use). The provision of good indoor environmental quality while achieving energy and cost efficient operation of the heating, ventilating and air-conditioning (HVAC) plants (devices) in buildings represents a multi variant problem (Omer, 2009a). The comfort of building occupants is dependent on many environmental parameters including air speed, temperature, relative humidity and air quality in addition to lighting and noise. The overall objective is to provide a high level of building performance (BP), which can be defined as indoor environmental quality (IEQ), energy efficiency (EE) cost efficiency (CE), environmental performance (EP), and the influence of the building to external environment.

- Indoor environmental quality is the perceived condition of comfort that building occupants experience due to the physical and psychological conditions to which they are exposed by their surroundings. The main physical parameters affecting IEQ are air speed, temperature, relative humidity and air quality.

Several definitions of sustainable development have been put forth, including the following common one: development that meets the needs of the present without compromising the ability of future generations to meet their own needs. The World Energy Council (WEC) study found that without any change in our current practice, the world energy demand in 2020 would be 50-80% higher than 1990 levels (Omer, 2009b). According to the US Department of Energy (DoE) report, annual energy demand will increase from a current capacity of 363 million kilowatts to 750 million kilowatts by 2020 (DOE, 2009). The world's energy consumption today is estimated to 22 billion kWh per year, 53 billion kWh by 2020 (Omer, 2009c). Such ever-increasing demand could place significant strain on the current energy infrastructure and potentially damage world environmental health by CO, CO_2, SO_2, NO_x effluent gas emissions and global warming (Omer, 2009d). Achieving solutions to environmental problems that we face today requires long-term potential actions for sustainable development. In this regards, renewable energy resources appear to be the one of the most efficient and effective solutions since the intimate relationship between renewable energy and sustainable development. More rational use of energy is an important bridge to help transition from today's fossil fuel dominated world to a world powered by non-polluting fuels and advanced technologies such as photovoltaics (PVs) and fuel cells (FCs) (Omer, 2008a). Some amount of renewable energy devices, e.g., biomass devices produce air pollution or greenhouse gases during life of energy devices and plants.

Energy security, economic growth and environment protection are the national energy policy drivers of any country of the world. As world populations grow, many faster than the growth rate 2%, the need for more and more energy is exacerbated (Figure 1). Enhanced lifestyle and energy demand rise together and the wealthy industrialised economics, which contain 25% of the world's population, consume 75% of the world's energy supply (Omer, 2008a). The world's energy consumption today is estimated to 22 billion kWh per year (Omer, 2008a). About 6.6 billion metric tons carbon equivalent of greenhouse gas (GHG) emission are released in the atmosphere to meet this energy demand (Omer, 2008b).

Approximately 80% is due to carbon emissions from the combustion of energy fuels (Omer, 2008b). At the current rate of usage, taking into consideration population increases and higher consumption of energy by developing countries, oil resources, natural gas and uranium will be depleted within a few decades. People could depend on new nuclear technologies that will enable much slower uranium depletion in the future. As for coal, it may take two centuries or so. Technological progress has dramatically changed the world in a variety of ways. It has, however, also led to developments, e.g., environmental problems, which threaten man and nature. Build-up of carbon dioxide and other GHGs is leading to global warming with unpredictable but potentially catastrophic consequences.

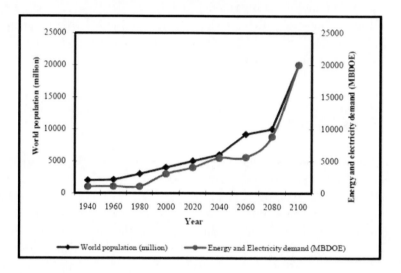

Figure 1. Annual and estimated world population and energy demand Million of barrels per day of oil equivalent (MBDOE).

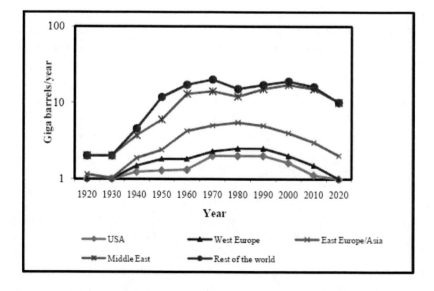

Figure 2. World oil productions in the next 10-20 years.

When fossil fuels burn, they emit toxic pollutants that damage the environment and people's health with over 700,000 deaths resulting each year, according to the World Bank review of 2000.

At the current rate of usage, taking into consideration population increases and higher consumption of energy by developing countries, oil resources, and natural gas will be depleted within a few decades. A Figure 2 shows the annual and estimated world population and energy demand, and Figure 3 the world oil productions in the next 10-20 years. As for coal, it may take two centuries or so. One must therefore endeavour to take precautions today for a viable world for coming generations.

Research into future alternatives has been and still being conducted aiming to solve the complex problems of this recent time, e.g., rising energy requirements of a rapidly and constantly growing world population and global environmental pollution.

Therefore, options for a long-term and environmentally friendly energy supply have to be developed leading to the use of renewable sources (water, sun, wind, biomass, geothermal, hydrogen production by electrolysis of water and fuel cells). Renewables could shield a nation from the negative effect in the energy supply, price and related environment concerns.

Hydrogen for fuel cells and the sun for the PV have been considered for many years as a likely and eventual substitute for oil, gas, and coal. The sun is the most abundant element in the universe.

The use of solar thermal energy or solar photovoltaics (PVs) for the everyday electricity needs has a distinct advantage: avoid consuming resources and degrading the environment through polluting emissions, oil spills and toxic by-products. A one-kilowatt PV system producing 150 kWh each month prevents 75 kg of fossil fuel from being mined (Omer, 2008c), and 150 kg of CO_2 from entering the atmosphere and keeps 473 litres of water from being consumed (WCED, 2009).

Electricity from fuel cells can be used in the same way as grid power: to run appliances and light bulbs and even to power cars since each gallon of gasoline produced and used in an internal combustion engine releases roughly 12 kg of CO_2, a GHG that contributes to global warming. Research into future alternatives has been and still being conducted aiming to solve the complex problems of this recent time, e.g., rising energy requirements of a rapidly and constantly growing world population and global environmental pollution.

Therefore, options for a long-term and environmentally friendly energy supply have to be developed leading to the use of renewable sources (water, sun, wind, biomass, geothermal, hydrogen) and fuel cells.

Renewables could shield a nation from the negative effect in the energy supply, price and related environment concerns. Hydrogen for fuel cells and the sun for the PV have been considered for many years as a likely and eventual substitute for oil, gas, coal and uranium. They are the most abundant elements in the universe.

2. ENERGY AND POPULATION GROWTH

Consequently, the focus has now shifted to non-commercial energy resources, which are renewable in nature. This is bound to have less environmental effects and also the availability is guaranteed.

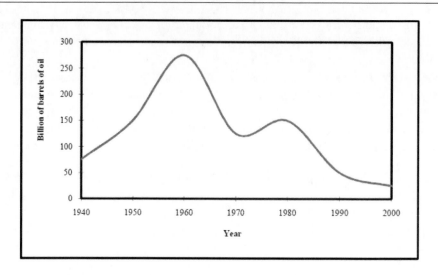

Figure 3. Volume of oil discovered worldwide.

However, even though the ideal situation will be to encourage people to use renewable energy resources, there are many practical difficulties, which need to be tackled.

Table 1. Global emissions of the top fourteen nations by total CO_2 volume (billion of tons) (EPA, 2008)

Rank	Nation	CO_2	Rank	Nation	CO_2	Rank	Nation	CO_2
1	US	1.36	6	India	0.19	11	Mexico	0.09
2	Russia	0.98	7	UK	0.16	12	Poland	0.08
3	China	0.69	8	Canada	0.11	13	S. Africa	0.08
4	Japan	0.30	9	Italy	0.11	14	S. Korea	0.07

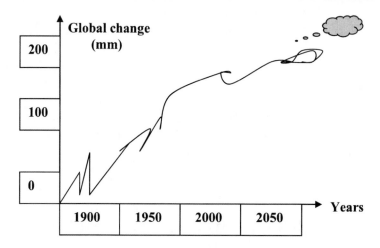

Figure 4. Change in global sea level.

2.1. Energy and Environmental Problems

Growing evidence of environmental problems is due to a combination of several factors since the environmental impact of human activities has grown dramatically because of the sheer increase of world population, consumption, industrial activity, etc., throughout the 1970s most environmental analysis and legal control instruments concentrated on conventional effluent gas pollutants such as SO_2, NO_x, particulates, and CO (Table 1).

The four more important types of harm from man's activities are global warming gases, ozone destroying gases, gaseous pollutants and microbiological hazards. Notably, human activities that emit carbon dioxide (CO_2), the most significant contributor to potential climate change, occur primarily from fossil fuel production.

Consequently, efforts to control CO_2 emissions could have serious, negative consequences for economic growth, employment, investment, trade and the standard of living of individuals everywhere. The earth is warmer due to the presence of gases but the global temperature is rising (Figure 4). This is a result of the increased concentration of carbon dioxide and other GHGs in the atmosphere as released by burning fossil fuels.

2.2. Environmental Transformations

To make sustainable water extraction economically viable, the sustainable policy has to break even (all costs are covered by revenues) while unsustainable policy has to be unprofitable (costs exceed revenues):

$$(1+r) \, vt_{-1} = 5y_t + v_t \tag{1}$$

where: r is the interest rate, t=year, y_t is the revenue, v_t is initial costs recovered by revenue, and vt_{-1} is all costs are covered by revenues.

Where: r is the interest rate, t=year, y is the revenue.

$$(1+r) \, vt_{-1} > 105y_t \tag{2}$$

$$(1+r) \, vt_{-1} < [105/(105-5)] \, v_t \tag{3}$$

The term $[105/(105-5)]$ is to define the natural productivity factor of the water resource as $(1+g) = [105/(105-5)]$; g is the natural productivity rate.

Rate g will be close to zero if the sustainable extraction level is much smaller than the unsustainable level. Using g, the equation can be as follows:

$$v_{t >} (1+r)/(1+g) \, v_{t-1} \tag{4}$$

Regulatory measures that prevent resource owners from adopting certain unsustainable extraction policies are a necessary pre-condition for the effective operation of a privatised natural resource sector.

Unregulated water privatisation would result in an inflationary dynamics whose distributional effects would threaten the long-term viability of the economy.

This inflationary dynamics is not due to any form of market imperfection but is a natural consequence of the competitive arbitrage behaviour of unregulated private resource owners.

2.3. Sustainability Concept

Absolute sustainability of electricity supply is a simple concept: no depletion of world resources and no ongoing accumulation of residues. Relative sustainability is a useful concept in comparing the sustainability of two or more generation technologies.

Therefore, only renewables are absolutely sustainable, and nuclear is more sustainable than fossil. Energy used to produce devices and plants for renewable use are not sustainable.

It is unlikely that consumers would tolerate any reduction in the quality of the service, even if this were the result of the adoption of otherwise benign generation technologies.

Renewables are generally weather-dependent and as such their likely output can be predicted but not controlled. The adoption of green or sustainable approaches to the way in which society is run is seen as an important strategy in finding a solution to the energy problem.

The only control possible is to reduce the output below that available from the resource at any given time. Therefore, to safeguard system stability and security, renewables must be used in conjunction with other, controllable, generation and with large-scale energy storage. There is a substantial cost associated with this provision.

2.4. Environmental Aspects

Environmental pollution is a major problem facing all nations of the world. People have caused air pollution since they learned to how to use fire, but man-made air pollution (anthropogenic air pollution) has rapidly increased since industrialisation began. Many volatile organic compounds and trace metals are emitted into the atmosphere by human activities. The pollutants emitted into the atmosphere do not remain confined to the area near the source of emission or to the local environment, and can be transported over long distances, and create regional and global environmental problems.

The privatisation and price liberalisation in energy fields has been secured to some extent (but not fully). Availability and adequate energy supplies to the major productive sectors. The result is that, the present situation of energy supplies is for better than ten years ago (Table 2).

3. WASTES MANAGEMENT

Lifecycle analysis of several ethanol feedstocks shows the emissions per ton of feedstock are highest for corn stover and switchgrass (about 0.65 tons of CO_2 per ton of feedstock) and lowest for corn (about 0.5 ton).

Table 2. Classifications of data requirements

	Plant data	**System data**
Existing data	Size Life Cost fixed and variation operation and maintenance (O and M) Forced outage Maintenance Efficiency Fuel Emissions	Peak load Load shape Capital costs Fuel costs Depreciation Rate of return Taxes
Future data	All of above, plus Capital costs Construction trajectory Date in service	System lead growth Fuel price growth Fuel import limits Inflation

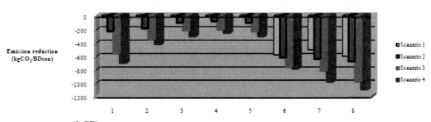

*1 Large steam power (LSP).
*2 Small steam power (SSP).
*3 Brayton cycle power (BCP).
*4 Bio-oil conversion power (B-CP).
*5 Gasification power (GP).
*6 Small steam CHP (SSCHP).
*7 Turboden cycle CHP (TCCHP).
*8 Entropic cycle CHP (ECCHP).

Figure 5. Comparison of thermal biomass usage options, CHP displacing natural gas as a heat source.

Emissions due to cultivation and harvesting of corn and wheat are higher than those for lignocellulosics, and although the latter have a far higher process energy requirement (Figure 5). The GHG emissions are lower because this energy is produced from biomass residue, which is carbon neutral.

On some climate change issues (such as global warming), there is no disagreement among the scientists. The greenhouse effect is unquestionably real; it is essential for life on earth. Water vapour is the most important GHG; next is carbon dioxide (CO_2).

Without a natural greenhouse effect, scientists estimate that the earth's average temperature would be $-18°C$ instead of its present $14°C$. There is also no scientific debate over the fact that human activity has increased the concentration of the GHGs in the atmosphere (especially CO_2 from combustion of coal, oil and gas). The greenhouse effect is also being amplified by increased concentrations of other gases, such as methane, nitrous oxide, and CFCs as a result of human emissions.

Most scientists predict that rising global temperatures will raise the sea level and increase the frequency of intense rain or snowstorms. Climate change scenarios sources of uncertainty and factors influencing the future climate are:

- The future emission rates of the GHGs.
- The effects of this increase in concentration on the energy balance of the atmosphere.
- The effect of these emissions on the GHGs concentrations in the atmosphere, and
- The effects of this change in energy balance on global and regional climate.

3.1. Particles

Some of the available control procedures for particles are summarised in Table 3. Figure 6 shows the variation of distribution factor with particle size. Emissions will also, of course, occur from petroleum-based or shale-based fuels, and in heavy consumption, such as in steam raising.

Table 3. Examples of SO$_2$ control procedures

Type of control	Fuel	Details
Pre-combustion	Fuels from crude oil	Alkali treatment of crude oil to convert thiols RSSR, disulphides; solvent removal of the disulphides
Post-combustion	Coal or fuel oil	Alkali scrubbing of the flue gases with CaCO$_3$/CaO
Combustion	Coal	Limestone, MgCO$_3$ and/or other metallic compounds used to fix the sulphur as sulphates

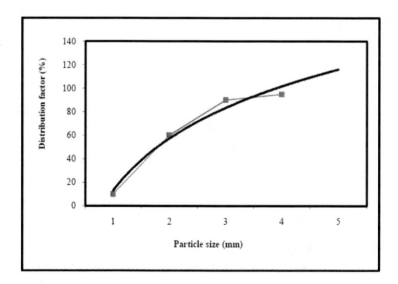

Figure 6. The variation of distribution factor against particle size for coal undersizes in a classifier. The sizes correspond to mid-point for ranges.

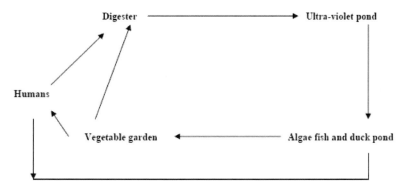

Figure 7. Biomass utilisation concept.

There will frequently be a need to control SO_2 emissions. There are, broadly speaking, three ways of achieving such control:

- Pre-combustion control: involves carrying out a degree of desulphurisation of the fuel.
- Combustion control: incorporating into the combustion system something capable of trapping SO_2.
- Post-combustion control: removing SO_2 from the flue gases before they are discharged into the atmosphere.

Figure 7 summaries biomass utilisation concept.

3.2. People, Power and Pollution

However, action must also be taken by developing countries to avoid future increases in emission levels as their economies develop and populations grow, as clearly captured by the Kyoto Protocol (Brain, and Mark, 2007).

4. GROUND SOURCE HEAT PUMPS

An advantage is gained from the necessity to provide filtered fresh air for ventilation purposes by providing every dwelling with a heat recovery mechanical ventilation system (Bos, My, Vu, and Bulatao, 1994).

Incorporation of a heating/cooling coil within the air-handling unit allows for active summertime cooling (i.e., collecting heat in summer), which along with the use of roof mounted solar panels to provide domestic hot water produces as well tempered and well engineered hybrid low energy scheme at very low carbon emissions.

As consumers in less-developed countries increase their capacity of electricity and green power, developed nations are starting to realise the benefits of using low-grade thermal energy for green heat applications that do not require high-grade electricity.

This shift will not only benefit renewable energies that are designed for space conditioning, but also will contribute to the global mix of green power and green heat capacity. Earth energy (also called geothermal or ground source heat pumps or GeoExchange), which transfers absorbed solar heat from the ground into a building for space heating or water heating. The same system can be reversed to reject heat from the interior into the ground in order to provide cooling (Figures 8-10).

A typical configuration buries polyethylene pipe below the frost line to serve as the head source (or sink), or it can use lake water and aquifers as the heat medium (Omer, 2008c).

4.1. GSHP Cost

The increased exploitation of renewable energy sources is central to any move towards sustainable development. However, casting renewable energy thus carries with it an inherent commitment to other basic tenets of sustainability, openness, democraticisations, etc. Due to increasing fossil fuel prices, the research in renewable energy technologies (RETs) utilisation has picked up a considerable momentum in the world. The present day energy arises has therefore resulted in the search for alternative energy resources in order to cope with the drastically changing energy picture of the world. The environmental sustainability of the current global energy systems is under serious question. A major transition away from fossil fuels to one based on energy efficiency and renewable energy is required. Alternatively energy sources can potentially help fulfill the acute energy demand and sustain economic growth in many regions of the world. The mitigation strategy of the country should be based primarily ongoing governmental programmes, which have originally been launched for other purposes, but may contribute to a relevant reduction of greenhouse gas emissions (energy-saving and afforestation programmes). Throughout the study several issues relating to renewable energies, environment and sustainable development are examined from both current and future perspectives. The exploitation of the energetic potential (solar and wind) for the production of electricity proves to be an adequate solution in isolated regions where the extension of the grid network would be a financial constraint. Figure 11 shows ground source heat pump system cost. Figure 12 the GSHPs extract solar heat stored in the upper layers of the earth.

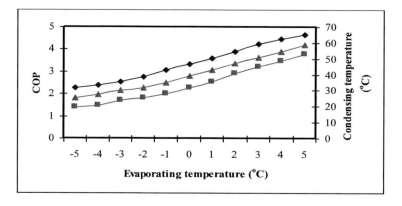

Figure 8. Coefficient of performance for cooling.

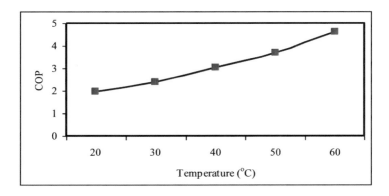

Figure 9. Coefficient of performance for heating.

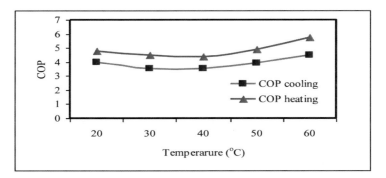

Figure 10. Coefficient of performance for heating and cooling.

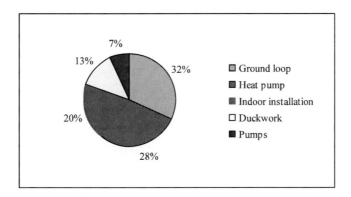

Figure 11. Ground source heat pump system cost.

Figure 13 conceptual illustration of an integrated energy system with thermal storage. Figure 14 measured data of soil thermal diffusivity. Figure 15 heat pump performance vs condensation temperature. Figure 16 heat pump performance vs evaporation temperature. Figure 17 measured data of soil thermal conductivity. Figure 18 experiments for saturated soil without groundwater flow (SS). Figure 19 experiments for saturated soil with groundwater flow (SSG). Figure 20 shows vertical closed-loop. Figure 21 shows horizontal closed-loop. Figure 22 shows pond loop.

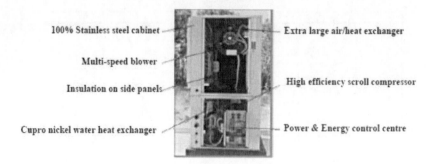

Figure 12. The GSHPs extract solar heat stored in the upper layers of the earth.

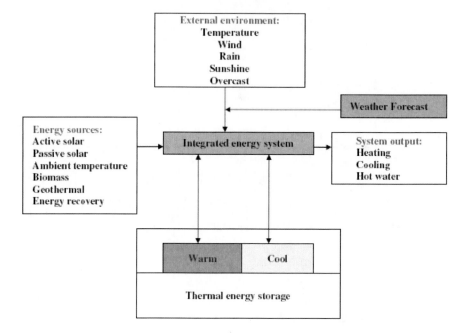

Figure 13. Conceptual illustration of an integrated energy system with thermal storage.

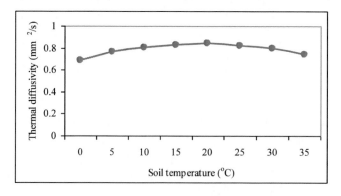

Figure 14. Measured data of soil thermal diffusivity.

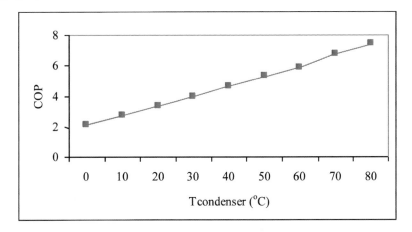

Figure 15. Heat pump performance vs condensation temperature.

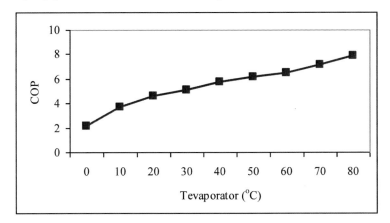

Figure 16. Heat pump performance vs evaporation temperature.

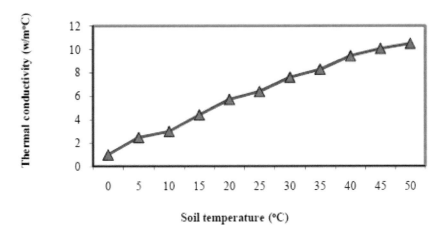

Figure 17. Measured data of soil thermal conductivity.

4.2. Biogas Technology

The requirements of gas for various purposes, and a comparison between biogas; and various commercial fuels in terms of calorific value, and thermal efficiency are presented successively in Table 4. The amount of biogas actually produced from a specific digester depends on the following factors: (1) Amount of material fed (2) Type of material (3) The carbon/nitrogen ratio and (4) Digestion time and temperature.

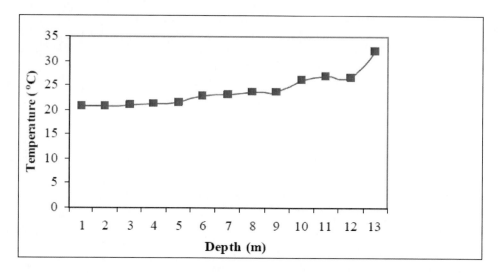

Figure 18. Experiments for saturated soil without groundwater flow (SS).

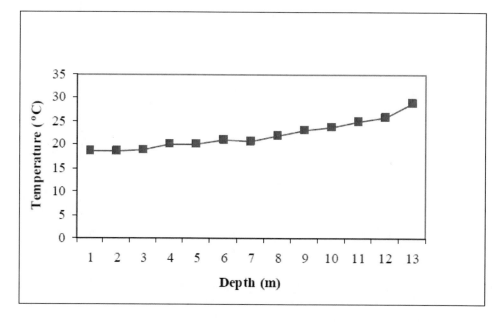

Figure 19. Experiments for saturated soil with groundwater flow (SSG).

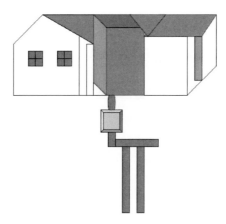

Figure 20. Vertical closed-loop.

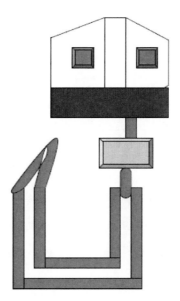

Figure 21. Horizontal closed-loop.

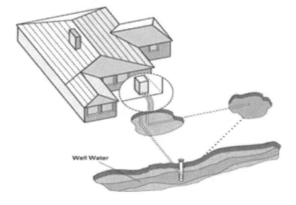

Figure 22. Pond loop.

Table 4. Biogas requirements for various purposes

Purpose	Specifications	Gas required (m³)
Cooking	per person	0.425/day
	stove 10 cm diameter	0.47
Lighting	200-candle power	0.1
	40-watt bulb	0.13
	2-mantle	0.14
Gasoline engine	Per HP	0.43
Diesel engine	Per HP	0.45
Refrigerator	Per m³	1.2
Incubator	Per m³	0.6
Table fan (indirectly)	30 cm diameter	0.17
Space heater	30 cm diameter	0.16

Table 5. Cost of construction materials 7 Cubic meters fixed dome digester in Sudanese Dinars (DS) (July 2010)

Construction material	Cost (DS)
1. Cement 23 bags (1 bag = 2500 Dinars)	57500
2. Sand 3 m³ (1 m³ = 2000 Dinars)	6000
3. Gravel 3 m³ (1 m³ = 2500 Dinars)	7500
4. Construction steel 1 m³ = 9000 Dinars	90000
5. Pipe 6'' x 3 m	23000
6. Pipe 8'' x 2 m	23000
7. Steel wire (1 kg = 400 Dinars)	400
8. Gas value 2 pieces (1 piece = 3500 Dinars)	7000
9. Rubber pipe 20-30 m (1 m = 150 Dinars)	4500
10. Burner	51250
Total	270150

In Sudan, people are requested to construct biogas plants by themselves in order to reduce costs. In remote areas, the costs for materials increase by about 15-20% due to transportation. The costs of construction of the fixed digester are given in Table 5.

In an economic analysis, many factors have to be considered, as outlined in Table 6. It is clear that many factors listed cannot be expressed in monetary terms.

The basis of evaluation of the gas produced is of significance. For example, in places where people use waste, but not kerosene as fuel, valuing the gas at the market price of kerosene equivalent is not correct since these over-estimates the benefits.

It has been observed that some of the present evaluations on biogas systems, while comparing the benefits with respect to existing practices, make the error of double accounting. For example, if the dung, which is already used as manure, is fed to a digester, only its incremental value can be taken into account.

Due to the lack of knowledge and awareness, villagers cannot be expected to understand the benefits of deforestation control, nutrient conservation, or health improvement. A poor rural peasant is very hesitant to enter a new venture. The negative attitude towards the use of nightsoil varies from place to place, but when it occurs, it is a major obstacle to the implementation of biogas technology.

Table 6. Factors to be considered in economic analysis

Economic Factors
• Interest on loan
• Current/future cost of alternative fuels
• Current/future cost of construction materials
• Saving of foreign currency
• Current/ future labour cost
• Inflation rate
• Costs of transport of feeding materials and effluents
Social factors
• Employment created
• Better lighting: more educational/cultural activities
• Less time consumed for fetching firewood and for cooking
• Improved facilities in villages; thus less migration to cities
• Less expense for buying alternative fuels
• More time for additional income earning activities
Technical factors
• Construction, maintenance and repairs of biogas plants
• Availability of materials and land required
• Suitability of local materials
Ecological/health factors
• Improved health
• Forest conservation (positive or negative)
• Environment pollution abatement
• Improvement in yields of agricultural products

4.3. Energy Consumption

Over the last decades, natural energy resources such as petroleum and coal have been consumed at high rates. The heavy reliance of the modern economy on these fuels is bound to end, due to their environmental impact, and the fact that conventional sources might eventually run out. The increasing price of oil and instabilities in the oil market led to search for energy substitute's way back in the early 1970s.

In addition to the drain on resources, such an increase in consumption consequences, together with the increased hazards of pollution and the safety problems associated with a large nuclear fission programmes. It would be equally unacceptable to suggest that the difference in energy between the developed and developing countries and prudent for the developed countries to move towards a way of life which, whilst maintaining or even increasing quality of life, reduce significantly the energy consumption per capita. Such savings can be achieved in a number of ways:

- Improved efficiency of energy use, for example better thermal insulation, energy recovery, and total energy.
- Conservation of energy resources by design for long life and recycling rather than the short life throwaway product.
- Systematic replanning of our way of life, for example in the field of transport.

Energy ratio is defined as the ratio of energy content of the food product/energy input to produce the food.

$$Er = Ec/Ei \qquad\qquad (5)$$

where Er is the energy ratio, Ec is the energy content of the food product, and Ei is the energy input to produce the food.

Currently the non-commercial fuelwood, crop residues and animal dung are used in large amounts in the rural areas of developing countries, principally for heating and cooking and the method of use is highly inefficient. The fossil fuels are currently of great importance in the developing countries (Sudan is not an exception).

Geothermal and tidal energy are less important though, of course, will have local significance where conditions are suitable. Nuclear energy sources are options for completeness, but are not likely to make any effective contribution in the rural areas. Figure 23 shows different renewable energy sources.

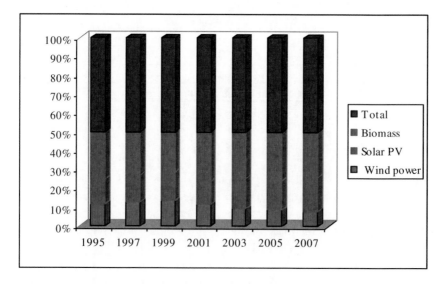

Figure 23. Different renewable energy sources.

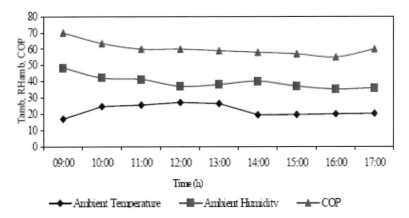

Figure 24. Temperature for different heights.

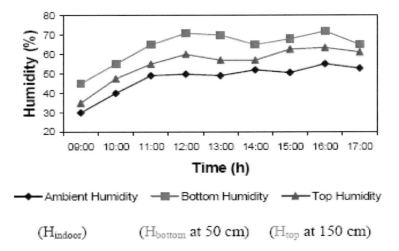

Figure 25. Humidity for different heights.

Figure 24 shows temperature for different heights. Figure 25 shows humidity for different heights. Figure 26 shows regional distribution of scientists. Figure 27 shows African global change scientists.

4.4. Wind Energy Potential

The use of wind as a source of power has a long history. Wind power has been used in the past for water pumping, corn grinding, and provision for power for small industries. In areas of low population density where implementation of a central power system would be uneconomical, the decentralised utilisation of wind energy can provide a substantial contribution to development (Omer, 1997; Omer, 1998).

The use of the wind machine is divided into two; one is the use of small-scale wind machines for water pumping or electricity generation, and the other is the use of large-scale wind machines for generating electricity (big wind machines or wind farms).

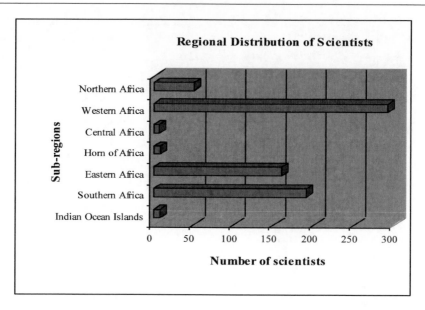

Figure 26. Regional distribution of scientists.

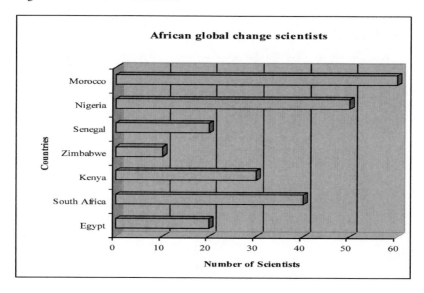

Figure 27. African global change scientists.

However, the wind machine can be used for pumping water, electricity generation or any other task. A programme of wind power for generating electricity as well as for pumping water appears to be attractive for rural development, e.g., lights, radios, televisions (Figure 28). Wind electric generators can be utilised to meet the power requirements of isolated settlements. Wind energy is found to match well with the demand pattern of the loads, high load during the day for illumination. Wind energy has considerable resources in Sudan where the annual average wind speeds exceeds 5 ms^{-1} (at 10 m height) in the most parts north latitude 12°'N (at the coastal area along the Red Sea), and along the Nile valley (from Wadi Halfa to Khartoum, and south of Khartoum covering the El Gezira area). The southern regions have the poorest potential because of the prevailing low wind speeds. Many designs of wind machines have been suggested and built in Sudan as shown in Table 7.

Table 7. Number of wind pumps installed for irrigation purpose in Sudan

Location	Number of pumps
Tuti island	2
Jebel Aulia	1
Soba	4
Shambat	4 (one was locally manufactured)
Toker (eastern Sudan)	2 (both locally manufactured)
Karima (northern Sudan)	2 (both locally manufactured)
Total	15

In Sudan, wind energy is today mainly used for water pumping. Wind has not yet been significantly exploited for power generation. Experience in wind energy in Sudan was started since 1950's, where 250 wind pumps from Australian government, had been installed in El Gezira Agricultural Scheme (Southern Cross Wind Pumps). These were gradually disappeared due to a lack of spare parts and maintenance skills combined with stiff competition from relatively cheep diesel pumps. However, the government has recently begun to recognise the need to reintroduce wind pump technology to reduce the country's dependence on foreign oil. This increases economic security, given high and/or fluctuating oil prices, and it helps to reduce the trade deficit. Using wind power also allows for pumping in rural areas where transportation of oil might be difficult (Figure 29).

In the last 15 years the Energy Research Institute (ERI) installed 15 Consultancy Services Wind Energy Developing Countries (CWD 5000 mm diameter) wind pumps around Khartoum area, Northern state, and Eastern state. Now ERI with cooperation of Sudanese Agricultural Bank (SAB) introduced 60 wind pumps to be use for water pumping in agricultural schemes, but not yet manufactured due to lack of financial support.

The maximum extractable monthly mean wind power per unit cross sectional area, P, is given by:

$$P = 0.3409 \ V^3 \tag{6}$$

where: P is the wind power Wm^{-2}; and V is the average wind speed ms^{-1}.

The amount of power extracted from the wind depends generally on the design of the wind rotor.

In practice the wind machine power will be lost by the aerodynamic affects of the rotor. An important problem with wind pump system is matching between the power of the rotor, and that of the pump. In general the wind pump system consists of the following items:

- The wind rotor
- Transmission
- The pump

The overall efficiency of the system is given by the multiplication of the rotor efficiency, transmission efficiency and the pump efficiency:

$$\eta_{overall} = \eta_{rotor} \times \eta_{transmission} \times \eta_{pump} \tag{7}$$

For wind pumps though efficiency is important, a more suitable definition is the number of gallons of water pumped per day per dollar. A sizing of wind pump for drinking and irrigation purposes usually requires an estimation of hourly, daily, weekly, and monthly average output.

The method for making such estimation is combining data on the wind pump at various hourly average wind speeds with data from a wind velocity distribution histogram (or numerical information on the number of hours in the month that wind blows within predefined speed). The result is given in Table 8, which gives the expected output of wind pump in various wind speeds, and the statistical average number of hours that the wind blows within each speed range.

Generally it is concluded that wind pump system have a potential to fulfill water lifting needs, both in Khartoum area and even in remote rural areas, both for irrigated agriculture and water supply for man and livestock. This conclusion is based on:

- Studies of several agencies dealing with the feasibility of wind pumps.
- The history of water pumping in the Gezira region for drinking purposes.
- The national policy of Sudan vis a vis wind energy.

Sudan is rich in wind; mean wind speed of 4.5 ms^{-1} are available over 50% of Sudan, which is well suited for water lifting and intermittent power requirements, while there is one region in the eastern part of Sudan that has a wind speed of 6 ms^{-1} which is suitable for power production. In areas where there is wind energy potential but no connection to the electric grid the challenge is simplicity of design, and higher efficiency (Abdeen, 2008). Because of this potential for fulfilment of rural water pumping needs, it is recommended to continue the development of wind pumping in Sudan.

Table 8. Wind speeds versus wind pump discharges

Wind speeds (ms^{-1})	Annual duration (h)	Output rate (m^{3}h^{-1})
3.0	600	0.3
3.5	500	1.4
4.0	500	2.3
4.5	400	3.0
5.0	500	3.7
5.5	450	4.3
6.0	450	4.7
6.5	300	5.2
7.0	300	5.7

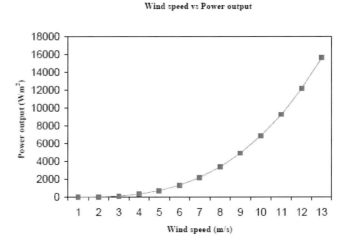

Figure 28. Wind speed vs power output.

The research and development in the field of wind machines should be directed towards utilising local skills and local available materials. Local production of wind machines should be encouraged in both public and private organisations.

The most obvious region to start with seems to be the northern regions because of a combination of:

- Favourable wind regime
- Shallow ground water level 5-10 meters depth
- Existing institutional infrastructures

4.5. Water Resources Management

Planning methodologies/processes for sustainable river basin management (Table 9) must include, support or promote:

- Integration of knowledge from all relevant disciplines.
- Handling of different kinds of uncertainty.
- Identification of most relevant value.
- Rational argumentation based on the identified values, relating them to alternative choices in the planning process.
- Inclusion of knowledge owned by relevant actors.
- Inclusion of the ideological orientations represented by relevant actors.
- Participation in the most critical phases of the process.
- A procedure for defining the actors that should be involved.
- Handling of power asymmetries.
- Learning.

- Procedures that ensure that ideological orientations are not suppressed (for consensus-based processes).

Water supply factor (WSF) = (Actual water supply/Planned water supply)   (8)

Diurnal stability factor (DSF) = (Standard deviation of diurnal flow rates
from an average daily flow rate/Average daily flow rate)   (9)

Water supply uniformity factor (WSUF) = (An absolute value of the difference
between WSF of individual off-take and WSF of whole canal/WSF of whole canal)   (10)

Technical efficiency factor (TEF) = (Water supply + transit flow + out flow) /
(Head water diversion + side inflow)   (11)

5. Discussions

Water is one of the most precious natural resources on earth. It is essential for human survival and development and cannot be replaced by any other resources. However, with rapid social and economic development, as well as explosive population growth, a water crisis has developed in Sudan during the 20th century. With economic development, and population growth the conflict between water supply and demand has become more and more acute in Sudan, and it has been aggravated further by the irrational utilisation of water resources.

As a result, the deterioration and destruction of the eco-environment have become increasingly serious. In order to effectively protect ecosystems and improve their ecological conditions, many developments on ecological and environmental water requirements have been carried out including rivers, vegetation, lakes, wetlands and groundwater.

Changes in the economy or in the population will have different impacts on the water resource availability, which in turn will impact economic output and population dynamics. Water conservation is a major challenge because of increasing competition between agricultural and non-agricultural use of water. Efficient use of water in agriculture is critical because of the large volumes of water used.

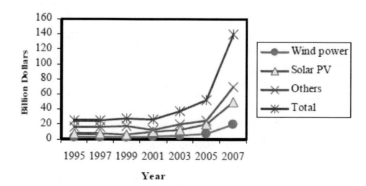

Figure 29. Different renewable sources.

Table 9. Alternative and complementary measures

Information based instruments	Targeted information provision Naming, shaping and faming Registration, labelling and certification
Private and voluntary regulation	Self-regulation Voluntary regulations Governments and negotiated agreements Private regulations Professional regulations Civic regulations
Support mechanisms and capacity	Research and knowledge generation Demonstration projects and knowledge diffusion Network building and joint problem solving

Figure 30 shows a schematic of the three sub-models and highlights their important linkages that include primary and secondary feedback loops between the components.

For example, changes in the economy or in the population will have different impacts on the water resource availability, which in turn will impact economic output and population dynamics. There is a need for some further development to suit local conditions, to minimise spares holdings, to maximise interchangeability both of engine parts and of the engine application.

According to the 1993 census 65 percent of the urban population have access to piped water (UNICEF, 2006), whereas only 20 of the rural population enjoy the same facility (Table 10).

Other sources give much lower figures. The Government National Programme of Action for Child Survival and Development sets availability of 20 litres of safe water per capita per day, within one kilometre from the users' dwelling, as the acceptable standard. A clear national definition of what a safe source of water is has yet to be adopted.

Urban populations theoretically consume 20 litres per capita per day while rural people get 8 litres at two and a half times the cost per litre. Variations are large from one place to another. People in Darfur consume less than one fourth of those in Khartoum. Population in pre-urban areas and in newly emerging urban centres are around 35 percent of the total urban population and generally do not have access to safe water. The quality of urban water supply is questionable (Omer, 2004). Surface water sources are contaminated and bear heavy loads of suspended silt. Waterworks are old and small and cannot cope with the escalating demand. Increased reliance on ground water in the past few years has partially released pressure off in the way of quantity but not quality. Septic tanks and soak-away wells are a constant threat of contamination.

Distribution networks are old and is a potential weak link taking up contaminants, in the case of the capital Khartoum for example. Increased reliance on booster pumps and storage tanks have contributed to the depletion of the major power supply as well as offering excellent breeding sites for mosquitoes. The worst-case scenario is the city of Port Sudan suffering from chronic water shortages on the one hand and the heavy incidence of malaria on the other.

Table 10. Percent population with access to sources of water

Sources of water	Northern States	Southern States	Rural	Urban
Pipe connection	19.62	65.81	35.33	7.02
Boreholes	56.12	20.25	43.93	55.16
River/ canal	14.09	2.25	10.05	30.01
Tanker	4.78	0.59	3.35	0.24
Haffirs/Fola	1.36	7.17	3.34	5.40
Others	3.97	3.82	3.92	1.74
Not stated	0.06	0.11	0.08	0.43
Total	100.0	100.0	100.0	100.0

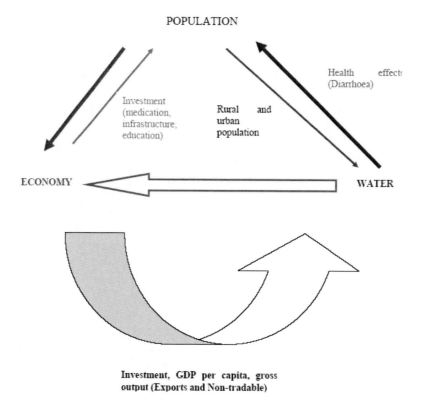

Figure 30. Water/economy/population relationship.

The celebrations of the city of El Obeid with the new source of water from Bara, was short lived. The old distribution system soon collapsed. Storm drains is another fiasco. It is a paradox that there is no clear link between the water supply and sanitation sectors. Even within the water sector there are poor linkages between the Federal and state water corporations. The water sector itself has experienced ten major institutional changes since independence.

The Sudanese National Plan of Action (NPA) sets the goal of universal access to safe drinking water and sanitary means of excreta disposal by the turn of the decade (Omer, 2004).

The Comprehensive National Strategy gives priority to the following strategies for achieving the set:

- Cost-effective utilisation and management of water resources.
- Introduction of low-cost appropriate technologies and encouragement of local production of equipment.
- Rehabilitation of deteriorating water sources and systems.
- An expanded programme of well-drilling and hand-pump installation, especially in priority rural areas.
- Training, capacity building and increased use of domestic technical resources, to increase cost efficiency and reduce dependence on external resources.
- Development and expansion of sanitation services.
- Increased community involvement in planning, execution and management of water supply and sanitation services encouraging cost sharing and self-help.
- Encouragement of research aimed at better water resources management, evaluation of existing schemes and identification of cost-effective alternative strategic elements.

The National Comprehensive Strategy gives priority to the rural sector, emphasising rehabilitation of existing water yards, expansion of low cost technology options (hand pumps) and improvement of surface water sources (water harvesting, haffirs, etc.). The priority in urban areas is to restore existing waterworks to their original designed output followed by rehabilitation of other sources and expansion of services especially in Khartoum and the sixteen state capitals recently upgraded from towns to cities (Table 11).

Table 11. Water sources for rural

Type	Number	Capacity (m³/ day)	Installed actual yield
Water yards	4,252	447,000	146,000
Hands pumps	8,000	48,000	32,000
Dug wells	7,000	56,000	44,800
Haffirs/ Dams	856	142,000	56,800
Slow sand filters	233	16,500	8,200
Total	20,441	710,000	287,800

Source: (UNICEF, 2006).

Only 40 percent of the urban population have access to safe water. Investment does not match growth in needs. The State Ministry of Housing and Public Works of Khartoum conceded, "The estimated deficit of clean water supply in Khartoum is about 40 thousand cubic metres per day". The unaccounted losses in the distribution system are as high as 40% (Table 12).

Globally and regionally a realisation is setting in that constraint of freshwater resource availability and their sharing could be a major impediment to security and subsequently to sustainable development of developing countries.

Scarcity of freshwater resources globally is fast becoming a major factor in limiting the development of regions and in causing conflict between regions sharing the same water resource. The Nile has many origins, draining 10 percent of the African continent's river water and flows 6825 km through three climatic zones and ten sovereign countries. The Nile flowing through Sudan is the confluence of two major rivers.

Table 12. Urban Demand for water

Region	Capacity m³/day	Demand/m³/day	Coverage (%)
Khartoum	350,000	700,000	50
Northern	30,000	64,000	47
Eastern	48,000	205,000	20
Central	80,000	184,000	39
Kordofan	23,000	114,000	17
Darfur	14,000	74,000	20
Southern	20,000	103,000	16
Total	565,000	1,445,400	39

Source: (UNICEF, 2006).

Table 13. Water consumption in different African countries

Country	Population (10⁶)	Total mean precipitation (km³)	Sectors (%)	Per capita consumption (litres/day)
Botswana	1.7	261.9	Domestic 25% Agriculture 37% Industry 25% Others 13%	221.5
Burundi	7.1	N.A.	Domestic 58.3% Agriculture 21.8% Industry 5.2% Others 14.7%	260.7
Congo	3.6	N.A.	N.A.	111.1
Egypt	73	80	N.A.	200
Eritrea	4	N.A.	Domestic 36.9% Industry 34.8% Others 26.9%	N.A.
Ethiopia	67.8	*N.A.*	N.A.	12.25
Kenya	33	N.A.	Domestic 27.6% Industry 33.2% Others 39.2%	143
Libya	5.3	43	Domestic 11% Agriculture 86% Industry 2% Others 1%	100
Rwanda	7.9	80	N.A.	43
Sudan	39	80	Domestic 20% Agriculture 70% Industry 5% Others 5%	20

Reproduced from (WAIG, 2006).

Concurrently, countries of the "Eastern Nile" and those of "Equatorial Lakes region" have identified joint projects (Subsidiary Action Programmes) (Table 13). Basically, the Nile Basin Initiative is a country and needs-driven programme. It should be stressed here that the Nile should have been treated as one indivisible whole (Jonathon, 1991; John, 1993).

Historically all control works and projects have been carried out where and when the need arose. Engineers and diplomats were the only players and other stakeholders were never involved. The not unexpected consequence is that six of the Nile Basin countries are among the poorest in the world (UNICEF, 2006).

Over the last decades, natural energy resources such as petroleum and coal have been consumed at high rates. The heavy reliance of the modern economy on these fuels is bound to end, due to their environmental impact, and the fact that conventional sources might eventually run out. The increasing price of oil and instabilities in the oil market led to search for energy substitutes. In addition to the drain on resources, such an increase in consumption consequences, together with the increased hazards of pollution and the safety problems associated with a large nuclear fission programmes. This is a disturbing prospect.

5.1. Environment

Biogas is a generic term for gases generated from the decomposition of organic material. As the material breaks down, methane (CH_4) is produced. Sources that generate biogas are numerous and varied.

These include landfill sites, wastewater treatment plants and anaerobic digesters (Brain, and Mark, 2007). Landfills and wastewater treatment plants emit biogas from decaying waste. To date, the waste industry has focused on controlling these emissions to our environment and in some cases, tapping this potential source of fuel to power gas turbines, thus generating electricity (Bos, My, Vu, and Bulatao, 1994; Achard, and Gicqquel, 1986). The primary components of landfill gas are methane (CH_4), carbon dioxide (CO_2), and nitrogen (N_2).

The average concentration of methane is ~45%, CO_2 is ~36% and nitrogen is ~18% (Omer, and Yemen, 2001). Other components in the gas are oxygen (O_2), water vapour and trace amounts of a wide range of non-methane organic compounds (NMOCs). Landfill gas-to-cogeneration projects present a win-win-win situation. Emissions of particularly damaging pollutant are avoided, electricity is generated from a free fuel and heat is available for use locally.

Presently, Sudan uses a significant amount of kerosene, diesel, firewood, and charcoal for cooking in many rural areas. Biogas technology was introduced to Sudan in 1973 when GTZ designed a unit as a side-work of a project for water hyacinth control in central Sudan. Anaerobic digesters producing biogas (methane) offer a sustainable alternative fuel for cooking that is appropriate and economic in rural areas. In Sudan, there are currently over 200 installed biogas units, covering a wide range of scales appropriate to family, community, or industrial uses.

The agricultural residues and animal wastes are the main sources of feedstock for larger scale biogas plants. There are in practice two main types of biogas plant that have been developed in Sudan; the fixed dome digester, which is commonly called the Chinese digester (120 units each with volumes 7-15 m^3).

The other type is with floating gasholder known as Indian digester (80 units each with volumes 5-10 m^3). The solid waste from biogas plants adds economic value by providing valuable fertiliser as by products. Table 14 shows growth in annual GHG emissions amount (10^6 Mt of CO$_2$) by which have increased since 2007.

Table 14. Growth in annual GHG emissions amount (10^6 Mt of CO$_2$) by which have increased since 2007 (EPA, 2008)

Item	10^6 Mt of CO$_2$
Electrical generation (coal)	436.5
Electrical generation (natural gas)	197.3
Transportation (petroleum)	403.7
Substitution of ozone depleting substances	108.0
All others	-94.1

5.2. Wave Power Conversion Devices

The patent literature is full of devices for extracting energy from waves, i.e., floats, ramps, and flaps, covering channels (Swift-Hook, et al., 1975). Small generators driven from air trapped by the rising and falling water in the chamber of a buoy are in use around the world (Swift-Hook, et al., 1975). Wave power is one possibility that has been selected. Sudan has potential of waves at the Red sea.

A wave power programme would make a significant contribution to energy resources within a relatively short time and with existing technology. Wave energy has also been in the news recently. There is about 140 megawatts per mile available round British coasts.

It could make a useful contribution people needs in the UK. Although very large amounts of power can be generated from the waves, it is important to consider how much power can be extracted. A few years ago only a few percent efficiency had been achieved.

Recently, however, several devices have been studied which have very high efficiencies. Some form of storage will be essential on a second-to-second and minute-to-minute basis to smooth the fluctuations of individual waves and wave's packets but storage from one day to the next will certainly not be economical. This is why provision must be made for adequate standby capacity.

A number of prospective areas have been identified by surveys and studies carried for exploration of mini-hydropower resources in Sudan. Mini and micro hydro can be utilised or being utilised in Sudan in two ways:

- Using the water falls from 1 m to 100 m; energy can be generated, and small power can be generated up to 100 kW.
- Using the current flow of the Nile water, i.e., the speed of the Nile water. The water speed can be used to run the river turbines (current river turbines), and then water can be pumped from the Nile to the riverside farms. There are more than 200 suitable sites for utilisation of current river turbines along the Blue Nile and the main Nile.

5.3. Sustainable Development

Like most African countries, Sudan is vulnerable to climate variability and change. Drought is one of the most important challenges.

The most vulnerable people are the farmers in the traditional rain-fed sector of western, central and eastern Sudan, where the severity of drought depends on the variability in amount, distribution and frequency of rainfall.

Three case studies were conducted in Sudan as part of the project. They examined the condition of available livelihood assets (natural, physical, financial, human and social) before and after the application of specific sustainable livelihood environmental management strategies, in order to assess the capacity of communities to adapt creased resilience through access to markets and income generating opportunities.

Figure 31 summarises oil production and consumption in Sudan.

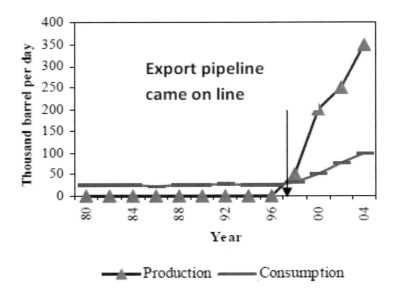

Figure 31. Sudan's oil production and consumption 1980-2005.

Table 15. Renewable applications

Systems	Applications
Water supply	Rain collection, purification, storage and recycling
Wastes disposal	Anaerobic digestion (CH_4)
Cooking	Methane
Food	Cultivate the 1 hectare plot and greenhouse for four people
Electrical demands	Wind generator
Space heating	Solar collectors
Water heating	Solar collectors and excess wind energy
Control system	Ultimately hardware
Building fabric	Integration of subsystems to cut costs

Table 15 presented some renewable applications. Considerations when selecting power plant include the following:

- Power level- whether continuous or discontinuous.
- Cost - initial cost, total running cost including fuel, maintenance and capital amortised over life.
- Complexity of operation.
- Maintenance and availability of spares.
- Life.
- Suitability for local manufacture.

The potential of the most important forms of renewable energy, such as solar, wind, biomass, and geothermal energies, is shown in Table 16. Existing renewable energy technologies could play a significant mitigating role, but the economic and political climate will have to change first. Climate change is real. It is happening now, and greenhouse gases produced by human activities are significantly contributing to it. The predicted global temperature increase of between 1.5 and 4.5 degrees C could lead to potentially catastrophic environmental impacts.

These include sea level rise, increased frequency of extreme weather events, floods, droughts, disease migration from various places and possible stalling of the Gulf Stream. This has led scientists to argue that climate change issues are not ones that politicians can afford to ignore, and policy makers tend to agree. However, reaching international agreements on climate change policies is no trivial task (Duchin, 1995).

Table 16. Potential, productive, end-uses of various energy sources and technologies

Energy source and technology	Productive end-uses and commercial activities
Solar	Lighting, water pumping, radio, TV, battery charging, refrigerators, cookers, etc. Dryers, cold stores for vegetables and fruits, water desalination, heaters, baking, etc.
Wind	Pumping water, grinding and provision for power for small industries
Hydro	Lighting, battery charging, food processing, irrigation, heating, cooling, cooking, etc.
Biomass	Sugar processing, food processing, water pumping, domestic use, power machinery, weaving, harvesting, sowing, etc.
Kerosene	Lighting, ignition fires, cooking, etc.
Dry cell batteries	Lighting, and small appliances
Diesel	Water pumping, irrigation, lighting, food processing, electricity generation, battery charging, etc.
Animal and human power	Transport, land preparation for farming, food preparation (threshing)

5.4. Renewable Energy Potential

There is strong scientific evidence that the average temperature of the earth's surface is rising. This is a result of the increased concentration of carbon dioxide and other greenhouse gases (GHGs) in the atmosphere released by burning fossil fuels. This global warming will eventually lead to substantial changes in the world's climate, which will, in turn, have a major impact on human life and the built environment. Therefore, effort has to be made to reduce fossil energy use and to promote green energies, particularly in the building sector. Energy use reductions can be achieved by minimising the energy demand, by rational energy use, for example, by recovering heat and the use of more green energies and green energy technologies. This study was a step towards achieving that goal. The adoption of green or sustainable approaches to the way in which society is run is seen as an important strategy in finding a solution to the energy problem. The key factors to reducing and controlling carbon dioxide (CO_2), which is the major contributor to global warming, are the use of alternative approaches to energy generation and the exploration of how these alternatives are used today and may be used in the future as green energy sources (Trevor, 2007). Even with modest assumptions about the availability of land, comprehensive fuel-wood farming programmes offer significant energy, economic and environmental benefits. These benefits would be dispersed in rural areas where they are greatly needed and can serve as linkages for further rural economic development. The developing nations as a whole would benefit from savings in foreign exchange, improved energy security, and socio-economic improvements. With a nine-fold increase in forest – plantation cover, a nation's resource base would be greatly improved. The international community would benefit from pollution reduction, climate mitigation, and the increased trading opportunities that arise from new income sources. The non-technical issues, which have recently gained attention, include: (1) Environmental and ecological factors, e.g., carbon sequestration, reforestation and revegetation. (2) Renewables as a CO_2 neutral replacement for fossil fuels. (3) Greater recognition of the importance of renewable energy, particularly modern biomass energy carriers, at the policy and planning levels. (4) Greater recognition of the difficulties of gathering good and reliable renewable energy data, and efforts to improve it. (5) Studies on the detrimental health efforts of biomass energy particularly from traditional energy users.

In Sudan, electricity reaches only about 30% of the population, mainly in urban areas. Hence, a major problem for rural people is the inadequate supply of power for lighting, heating, cooking, cooling, water pumping, radio or TV communications and security services. Petroleum product supplies, including diesel, kerosene and LPG are irregular and often subject to sudden price increases. Because of the inadequate supply of these fuels, women trek great distances into the forest to collect fuelwood, charcoal and biomass residues from animal and agriculture, which account for more than half of total energy consumption. Most of this is utilised for cooking and heating water in rural and semi urban areas and by the urban poor.

It is a need to provide alternative renewable energy sources to enhance women's participation in, and benefit from development. Household energy was the first energy sector that paid explicit attention to women and their energy needs. The contribution of women to environmental policy is largely ignored. Decision-making and policy formulation at all environmental levels, i.e., conservation, protection and rehabilitation and environmental management are more or less a male preserve. Women have been involved in promotion of appropriate energy technologies, primarily for rural populations over the past 15 years. Currently, distribution networks generally differ greatly from transmission networks, mainly in terms of role, structure (radial against meshed) and consequent planning and operation philosophies.

The increased availability of reliable and efficient energy services stimulates new development alternatives. Anticipated patterns of future energy use and consequent environmental impacts (acid precipitation, ozone depletion and the greenhouse effect or global warming) are comprehensively discussed in this study. Throughout the theme several issues relating to renewable energies, environment and sustainable development are examined from both current and future perspectives.

6. MEROWE HIGH DAM

The Merowe High Dam, also known as Merowe Multi-Purpose Hydro Project or Hamdab Dam, is a large construction project in Merowe town in northern Sudan, about 350 km north of the capital Khartoum. It is situated on the river Nile, close to the 4th Cataract where the river divides into multiple smaller branches with large islands in between.

Merowe is a city about 40 km downstream from the construction site at Hamdab. The main purpose of the dam is the generation of electricity. Its dimensions make it the largest contemporary hydropower project in Africa (Figure 32). The dam is designed to have a length of about 9 km and a crest height of up to 67 m. It will consist of polystyrene-faced rockfill dams on each river bank, an earth-rock dam with a pepper core in the left river channel and a live water section in the right river channel (sluices, spillway and power intake dam with turbine housings). It contains a reservoir of 12.5 km³, or about 20% of the Nile's annual flow (Habeballa, 1991). The reservoir lake is planned to extend 174 km upstream. The powerhouse equipped with ten 125 MW Francis turbines, each one designed for a nominal discharge rate of 300 m³/s, and each one driving a 150 MVA, and 15 kV synchronous generator. The expected annual electricity yield is 5.5 TWh, corresponding to an average load of 625 MW, or 50% of the rated load. To utilise the extra generation capacity, the Sudanese power grid will be upgraded and extended as part of the project. It is planned to build about 500 km of new 500 kV aerial transmission line across the Bayudah desert to Atbara, continuing to Omdurman /Khartoum, as well as about 1000 km of 220 kV lines eastwards to Port Sudan and westwards along the Nile, connecting to Merowe, Dabba and Dongola.

6.1. Planning and Construction

The idea of a Nile dam at the 4th cataract is quite old. The authorities of the Anglo-Egyptian Sudan proposed it several times during the first half of the 20th century. It was supposed to equalise the large annual Nile flow fluctuations, create the possibility of growing cotton and provide flood protection for the lower Nile valley. After Sudan achieved independence in 1956, Egypt finally decided to control the Nile floods with a dam and reservoir on its own territory - the Aswan High Dam and Nasser Lake. However, insufficient funding and lack of investor interest effectively stalled the project at the planning stage. This appears to have changed fundamentally since the country started exporting oil in commercial quantities in the years 1999/2000. A greatly improved creditworthiness brought an influx of foreign investment, and the contracts for the construction of what is now known as the Merowe Dam project were signed in 2002 and 2003. The main contractors are:

- China International Water and Electric Corporation, China National Water Resources and Hydropower Engineering Corp. (construction of dam, and hydromechanical works).
- Lahmeyer International (Germany - planning, project management, and civil engineering).
- Alstom (France - generators, and turbines).
- Harbin Power Engineering Company, Jilin Province Transmission and Substation Project Company (both China - transmission system extension).

By the time the contracts were signed, the Merowe Dam had been the largest international project the Chinese industry ever participated in. River diversion and work on the concrete dams began in early 2004. The project timeline schedules the reservoir impounding to start in mid-2006 and the first generating unit went on-line in mid-2007.

The work finished when the water level in the reservoir will have reached 300 m above sea level and all ten generating units will be operational, scheduled for 2009. The dam was inaugurated on March 3, 2009.

Source: (Merowe Dam from air, 2007).

Figure 32. Merowe high dam locations.

6.2. Financing

The total project cost is reported to be Euro 1200 million. This can be subdivided into partial amounts for the construction work on the dam itself (ca. 45%), its technical equipment (ca. 25%) and the necessary upgrade of the power transmission system (ca. 30%). The project receives funding from:

- China Import Export Bank - approximately Euro 240 million.
- Arab Fund for Economic and Social Development - approximately Euro 130 million.
- Saudi Fund for Development - approximately Euro 130 million.
- Oman Fund for Development - approximately Euro 130 million.
- Abu Dhabi Fund for Development - approximately Euro 85 million.
- Kuwait Fund for Arab Economic Development - approximately Euro 85 million.

The remaining cost - approximately Euro 400 million - is supposed to be covered by the Sudanese government.

6.3. Benefits

The electrification level in Sudan is very low, even by the standards of the region. In 2002, the average Sudanese consumed 58 kWh of electricity per year (UN, 2007a), i.e., about one fifteenth of their Egyptian neighbours to the north, and less than one hundredth of the Organisation for Economic Co-operation and Development (OECD) average.

The capital Khartoum and a few large plantations account for more than two thirds of the country's electric power demand, while most of the rural areas are not connected to the national grid.

Many villages use the option of connecting small generators to the ubiquitous diesel-powered irrigation pumps. This way of generating electricity is rather inefficient and expensive.

The combined grid-connected generating capacity in Sudan was 728 MW in 2002, about 45% hydroelectricity and 55% oil-fired thermal plants. However, the effective capacity has always been a lot lower. The two main facilities, the Sennar (constructed in 1925) and Roseires (1966) dams on the Blue Nile were originally designed for irrigation purposes rather than power production. Generating units were added during the 1960s and 1970s when the demand for electric power increased, but their power production is often heavily restricted by irrigation needs.

The government in Khartoum has announced plans to raise the country's electrification level from an estimated 30% to about 90% in the mid-term (UN, 2007b).

Large investments into the medium and low voltage distribution grids will be necessary but not sufficient to reach this ambitious goal: First and foremost, the foreseeable increase in power consumption would require the addition of generating capacity.

During the 1990s, Sudanese electricity customers have already been plagued by frequent blackouts and brownouts due to insufficient generation. Three new thermal power plants went into operation in the Khartoum area in 2004, increasing the installed capacity to 1315 MW.

The Merowe dam with its peak output of 1250 MW would almost double this capacity once it comes online.

6.4. Human Impact

The following are human impacts:

6.4.1. Resettlement

Before the construction began, an estimated 55,000 to 70,000 people were residents of the area which will be covered by the reservoir lake, mainly belonging to the Manasir, Hamadab and Amri tribes. They settle in small farming villages along the banks of the Nile and on the islands in the cataract. The whole region is relatively isolated, without paved roads or other infrastructure, and the communities are widely self-sufficient. Except for beans and millets the farmers grow vegetables, both for their own consumption and for trading at the weekly regional markets.

However, their main sources of income - and their most valuable possession - are the groves of date palms growing in the fertile silt on the river banks.

Once the dam is completed and the reservoir impounding begins, the whole population will be forced to move. While the majority of the farmers would prefer to stay as close to their old grounds as possible and build themselves a new existence at the shores of the new lake, the government has decided otherwise and pointed out three resettlement sites: Al-Multaqah, Al-Makabrab and Al-Muqadam.

At these locations, farmers will receive plots of land relative in size to their former possessions, in addition to financial compensation for lost assets - houses and date palms.

Though government officials claim there are improved living conditions at the resettlement areas, with relatively modern buildings and infrastructure, the affected people reject the compensation plans. Their main objections are:

- The soil at the resettlement areas is sandy, and its quality is extremely poor, particularly if compared to the excellent farmland by the Nile. It would take much effort and a long time—probably decades—until it became fertile enough for growing vegetables and other marketable produce.
- The government has announced that it will provide free water supply, sand removal and fertiliser during the first two years after the resettlement. After this period, the farmers will have to pay the full price for these services, none of which had to be used at the old site.
- Compensation for a date palm amounts to about four years harvest, while a good palm tree can bear fruit for a hundred years. Compensation for vegetable gardens is very low, and only married men will receive compensation for their houses and date palms.

About 6,000 people have been resettled to the Al-Multaqah site in the Nubian Desert during 2003 and 2004. Their villages were the closest to the dam construction site near Hamdab.

According to a survey conducted in early 2005, (Marshall, et al., 2006) stated the poverty rate has shot up dramatically since, because the farmers are not able to produce anything they could sell in the local markets.

A few years ago only a few percent efficiency had been achieved. Recently, however, several devices have been studied which have very high efficiencies.

6.4.2. Nomads

A significant fraction of the Manasir tribe inhabits the desert regions close to the Nile valley. The exact size of this nomadic population is unknown, but estimated to be of the same order of magnitude as that of the resident farmers, i.e., tens of thousands. Both groups maintain tight cultural interchanges and trade relations with each other.

Only the owners of real estate purportedly are covered under the compensation scheme, although reports are that families have been displaced without compensation or adequate provisions for relocation. Nomadic families will not receive any compensation, even though the resettlement of the farming Manasir will deprive them of their symbiotic partners. The consequences for their ability to sustain their lives in a harsh environment remain to be assessed.

6.4.3. Human Rights Concerns and Fears of Another War

The UN Special Rapporteur on Adequate Housing Miloon Kothari issued a statement on 27th of August 2007, calling for a halt to dam construction at Merowe until an independent assessment of the dam's impacts on the more than 60,000 people who stand to be displaced by the dams at Merowe and Kajbar. Kothari stated he has "received reports that the Merowe reservoir's water levels have already risen, destroying dozens of homes in the area and putting many more at risk (UN, 2007b). Kothari announced, "The affected people have claimed that they received no warning that water levels would be raised and that no assistance from government authorities has been forthcoming since their houses were destroyed". According to reports, the Government of Sudan has not honoured its promises to those who have been displaced. Kothari noted that, "thousands of people in the same area were relocated in similar circumstances that left many temporarily without food or shelter, and that some of those people remain homeless today" (UN, 2007b). Kothari called upon the Sudanese government to ensure safety and adequate housing to all those affected by the dams and warned the projects "would lead to large-scale forced evictions and further violence".

6.4.4. Archaeology

The following are discussed:

6.4.4.1. Kingdom of Kush

The fertile Nile valley has been attracting human settlement for thousands of years. The section between the 4th and 5th cataract - a significant portion of which will be inundated by the reservoir lake - has been densely populated through nearly all periods of (pre) history, but very little archaeological work has ever been conducted in this particular region. Recent surveys have confirmed the richness and diversity of traceable remains, from the Stone Age to the Islamic period.

Several foreign institutions have been recently or are currently involved in salvage archaeology in the region, among them the ACACIA project University of Cologne, Gdańsk Archaeological Museum Expedition (GAME), Polish Academy of Sciences, Humboldt University of Berlin, the Italian Institute for Africa and the Orient (IsIAO), the University College London, the Sudan Archaeological Research Society, the Hungarian Merowe Foundation, University of California at Santa Barbara - Arizona State University consortium, and the Oriental Institute Museum of the University of Chicago.

Their main problems are the shortness of the remaining time and limited funding.

Unlike the large UNESCO campaign conducted in Egypt before the completion of the Aswan High Dam, when more than a thousand archaeological sites could be documented and complete buildings were moved to prevent them from drowning in Lake Nasser's floods, work at the 4th cataract is much more restricted.

6.4.4.2. Political Impact

Usage rights to the waters of the Nile are fixed in the Nile Waters Treaty (Daniel, 2005), negotiated by the British in 1959. It allots 82 percent of the water volume to Egypt, while Sudan is granted the rights to the remaining 18 percent.

None of the riparian countries further upstream in the Nile basin - Ethiopia, Uganda, Rwanda, Burundi, Kenya and Tanzania - are entitled to any significant use of the water, be it for irrigation (of particular interest to Ethiopia and Kenya) or hydropower (Rwanda, Burundi, Uganda).

As Sudan now pushes forward to make use of its water allotment, those countries have begun to call for a revision of the treaty, arguing that - with the exception of Ethiopia - they had all been under colonial rule at the time the negotiations took place, and had not been represented in their best interest. Moreover, the decision of distribution of water was made without any negotiations with Ethiopia, which had rejected the agreement and is the source of 90% of the water and 96% of transported sediment of the Nile (Daniel, 2005; Marshall, et al., 2006).

6.4.4.3. Domestic

While a peace treaty seems to have stopped the fighting in southern Sudan after almost 20 years, there is no end in sight yet for the civil war in the western Darfur. More recently, unrest in Nubia as a direct result of the dams and the forced permanent displacement of Nubians from their homelands threatens to erupt into war. A group calling itself the Nubian Liberation Front is threatening armed resistance in order to thwart the series of dams along the Nile, and particularly at Kajbar.

6.4.4.4. Health

The resettlement area is a vast area with an expected 50,000–70,000 inhabitants who would be going through a transitional period for a few years before the get acclimatised and psychologically adapted to the new-life ahead. Governing by the two eminent health impact experiences of New Halfa resettlement projects and Aswan Dam in Egypt, strategic health planning ought to start early to foresee what water born diseases and other ecological health problems (such as bilharzias, malaria, etc.) are likely to prevail and to plan how to guard against that.

6.4.4.5. Evaporation

The creation of the reservoir lake will increase the surface area of the Nile by about 700 km². Under the climatic conditions at the site, additional evaporation losses of up to 1,500,000,000 m³ per year can be expected. This corresponds to about 8% of the total amount of water allocated to Sudan in the Nile Waters Treaty.

6.5. Ventilation

The increased exploitation of renewable energy sources is central to any move towards sustainable development. However, casting renewable energy thus carries with it an inherent commitment to other basic tenets of sustainability, openness, democraticisations, etc. Due to increasing fossil fuel prices, and the research in renewable energy technologies (RETs) utilisation has picked up a considerable momentum in the world. The present day energy arises has therefore resulted in the search for alternative energy resources in order to cope with the drastically changing energy picture of the world.

As fan speed is reduced on a fixed capacity compressor, so is energy efficiency. Reducing air volume on a fixed capacity compressor increases latent cooling capacity and increases the risk of coil freeze. Digital scroll compressors do not carry that risk. Figure 33 shows that energy efficiency (coefficient of performance) goes down as the air volume is reduced. Figure 34 peak power demands for conventional and radiant cooling. Figure 35 radiantly cooled slab with under floor ventilation.

The heating strategy includes four concepts (Figure 36):

- Solar collection: collection of the sun's heat through the building envelope.
- Heat storage: storage of the heat in the mass of the walls and floors.
- Heat distribution: distribution of collected heat to the different spaces, which require heating.
- Heat conservation: retention of heat within the building.

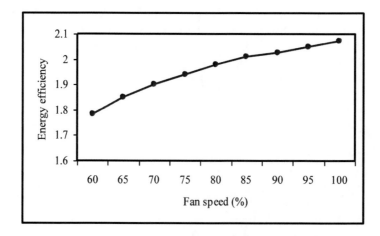

Figure 33. Lower air volume and lower evaporator temperature increase the risk of coil freeze in fixed compressor systems.

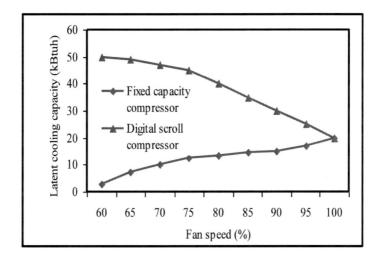

Figure 34. Digital scroll compressors reduce the need for rehumidification when the fan operates at lower speeds.

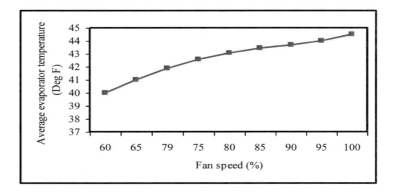

Figure 35. As fan speed is reduced on a fixed capacity compressor, so is energy efficiency.

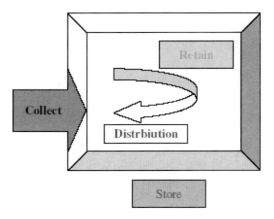

Figure 36. Heating strategy.

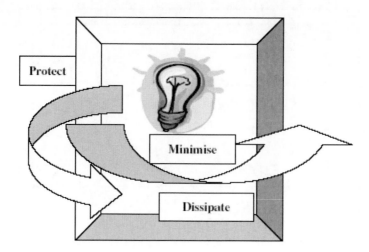

Figure 37. Daylighting strategies.

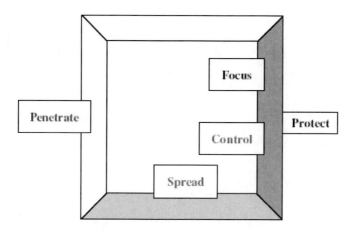

Figure 38. Cooling strategies.

The daylighting strategy includes four concepts (Figure 37):

- Penetration: collection of natural light inside the building.
- Distribution: homogeneous spreading of light into the spaces or focusing.
- Protect: reducing by external shading devices the sun's rays penetration into the building.
- Control: control light penetration by movable screens to avoid discomfort.

The cooling strategies include five concepts (Figure 38):

- Solar control: protection of the building from direct solar radiation (Figure 39).
- Ventilation: expelling and replacing unwanted hot air.

- Internal gains minimisation: reducing heat from occupants, equipments and artificial lighting.
- External gains avoidance: protection from unwanted heat by infiltration or conduction through the envelope (hot climates).
- Natural cooling: improving natural ventilation by acting on the external air (hot climates).

The environmental sustainability of the current global energy systems is under serious question.

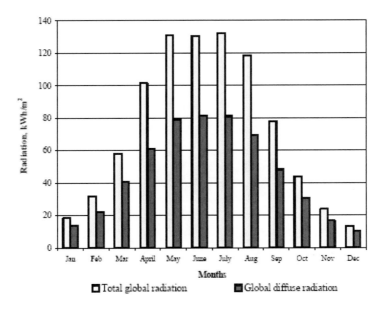

Figure 39. Average monthly global and diffuse radiations over Nottingham.

A major transition away from fossil fuels to one based on energy efficiency and renewable energy is required. Alternatively energy sources can potentially help fulfill the acute energy demand and sustain economic growth in many regions of the world. The mitigation strategy of the country should be based primarily ongoing governmental programmes, which have originally been launched for other purposes, but may contribute to a relevant reduction of greenhouse gas emissions (energy-saving and afforestation programmes).

Throughout the study several issues relating to renewable energies, environment and sustainable development are examined from both current and future perspectives. The exploitation of the energetic potential (solar and wind) for the production of electricity proves to be an adequate solution in isolated regions where the extension of the grid network would be a financial constraint. The provision of good indoor environmental quality while achieving energy and cost efficient operation of the heating, ventilating and air-conditioning (HVAC) plants in buildings represents a multi variant problem. The comfort of building occupants is dependent on many environmental parameters including air speed, temperature, relative humidity and quality in addition to lighting and noise. Renewable energy technologies

(RETs) and its role for harnessing for the society are well known. A lot of research has been conducted around the globe to make the utilisation of the RETs easy and simple through the development and demonstration activities.

The overall objective is to provide a high level of building performance (BP), which can be defined as indoor environmental quality (IEQ), energy efficiency (EE) and cost efficiency (CE).

- Indoor environmental quality is the perceived condition of comfort that building occupants experience due to the physical and psychological conditions to which they are exposed by their surroundings. The main physical parameters affecting IEQ are air speed, temperature, relative humidity and quality.
- Energy efficiency is related to the provision of the desired environmental conditions while consuming the minimal quantity of energy.
- Cost efficiency is the financial expenditure on energy relative to the level of environmental comfort and productivity that the building occupants attained. The overall cost efficiency can be improved by improving the indoor environmental quality and the energy efficiency of a building.

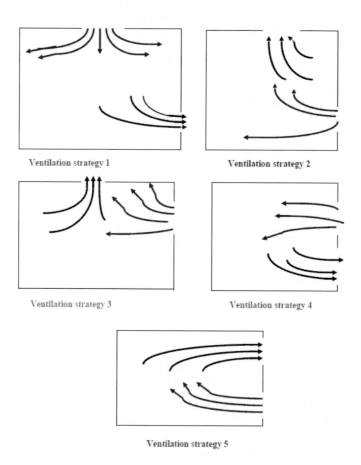

Figure 40. Different ventilation strategies.

The concentration of indoor aerosol particles can be reduced by using different ventilation strategies (Figure 40) such as displacement and perfect mixing. However, there are insufficient data to quantify the effectiveness of these methods, as removal of particles is influenced by particle deposition rate, particle type, size, source and concentrations (Rawlings, 1999).

6.6. Effects of Urban Density

Compact development patterns of buildings can reduce infrastructure demands and the need to travel by car. As population density increases, transportation options multiply and dependence areas, per capita fuel consumption is much lower in densely populated areas because people drive much less (Erlich, 1991; Herath, 1985; IEA, 2007).

Table 17. Summary of material recycling practices in the construction sector

Construction and demolition material	Recycling technology options	Recycling product
Asphalt	Cold recycling: heat generation; Minnesota process; parallel drum process; elongated drum; microwave asphalt recycling system; finfalt; surface regeneration	Recycling asphalt; asphalt aggregate
Brick	Burn to ash, crush into aggregate	Slime burn ash; filling material; hardcore
Concrete	Crush into aggregate	Recycling aggregate; cement replacement; protection of levee; backfilling; filter
Ferrous metal	Melt; reuse directly	Recycled steel scrap
Glass	Reuse directly; grind to powder; polishing; crush into aggregate; burn to ash	Recycled window unit; glass fibre; filling material; tile; paving block; asphalt; recycled aggregate; cement replacement; manmade soil
Masonry	Crush into aggregate; heat to 900°C to ash	Thermal insulating concrete; traditional clay
Non-ferrous metal	Melt	Recycled metal
Paper and cardboard	Purification	Recycled paper
Plastic	Convert to powder by cryogenic milling; clopping; crush into aggregate; burn to ash	Panel; recycled plastic; plastic lumber; recycled aggregate; landfill drainage; asphalt; manmade soil
Timber	Reuse directly; cut into aggregate; blast furnace deoxidisation; gasification or pyrolysis; chipping; moulding by pressurising timber chip under steam and water	Whole timber; furniture and kitchen utensils; lightweight recycled aggregate; source of energy; chemical production; wood-based panel; plastic lumber; geofibre; insulation board

6.7. Policy Recommendations for a Sustainable Energy Future

Sustainability is regarded as a major consideration for both urban and rural development. People have been exploiting the natural resources with no consideration to the effects, both short-term (environmental) and long-term (resources crunch).

It is also felt that knowledge and technology have not been used effectively in utilising energy resources. Energy is the vital input for economic and social development of any country. Its sustainability is an important factor to be considered. The urban areas depend, to a large extent, on commercial energy sources. The rural areas use non-commercial sources like firewood and agricultural wastes (Robinson, 2007; Sims, 2007). A review of the potential range of recyclables is presented in Table 17.

Sustainability is regarded as a major consideration for both urban and rural development. People have been exploiting the natural resources with no consideration to the effects, both short-term (environmental) and long-term (resources crunch).

It is also felt that knowledge and technology have not been used effectively in utilising energy resources. Energy is the vital input for economic and social development of any country. Its sustainability is an important factor to be considered. The urban areas depend, to a large extent, on commercial energy sources.

The rural areas use non-commercial sources like firewood and agricultural wastes. With the present day trends for improving the quality of life and sustenance of mankind, environmental issues are considered highly important.

In this context, the term energy loss has no significant technical meaning. Instead, the exergy loss has to be considered, as destruction of exergy is possible.

Hence, exergy loss minimisation will help in sustainability.

The development of a renewable energy in a country depends on many factors. Those important to success are listed below:

1. Motivation of the population
   The population should be motivated towards awareness of high environmental issues, rational use of energy in order to reduce cost.
   Subsidy programme should be implemented as incentives to install biomass energy plants.
   In addition, image campaigns to raise awareness of renewable energy technology.
2. Technical product development
   To achieve technical development of biomass energy technologies the following should be addressed:
   • Increasing the longevity and reliability of renewable energy technology.
   • Adapting renewable energy technology to household technology (hot water supply).
   • Integration of renewable energy technology in heating technology.
   • Integration of renewable energy technology in architecture, e.g., in the roof or façade.
   • Development of new applications, e.g., solar cooling.
   • Cost reduction.
3. Distribution and sales

Commercialisation of biomass energy technology requires:

- Inclusion of renewable energy technology in the product range of heating trades at all levels of the distribution process (wholesale and retail).
- Building distribution nets for renewable energy technology.
- Training of personnel in distribution and sales.
- Training of field sales force.

It represents an excellent opportunity to offer a higher standard of living to local people and will save local and regional resources. Implementation of greenhouses offers a chance for maintenance and repair services.

4. Consumer consultation and installation

To encourage all sectors of the population to participate in adoption of biomass energy technologies, the following has to be realised:

- Acceptance by craftspeople, and marketing by them.
- Technical training of craftspeople, and initial and follow-up training programmes.
- Sales training for craftspeople.
- Information material to be made available to craftspeople for consumer consultation.

5. Projecting and planning

Successful application of biomass technologies also require:

- Acceptance by decision makers in the building sector (architects, house technology planners, etc.).
- Integration of renewable energy technology in training.
- Demonstration projects/architecture competitions.
- Biomass energy project developers should prepare to participate in the carbon market by:
  - Ensuring that renewable energy projects comply with Kyoto Protocol requirements.
  - Quantifying the expected avoided emissions.
  - Registering the project with the required offices.
  - Contractually allocating the right to this revenue stream.
- Other ecological measures employed on the development include:
  - Simplified building details.
  - Reduced number of materials.
  - Materials that can be recycled or reused.
  - Materials easily maintained and repaired.
  - Materials that do not have a bad influence on the indoor climate (i.e., non-toxic).
  - Local cleaning of grey water.
  - Collecting and use of rainwater for outdoor purposes and park elements.
  - Building volumes designed to give maximum access to neighbouring park areas.
  - All apartments have visual access to both backyard and park.

6. Energy saving measures

The following energy saving measures should also be considered:

- Building integrated solar PV system.
- Day-lighting.
- Ecological insulation materials.
- Natural/hybrid ventilation.
- Passive cooling.
- Passive solar heating.
- Solar heating of domestic hot water.
- Utilisation of rainwater for flushing.

Energy efficiency and renewable energy programmes could be more sustainable and pilot studies more effective and pulse releasing if the entire policy and implementation process was considered and redesigned from the outset. New financing and implementation processes are needed which allow reallocating financial resources and thus enabling countries themselves to achieve a sustainable energy infrastructure.

The links between the energy policy framework, financing and implementation of renewable energy and energy efficiency projects have to be strengthened and capacity building efforts are required. Figure 41 shows solar photovoltaic arrays.

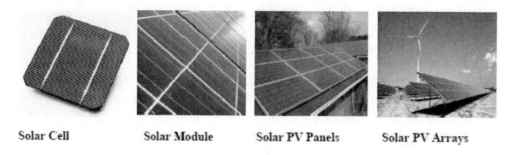

Solar Cell   Solar Module   Solar PV Panels   Solar PV Arrays

Figure 41. Solar photovoltaic arrays.

Improving access for rural and urban low-income areas in developing countries must be through energy efficiency and renewable energies. Sustainable energy is a prerequisite for development. Energy-based living standards in developing countries, however, are clearly below standards in developed countries.

Low levels of access to affordable and environmentally sound energy in both rural and urban low-income areas are therefore a predominant issue in developing countries.

In recent years many programmes for development aid or technical assistance have been focusing on improving access to sustainable energy, many of them with impressive results. Apart from success stories, however, experience also shows that positive appraisals of many projects evaporate after completion and vanishing of the implementation expert team.

Altogether, the diffusion of sustainable technologies such as energy efficiency and renewable energies for cooking, heating, lighting, electrical appliances and building insulation in developing countries has been slow.

CONCLUSION

Two of the most essential natural resources for all life on the earth and for man's survival are sunlight and water. Sunlight is the driving force behind many of the renewable energy technologies. The worldwide potential for utilising this resource, both directly by means of the solar technologies and indirectly by means of biofuels, wind and hydro technologies is vast. During the last decade interest has been refocused on renewable energy sources due to the increasing prices and fore-seeable exhaustion of presently used commercial energy sources. Plants, like human beings, need tender loving care in the form of optimum settings of light, sunshine, nourishment and water. Hence, the control of sunlight, air humidity and temperatures in greenhouses are the key to successful greenhouse gardening. The mop fan is a simple and novel air humidifier; which is capable of removing particulate and gaseous pollutants while providing ventilation. It is a device ideally suited to greenhouse applications, which require robustness, low cost, minimum maintenance and high efficiency.

A device meeting these requirements is not yet available to the farming community. Hence, implementing mop fans aides sustainable development through using a clean, environmentally friendly device that decreases load in the greenhouse and reduces energy consumption.

REFERENCES

Abdeen, MO. 2008. Chapter 10: Development of integrated bioenergy for improvement of quality of life of poor people in developing countries, In: *Energy in Europe: Economics, Policy and Strategy - IB*, Editors: Flip L. Magnusson and Oscar W. Bengtsson, 2008 NOVA Science Publishers, Inc., 341-373, New York, US.

Achard, P; Gicqquel, R. 1986. *European passive solar handbook*. Brussels: Commission of the European Communities. 1986.

Bos, E; My, T; Vu, E; and Bulatao, R. 1994. *World population projection: 1994-95*. Edition, published for the World Bank by the John Hopkins University Press. Baltimore and London.

Brain, G; Mark, S. 2007. Garbage in, energy out: landfill gas opportunities for CHP projects. *Cogeneration and On-Site Power*, 8 (5), 37-45. 2007.

Daniel, K. 2005. *The Five Dimensions of the Eritrean Conflict 1941 – 2004: Deciphering the Geo-Political Puzzle*. US: Signature Book Printing, Inc., 2005, 198.

Department of the Environment (DOE). (2009). *Energy consumption guide No. 19*, DOE, Washington, DC.

Duchin, F. 1995. *Global scenarios about lifestyle and technology, the sustainable future of the global system*. United Nations University. Tokyo.

Environmental Protection Agency (EPA). 2008. Inventory of US GHG emissions and sinks: 1990-2007.

Erlich, P. 1991. Forward facing up to climate change, In: *Global Climate Change and Life on Earth*. R. C. Wyman (Ed), Chapman and Hall, London. 1991.

Habeballa, H. 1991. The Treatability of the Blue and White Nile Waters for Drinking Purposes in Khartoum, Institute of Environmental Studies, University of Khartoum.

Herath, G. 1985. The green revolution in Asia: productivity, employment and the role of policies. *Oxford Agrarian Studies*, 14, 52-71.

International Energy Agency (IEA). (2007). *Indicators for Industrial Energy Efficiency and CO_2 Emissions: A Technology Perspective*. 2007.

Jonathon, E. 1991. *Greenhouse gardening*. The Crowood Press Ltd. UK.

John, W. 1993. *The glasshouse garden*. The Royal Horticultural Society Collection. UK.

Marshall, et al. 2006. Late Pleistocene and Holocene environmental and climatic change from Lake Tana, source of the Blue Nile. (247KiB), 2006.

Omer, AM. 1997. Compilation and evaluation of solar and wind energy resources in Sudan. *Renewable Energy*, 12 (1), 39-69.

Omer, AM. 1998. Horizons of using wind energy and establishing wind stations in Sudan. *Dirasat*, 25 (3), 545-552.

Omer, AM; Yemen, D. 2001. Biogas an appropriate technology. *Proceedings of the 7th Arab International Solar Energy Conference*, pp.417, Sharjah, UAE, 19-22 February 2001.

Omer, AM. 2004. Water resources development and management in the Republic of the Sudan. *Water and Energy International*, 61(4), 27-39.

Omer, AM. (2008a). *Energy demand for heating and cooling equipment systems and technology advancements*. In: *Natural Resources: Economics, Management and Policy*, 131-165.

Omer, AM. 2008b. Green energies and the environment. *Renewable and Sustainable Energy Reviews*, 12, 1789-1821.

Omer, AM. 2008c. Ground-source heat pumps systems and applications. *Renewable and Sustainable Energy Reviews*, 12, 344-371.

Omer, AM. 2009a. Environmental and socio-economic aspect of possible development in renewable energy use, In: *Proceedings of the 4th International Symposium on Environment,* Athens, Greece, 21-24 May 2009.

Omer, AM. 2009b. Energy use, consumption, environment and sustainable development, In: *Proceedings of the 3rd International Conference on Sustainable Energy and Environmental Protection* (SEEP 2009), Paper No.1011, Dublin, Republic of Ireland, 12-15 August 2009.

Omer, AM. 2009c. Energy use and environmental impacts: a general review, *Journal of Renewable and Sustainable Energy*, Vol.1, No.053101, 1-29, United State of America, September 2009.

Omer, AM. 2009d. Chapter 3: Energy use, environment and sustainable development, In: *Environmental Cost Management*, Editors: Randi Taylor Mancuso, 2009 NOVA Science Publishers, Inc., 129-166, New York, US, 2009.

Rawlings, RHD. 1999. *Technical Note TN 18/99 – Ground Source Heat Pumps: A Technology Review*. Bracknell. The Building Services Research and Information Association. 1999.

Robinson, G. 2007. Changes in construction waste management. *Waste Management World*, 43-49. 2007.

Sims, RH. 2007. Not too late: IPCC identifies renewable energy as a key measure to limit climate change. *Renewable Energy World*, 10 (4), 31-39. 2007.

Swift-Hook, DT; et al. 1975. Characteristics of a rocking wave power devices. *Nature*, 254, 504. 1975.

Trevor, T. 2007. Fridge recycling: bringing agents in from the cold. *Waste Management World*, 5, 43-47. 2007.

UNICEF. 2006. Situation Analysis of Children and Women in the Sudan. Khartoum: Sudan.

UN. 2007a. UN rights expert urges suspension to dam projects in northern Sudan. UN News Centre. *September 9, 2007.* http://www.un.org/apps/news/story.asp?NewsID= 23617andCr=sudanandCr1.

UN. 2007b. UN rights expert urges suspension to dam projects in northern Sudan. UN News Centre. *September 9, 2007.* http://en.wikipedia.org/w/ index.php?title=Merowe_ Damandaction=edit.

2006 World Atlas and Industry Guide (WAIG). (2006). Hydropower and sustainable water resources management in the world. *The International Journal on Hydropower and Dams*, United Kingdom.

World Commission on Environment and Development (WCED). (2009*). Our common future.* New York. Oxford University Press.

INDEX

D

F

G

S

T

U

V

W

Y

Z